W9-DES-919

MAY 2008

Management for Engineers, Technologists and Scientists

CD ROM IN
BACK POCKET

Wilhelm Nel

EDITOR

JUTA
AND COMPANY LTD

First published 2000
Reprinted 2004 (twice)
Second edition 2006
Reprinted 2007
by Juta & Co.
Mercury Crescent
Wetton, 7780
Cape Town, South Africa

© 2006 Juta & Co. Ltd

ISBN 10: 0 7021 7161 1

ISBN 13: 978 0 7021 7161 1

Typeset in 10 on 13 pt Baskerville

Project Manager: Sarah O'Neill
Editor: Alex Potter of FPP Productions
Proofreader: Cecily van Gend
Indexer: Cecily van Gend
Design and typesetting: Martingraphix
Cover designer: Pumphaus Design Studio
Printed in South Africa by Shumani Printers

658.0024

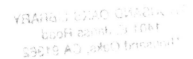

Contents

Chapter 4: The Impact of Employment Relations and Labour Legislation on an Organisation (*Marinda Conradie*)

Chapter 5: Managing People and Teams (*Marius Meyer*)

Chapter 6: Engineering Contracts and Law (*Alistair Glendinning*)

Chapter 7: Operations Management (*David Kruger*)

Chapter 8: Total Quality Management (*Winton Myers*)

Chapter 9: An Introduction to Safety Management (*Carl Marx*)

Chapter 10: Maintenance Management (*Krige Visser*)

Chapter 11: Marketing for Technical People (*Darryl Aberdein, Danie Petzer & Wilhelm Nel*)

Chapter 12: The Engineer, User of Information and Communication Systems (*Wilhelm Nel*)

Chapter 13: Principles of Project Management (*Ad Sparrius*)

Chapter 14: Introduction to Accounting, Economics, Financial Management and Budgeting (*Wilhelm Nel*)

Chapter 15: Cost Estimating, Cost Engineering and Cost Management (*Wilhelm Nel*)

Chapter 16: Introduction to Time Value of Money and Project Selection (*Wilhelm Nel*)

Chapter 17: Business and Technology Strategy (*Jannie Lourens*)

Chapter 18: Managing Technology and Innovation (*Darryl Aberdein*)

Chapter 19: An Overview of Environmental Management and Sustainable Development Concepts for Management Practices (*Alan Brent*)

Chapter 20: Entrepreneurship (*Jopie Coetzee*)

Preface

This book is primarily focused on the educational needs of technical people such as engineers, technologists, scientists and technicians. Young engineers often find themselves in control of people and resources soon after they start their careers. Some may find themselves in a supervisory/management position or in an entrepreneurial role much sooner than they would have imagined. Most engineers cannot escape involvement in commercial or administrative dealings with customers, suppliers, contractors, accountants and managers. For these and many other reasons, some engineering programmes include subjects such as communication, administration, supervision and management. Such subjects, however, usually form a small percentage of engineering or natural sciences programmes and are, in some cases, optional or excluded. Such an approach may be very tempting if one considers the pace at which technology advances and the vast amount of technical knowledge that learners have to absorb. Not to introduce engineering students to the business environment is, however, an irresponsible approach. This is further explained in chapter 1 of this book.

As a lecturer at the Unisa School of Engineering, I identified the need for a management textbook that targets the specific needs of engineers and scientists at an introductory level. Most of the existing management textbooks have been written for learners who are enrolled for courses in management and commerce. Course material and textbooks that are used in such programmes fill many volumes and in my opinion contain too much detail to be used, in their entirety, in engineering programmes, where the focus is on technical subjects and only a small part of the programme may be dedicated to management. Textbooks that were written for B.Com and N.Dip. Management students also do not focus on the content that is of particular importance to the technical person. The present book tries to rectify this. The syllabuses of several South African technikon and university engineering and technology management subjects were studied and an attempt was made to cover most of what was found in these syllabuses.

The book introduces technical people to various management and related issues and should make readers aware of issues that will help them to become better engineers, supervisors and engineering managers. To keep the content of chapters as brief and as flowing as possible, further definitions are provided in the glossary at the end of the book. The book was written according to the principles of Outcomes Based Education (OBE). Study objectives at the beginning of each chapter can be used by the reader to determine his/her level of knowledge of the chapter content and should be useful in determining how intensively a particular chapter or section should be studied. Chapters do not have to be studied in the sequence in which they appear in the book. I hope that readers who have English as a second or even a third language will find the book easy to read.

At the back of the book you will find a CD that contains related software and other useful content. You will also find instructions at the back of the book on how to use this CD.

Convention: Please note that the term 'engineer' is used frequently for practical reasons in place of 'engineers, technologists, scientists and technicians'.

THE SECOND EDITION

The new chapter entitled Environmental Management and Sustainable Development improves the 'greenness' of this second edition. In addition to the new chapter, you will also find a number of other changes and improvements. An effort was again made in this edition to include information that may be useful to technical people in small and medium as well as large organisations. Most of the mini-projects that were added to some chapters will require learners to consult additional sources.

I thank the staff at Juta and the various authors for the contributions they have made to the development and publishing of this book. I also thank all the people who read some of the chapters and provided me and other authors with useful information on how to improve the book. Some of the people who made such contributions are John Grove, George Lilford, Dr Dawie Roos, Jo Schwenke, Peet Snyman, Wim van Steenderen and Mariette Vercuil.

I hope that you will enjoy reading and studying this book. Please contact me at wnel@unisa.ac.za if you have any suggestions on how this book can be improved further.

See www.engmanage.co.za for more information on authors, additional questions, updates and an extended glossary.

Wilhelm Nel

About the Authors

Darryl Aberdein is a partner in the technology and innovation management consultancy, BrainWorks Management. He specialises in product development and technology marketing strategies.

Dr Alan Brent is Professor at the University of Pretoria. Since its inception in April 2002, he has held the AIDC (Automotive Industry Development Centre) Chair of Life Cycle Engineering, situated in the Department of Engineering and Technology Management at the UP.

Jopie Coetzee is a senior lecturer at Unisa's Graduate School of Business Leadership. His lecturing and research areas are Strategy Implementation, Executive Project Management and International Business. See www.sblunisa.ac.za for a profile of him.

Marinda Conradie is a senior lecturer in Industrial Relations at Unisa.

David Kruger is a lecturer in Operations Management at Unisa.

Jannie Lourens holds a business development position in a large international oil and gas company.

Alistair Glendinning is a qualified engineer and is registered as a professional construction project manager. He runs his own consulting company specialising in construction and contractual-related issues. In addition, he is an advocate specialising in labour, engineering and construction matters.

Carl Marx is from a well-known SHERQ company supplying risk-based solutions globally.

Marius Meyer is a senior lecturer in Human Resource Development at Unisa. He is a board member of the American Society for Training and Development Global Network South Africa and is registered as a mentor and master's human resource practitioner with the South African Board of Personnel Practice.

Winton Myers is principal of Viewpoint Training and Consulting, where he works as a management consultant and trainer in process improvement and operations management.

Wilhelm (Willie) Nel is a senior lecturer at the School of Engineering, Unisa. He lectures in the areas of Engineering and Technology Management, Fuel Cell Technology and Systems, and Mining Engineering.

Danie Petzer is a senior lecturer in the Department of Marketing Management at the University of Johannesburg. He teaches marketing research, strategic marketing, distribution management and pricing decisions to undergraduate students, and international marketing at a post-graduate level.

Ad Sparrius is a consultant and extraordinary professor at the Graduate School of Business Leadership at Unisa.

Krige Visser is Professor in the Department of Engineering and Technology Management at the University of Pretoria.

1 The Environment in which Technical People Work

Study objectives

After studying this chapter, you should be able to:

- define the terms technology, technologist, scientist and engineer
- describe the role of engineers in modern society
- list the different skills required by a successful engineer
- explain why engineers have to appreciate the needs of business
- explain why engineers need to have management skills
- discuss the life cycle of an organisation and a facility

The aim of this chapter is to provide information on the environment in which engineers, technologists, scientists and technicians practise their professions. It also provides a framework that may assist the reader in putting the content of the rest of the book in context.

The story of civilisation is, in a sense, the story of engineering – that long and arduous struggle to make the forces of nature work for man's good and for man's advancement. Engineers are the men and women who, down the ages, have learned to exploit the properties of matter and the sources of power for the benefit of mankind (L. Sprague de Camp)

1.1 THE ENGINEER, TECHNOLOGIST AND SCIENTIST DEFINED

It is not easy to define or explain an engineer's role in an organisation or in society in general. This task is complicated by the broad use of the word 'engineering' – e.g. consider its use in the terms social engineering, genetic engineering, business process re-engineering, software engineering and all the other different types of engineering. Defining an engineer is further complicated by terms such as 'technician', 'technologist' and 'scientist'. One method of defining an engineer is to describe what that person does, how he/she does it, where he/she works and to explain his/her contribution to society.

The Engineering Council of the United Kingdom (UK) defines an engineer as: 'one who has and uses scientific, technical and other pertinent knowledge, understanding and skills to create, enhance, operate or maintain safe, efficient systems, structures, plant, processes or devices of practical and economic value ...'.

1

Dillon (1998:188) says that engineering is directed to developing and providing infrastructure, goods and services for industry and the community. De Camp (1963:15) defines an engineer as somebody who *designs* some structure or machine, or who *directs* the building of it, or who *operates* and *maintains* it.

Many different types of engineers exist today. Examples are civil, mechanical, electrical, computer, chemical, aerospace, nuclear, mining, biochemical, agricultural, biomedical, materials, metallurgical, petroleum, marine and industrial engineers. Although a certain component of all engineers' training may be common, some degree of specialisation in a specific discipline is required to manage the expanding engineering knowledge base.

A technologist can be described as somebody who uses or applies technology. Technology is needs driven, whereas curiosity and speculation about the natural world usually drive science. A definition of technology is as follows.

> Technology refers to the theoretical and practical knowledge, skills and artifacts that can be used to develop products and services as well as their production and delivery systems. Technology can therefore be embedded in people, materials, cognitive and physical processes, plant, equipment and tools (Burgelman *et al.*, 2004:2).

The aim of technology is to create and improve artefacts and systems that satisfy human needs or aspirations. Success is judged in terms of, for example, performance, efficiency, reliability, durability, cost of production, ecological impact, safety and end-of-life disposability. Whereas the output from science may be a published paper, that from technology may be a patent (Microsoft *Encarta 98*). Techno-economic problems may have more than one solution due to the variables mentioned. Various solutions to problems are often analysed and adjusted before an optimal solution is achieved.

Scientists generally aim to produce theories that can be tested experimentally in the public domain and valued according to criteria such as simplicity, elegance, comprehensiveness and range of explanatory power. Such scientists try to advance human knowledge. Other scientists may work in industry on practical problems.

The practice of technology is much older than that of science and (modern) engineering. Humankind was developing and using technology very early in its evolution when basic materials such as stone and wood were used to improve such skills as hunting, food gathering, food preparation and self-defence. Over the years, archaeologists have uncovered many artifacts and evidence of mining activities. At the early stages of humankind's existence, there was very little understanding of science. There may have been some basic, intuitive understanding of the various laws of nature that had existed since the beginning of the universe, but these laws were only very recently (relative to the much longer existence of humankind) discovered by people such as Newton. Humankind established a technology base by means of experimentation, the adoption of best practice and by incorporating such practice into its culture. Various cultural practices resulted (and still result) in the transfer of knowledge from one generation to another and thereby humankind's knowledge base continues to be built and expanded, something that distinguishes modern humans from most animals.

Today's technical challenges are seldom solved by one person. A team of people is usually involved when new products, structures, processes or technologies are developed, tested and subsequently maintained. Such teams may consist of scientists, engineers, technologists and a further two groupings called technicians and artisans.

Technologists, as defined previously, apply science and mathematics to well-defined problems. Technicians, however, accomplish specific tasks such as drafting, laboratory procedures and model building. Artisans in turn have very practical skills (e.g. welding, machining and carpentry), which they use to construct devices specified by the scientists, engineers, technologists and technicians (Holtzapple & Reece, 2003:6).

1.2   THE ROLE OF ENGINEERS IN SOCIETY

Many roles can be assigned to engineers. They are seen as the guardians of the new, the innovators and the creative bringers of technological change. Some describe engineering as the driving force behind the prosperity of nations. Engineers raise living standards and bring benefits to society as a whole (Dillon, 1998:189).

In *The Ancient Engineers,* De Camp explains that civilisation has only risen during the last 1 per cent of humankind's time on Earth, when we discovered how to raise crops and tame animals. Soon after this, humans discovered how to support from 20 to 200 people on a square kilometre of fertile land. When farmers learned to raise more food than they needed for themselves and their own families, it was no longer necessary for a large percentage of the population to be involved in farming activities. The rest of the population who were not farming spent their time making things to exchange for various goods. As specialisation increased, merchants, physicians, smiths and other craftspeople of many kinds came into being. Instead of building their own houses or making their own carts and boats, people began to buy them from workpeople skilled in these arts (De Camp, 1963:4).

As wealth and experience accumulated, man undertook projects too large for a single craftsman. Such projects called for the work of hundreds of men, organised and directed towards a common goal. Hence arose a new class of men: the technicians or engineers, who could negotiate with a king or priesthood for building a public work, plan the details, and direct the workmen. These men combined practical experience with knowledge of general, theoretical principles. Sometimes they were inventors as well as contractors, designers and foremen, but all were men who could imagine something new and transform a mental picture into physical reality (De Camp, 1963:3–4).

The first engineers were mining and military engineers. Mining activities expanded the range of metals and other natural products that humankind had at its disposal for the manufacturing of products. Military engineers had to prepare topographical maps; locate, design and construct roads and bridges; and build forts and docks. In the 18th century, the term 'civil engineering' was used to describe engineering work done by civilians for non-military purposes. With the increasing use of machinery in the 19th century, mechanical engineering was recognised as a specific branch of engineering. The other branches of engineering followed as technology advanced and the demands of the socio-economic environment changed. Dillon (1998:189) explains the role of engineers in society as follows.

Engineers are entrusted with difficult and dangerous projects; they are called on to make ethical decisions; they are providers of needs, fulfillers of aspirations, generators of the driving force behind prosperity, they are builders of the dreams and fantasies of a nation; they are potential saviours of the planet.

1.3   ENGINEERING SKILLS AND KNOWLEDGE

Most people want to be successful at what they do. In this book the focus is firstly on those non-technical skills and knowledge that engineers require for success in the business environment and secondly, to provide the context in which many technical skills are used. The wide range of skills used by engineers is one of the reasons why engineering is such an exciting and challenging profession. Engineers often have to combine technical skills with various non-technical skills and knowledge. Industry and society require engineers to possess knowledge, skills and aptitudes for activities like research, design, development of new products, production, maintenance, testing, construction, operations, sales, management, consulting and teaching (John, 1995:101; Holtzappel & Reece, 2003:15). An engineer or engineering team has to consider many factors and variables when designing a new product, plant, shaft system and factory including:

- ease of maintenance
- aesthetics (appearance)
- quality
- environmental concerns
- production/construction method(s)
- cost and economic viability
- safety
- materials used
- moral and ethical concerns.

Further information on the requirements that products should meet can be found in chapter 11 ('Marketing for Technical People'). One important requirement is that products and services need to be internationally competitive. Many of the larger engineering firms operate internationally, tendering for projects and competing for markets world-wide. The implication is that engineering graduates need to have a good knowledge of both the language and culture of at least one country other than their own (John, 1995:100).

The list of skills, knowledge and aptitudes that engineers and engineering managers should have seems endless. Parnaby (1998:182) states that graduate engineers are required to have a greater variety of skills and to perform a greater variety of tasks than they are currently equipped to do.

It is impossible to predict the nature of the jobs they will hold in the course of their careers. Engineering has become increasingly multidisciplinary; an appreciation of marketing, business cash-flow management, project management, contract bidding, technical management of a new type of process, team organisation, design for quality economic manufacture, supply chain control and manufacturing systems engineering is essential for career success (Parnaby, 1998:183).

The correct mix of these skills, knowledge areas and aptitudes that engineers require depends on the company, industry and business ecosystem in which they will apply their knowledge. A relatively small percentage of engineers are required for specialist positions in research, design and development, whereas small and medium-sized industrial companies essentially require engineers with a broad range of skills. The aim of this book is to focus on the skills engineers should have to appreciate business. Technical people who do not appreciate business may be partially responsible for the phenomenon that high-quality and successful research and development projects often fail to lead to the creation of new products, new businesses or even new industries. This phenomenon is often called the 'innovation chasm'. Some institutions – e.g. the Institution of Structural Engineers – recommend that approximately 25 per cent of the total study time of engineers should be used for studying non-technical topics (John, 1995:102). Figure 1.1 gives examples of non-technical skills that an engineer should possess.

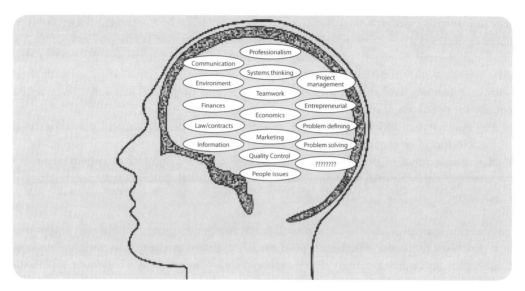

Figure 1.1 Non-technical skills and knowledge areas that an engineer should acquire (adapted from Redmill, 1995:133)

Whether or not engineers need the skills in figure 1.1 is debatable. Some people may think that engineers do not need entrepreneurial or intrapreneurial skills. Others may believe that engineers will fail to be the technical experts that they should be if most of their education is not spent on acquiring technical skills. Still others may think that too narrow a concern with technicalities encourages a view that only technicalities matter. Technical skills and creative talent are, however, not enough for an engineering practitioner to survive in today's business environment.

1.4 ENGINEERS AS (PROJECT) MANAGERS, EXECUTIVES AND SUPERVISORS

There is no hard-and-fast division between the skills required by an engineer and an engineering manager. Engineers not only generate ideas and design objects – they plan projects, administer budgets and direct others to implement their ideas. It is especially for the execution aspect of their ideas that engineers need management knowledge and skills. Many engineers spend much of their working lives managing rather than directly practising their engineering specialisation. Within ten years of starting their careers, many engineers take on leadership and management roles.

White (in Lock, 1993:15) suggests that engineering managers involved in the manufacture of products face six main challenges:
- getting the right people;
- aiming at the right targets;
- continuing action (monitoring, checking quality, etc.) throughout the product realisation process;
- creating a climate that encourages innovation;

- developing awareness of the social and other consequences of new technology; and
- dealing with increasing international competition.

It is important to understand that some managers should have a technical background in order to succeed in certain business environments. According to White (in Lock, 1993:17), good engineering managers are distinguished from other good managers by the fact that they simultaneously use their abilities to apply engineering principles together with their skills in organising and directing resources, people and projects. Thus they are qualified equally for two types of roles:

- the management of functions such as research, design, production and operations in a very technical environment; and
- the management of broader functions such as marketing, project management or top management in the high-technology enterprise that is usually subject to rapid technological change.

Engineering managers have become prominent in companies using sophisticated equipment and processes to produce high-tech products. The ability of companies to generate new business has become increasingly dependent on technical knowledge. Therefore people with both technical and managerial knowledge together with the ability to think strategically about the use of technology are increasingly appointed at board level. Engineers that manage scientists who are employed by organisations that thrive on research, development and technical consulting are included in this category. Engineering managers should understand the nature and culture of such organisations and be able to motivate staff. In high-technology enterprises, functions such as research, engineering design and quality assurance may enjoy equal status with functions such as marketing and finance.

1.5 THE ORGANISATIONAL AND OTHER LIFE CYCLES

All successful organisations are started at some point and tend to become larger and more complex over time. A number of researchers have proposed that the changes that take place during the life of an organisation follow a certain pattern, called the organisational life cycle. The stages in the organisational life cycle of an organisation are the inception (or birth) stage, high-growth stage, maturity stage and decline stage. The activities that take place during the inception stage, such as the development of a business plan, are discussed in the last chapter of this book, entitled 'Entrepreneurship'. Technology and products follow analogous patterns to that of the organisational life cycle and these patterns are called the technology life cycle and the product life cycle.

An organisation such as a mining company will have various assets and may have developed various facilities such as shafts and beneficiation plants during its life. In general, the lives of all facilities such as hospitals, universities, shaft systems, factories and plants can be divided into an initial project stage and an operational stage. During the initial project stage, money is spent (invested) to establish the infrastructure that will produce products and provide services during the operational stage. In chapter 13 ('Project Management'), the project stage is discussed. It is important to note that the same methods and knowledge (project management body of knowledge – PMBOK) that is used to manage the initial project for establishing a facility such as a plant or factory can also be applied to manage some of the activities that take place during the operational stages of such facilities. Examples of projects that may be initiated during the operational stage of a facility's life are maintenance projects; projects to upgrade equipment, processes and systems; projects to expand the capacity of the facility; and the disposal project at the end of the life of such a facility. These are illustrated in figure 1.2.

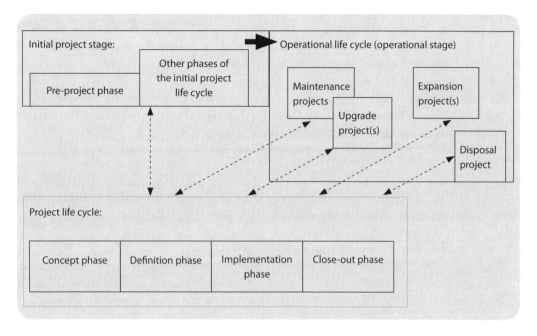

Figure 1.2 The life cycle of a shaft, plant, factory or any other facility (adapted from Burke, 2003:39)

Table 1.1 provides references to other chapters in the book where you will find more information on the life cycles mentioned in this section.

Table 1.1 Quick references to other chapters that contain information on life cycles

Life cycle discussed	Chapter
Operational life cycle/operational stage	7 'Operations Management'
Product life cycle (from a marketing management perspective)	11 'Marketing for Technical People'
Product life cycle (from an environmental management perspective)	19 'An Overview of Environmental Management and Sustainable Development Concepts for Management Practices'
Project life cycle/project stage	13 'Principles of Project Management'
Technology life cycle	18 'Managing Technology and Innovation'
Project, asset and product life cycles (from an environmental management perspective)	19 'An Overview of Environmental Management and Sustainable Development Concepts for Management Practices'

Most of this book's content applies to both the initial project stage to establish a new facility and the operational stage of such a facility.

1.6   THE WORLD VIEWS OF ENGINEERS, TECHNOLOGISTS AND SCIENTISTS

A world view can be defined as the comprehensive framework of one's basic belief about things (you will find other definitions in the 'Glossary' at the back of the book). It therefore serves as a framework and filter through which we make sense of what happens in the world around us. The world view of one human being is often quite different from that of another. We see it for example in the different beliefs about:

- how the universe and living things came into being (e.g. creationism and evolutionism);
- how the needs of society should be met at a macro-level (e.g. by means of free-market mechanisms or through a centrally controlled system); and
- how one ethnic group or nation should treat another; etc.

World views may therefore be incorrect or incomplete and result in narrow-mindedness and an inability to see (or want to see) things through the eyes of others. Because of this, scientists often have to question their own world views in an attempt to be objective in all they do.

This was described by Richard Feynman as follows:

The only way to have real success in science ... is to describe the evidence very carefully without regard to the way you feel it should be. If you have a theory, you must try to explain what's good about it and what's bad about it equally. In science you learn a kind of standard integrity and honesty.

Similarly, James Hansen has stated that:

The fun in science is to explore a topic from all angles and figure out how something works. To do this well, a scientist learns to be open-minded, ignoring prejudices that might be imposed by religious, political or other tendencies (Galileo being a model of excellence). Indeed, science thrives on repeated challenge of any interpretation, and there is even special pleasure in trying to find something wrong with well-accepted theory. Such challenges eventually strengthen our understanding of the subject, but it is a never-ending process as answers (often) raise more questions to be pursued in order to further refine our knowledge. Scepticism thus plays an essential role in scientific research, and, far from trying to silence sceptics, science invites their contributions.

The engineer and technologist should also be familiar with this scientific approach and thus value objectivity. Most engineers and technologists like to apply their creative talents at a practical level, e.g. to build something that will address a specific societal need. The final result of the successful execution of the engineer's and engineering manager's ideas and plans is usually something physical, such as a structure or product. The origins of these ideas and plans are, however, in the idea worlds and world views of the engineer. Expanding and pursuing these ideas could have major effects on society, as they may result in new technological breakthroughs. Such breakthroughs can be viewed as good or bad (ethical or unethical). It can be argued that technologies such as nuclear fission can be used constructively, e.g. to generate electricity, or destructively, e.g. to develop nuclear weapons that may be used to destroy life and habitats. Concepts, information and goals have causal effects in the material world that consists of forces and particles (Ellis, ND). Through top-down action, the plans of the civil engineer result in the movement of millions of atoms, structured in a specific way, to emerge as a bridge or highway. This ability to have an impact on the physical world points to the creative side of the engineer and probably the associated world view that although we as human beings are also just made up of atoms, there is more than that to us and as a result it is possible for us to 'manipulate', 'instruct' or even 'conquer' other atoms in the universe. A large burden therefore rests on the shoulders of humankind to use its technologies in constructive and sustainable ways. Many engineers believe that nature can be used for humankind's benefit without having to destroy or damage the ability of nature to sustain life on Earth.

The above may have illustrated why it is important for engineers to act in the best interest of clients and society, i.e. ethically. You will find more on ethics in the chapter on 'Safety Management' and more on sustainability in chapter 19 ('Environmental Management and Sustainable Development').

1.7   CONCLUSION

Engineers and engineering managers need both technical and non-technical skills. Technical skills are self-evident and (some of) these non-technical skills are listed and discussed in this chapter.

Engineering is a holistic discipline that does not focus on a single objective, such as the design and construction of a technical workable product. It should also focus on whether that product fulfils societal needs. It can therefore be argued that a holistic approach should be followed in the education of engineers, while at the same time taking educational and time constraints into consideration.

The reasons and causes why engineers act in certain ways are to be found not only in technical disciplines, but also in disciplines such as management, engineering economics and quality control.

Whereas the earliest engineers were employed by governments, kings and the wealthy, engineers today are employed by a variety of businesses. Many engineers are also self-employed as consultants, contractors and business owners.

This book seeks to provide a sound introduction to the non-technical issues that engineers, technologists and scientists may encounter in the workplace.

One of the objectives of this book is also to include information that should broaden the world view of the engineer, technologist and scientist.

Self-assessment

1.1   Define technology, technologist, scientist, engineer, 'innovation chasm' and 'engineering manager'.

1.2   List factors that an engineer or engineering team has to consider when designing a new product.

1.3   Describe the role of engineers in modern society.

1.4   Critically evaluate the following statement:
      'Without ancient technologists and engineers, humankind would still be living today as small bands of hunters and food-gathering tribespeople.'

1.5 Discuss whether engineers in training should take engineering and technology management as subjects.

1.6 List the various skills that technical people should have.

1.7 The successful engineer of today needs more than merely technical skills and knowledge. Discuss the non-technical skills and knowledge areas that an engineer should acquire.

1.8 Briefly discuss the life cycle of a shaft, plant, factory or any other facility.

1.9 List the type of projects that may be initiated during the life of a shaft, plant, factory or any other facility.

References

Bissel, C. 1996. 'Revitalising the engineering curriculum: the role of information technology.' *Engineering Science and Education Journal*, June.

Burgelman, R.A., Christensen, C.M. & Wheelwright, S.C. 2004. *Strategic Management of Technology and Innovation*. Boston: McGraw-Hill.

Burke, R. 2003. *Project Management: Planning and Control Techniques*. 4th edition. Cape Town: Burke Publishing.

Conelly, J.D. & Middleton, J.C.R. 1996. 'Personal and professional skills for engineers: One industry's perspective.' *Science and Education Journal*, June, pp. 139–42.

De Camp, L.S. 1963. *The Ancient Engineers*. New York: Balantine.

Dillon, C. 1998. 'Engineering education: Time for some new stories.' *Engineering Science and Education Journal*, August, pp. 188–92.

Dring, D. & O'Farrel, T. 1997. 'Concordat towards better researchers.' *Engineering Science and Education Journal*, August, pp. 167–70.

Duffy, M. 1996. 'The changing nature of engineering.' *Engineering Science and Education Journal*, October, pp. 231–9.

Ellis, G.F.R. n.d. *On the Nature of Emergent Reality*. Available at www..mth.uct.ac.za/~ellis/

Engineering Council of SA. n.d. *Standards and Procedures Initiative*. Available at www.ecsa.co.za

Holtzapple, M.T. & Reece, W.D. 2003. *Foundations of Engineering*. 2nd edition. Boston: McGraw-Hill.

John, V. 1995. 'A future path for engineering education.' *Engineering Science and Education Journal*, June, pp. 99–103.

Lock, D. (ed.). 1993. *Handbook of Engineering Management*. 2nd edition. Oxford: Butterworth-Heinemann.

Mayor. J. 1998. 'The evolution of engineering management.' *Engineering Science and Education Journal*, April, pp. 50, 51.

McDonald, M. 1997. 'The role of marketing in creating customer value.' *Engineering Science and Education Journal*, August, pp. 177–81.

Meyer, T. 1996. *Creating Competitiveness through ... Competencies: Currency for the 21st Century*. Randburg: Knowledge Resources.

Nel, W.P. 2005. 'Technology and OD.' In Botha, E. & Meyer, M. (eds). *Organisational Development and Transformation*. Durban: Butterworths.

Parnaby, J. 1998. 'The requirements for engineering degree courses and graduate engineers: An industrial viewpoint.' *Engineering Science and Education Journal*, August, pp. 181–7.

Redmill, F. 1995. 'Some suggestions for the (software) engineering degree curriculum.' *Engineering Science and Education Journal*, June, pp. 131–6.

Richardson, P. 1997. 'Business success, marketing and the engineer.' *Engineering Science and Education Journal*, August, pp. 171–6.

Shaw, M.C. 2001. *Engineering Problem Solving: A Classical Perspective*. Norwich: Andrew Publishing & Noyes Publications.

2 Principles of General Management

Jannie Lourens

Study objectives

After studying this chapter, you should be able to:
- distinguish the different management functions
- understand each function's objectives and mechanisms
- apply the functions in your work environment
- understand how the functions interrelate
- understand how leadership as a function has changed
- understand self-managing work teams (SMWT) and their significance in contemporary occupations
- understand the importance the SMWT concept has on daily scientific and engineering work
- note the differences among the classical application of planning, organising, leading and controlling, and how they are applied in the new types of work organisation such as SMWTs
- explain the application of general management principles when managing projects
- note how the general management principles are applied in other chapters of this book

The aim of this chapter is to introduce readers to principles of general management.

2.1 MANAGEMENT

What do managers really do? What is management about? The answers to these questions are not straightforward. Besides, the work of an engineering team leader differs dramatically from the work of a company director or a research leader – or does it? While management is recognised as a subject for systematic study, it is not regarded as a profession or discipline in itself. It incorporates many conventional disciplines and is an applied activity, regardless of the field of endeavour, be it research, engineering, marketing or health care.

Generally, all managers perform the same functions, activities and roles, to a lesser or greater extent. The content and context of the work will differ depending on the manager's level, the kind of business or service, the organisation's structure and its co-ordination systems. The functions of management are universal in the sense that they are applicable to a one-person business and to large multinational conglomerates. The same principles apply and the same kinds of activities are carried out, although at different levels and with different complexities.

2.1.1 The process of management

The process of management is illustrated in figure 2.1. It can be described as a systematic and rational process that directs all the efforts within an organisation that support the process of converting inputs into outputs and influences the organisation in order to produce valuable outcomes. The outcomes may be products, services or specific pre-determined objectives. Traditionally, the management functions necessary to perform this process are planning, organising, leading and controlling. Leading or leadership is the all-encompassing activity that directs the organisation and the management process towards their objectives. These management functions are dealt with in detail in this chapter.

In many contemporary, high-performance, team-based organisations, the classic functions of planning, controlling and organising are replaced by facilitation, mentoring, coaching and empowering. The teams handle the classical managerial functions themselves. You will find more about high-performance work teams (HPWTs) or self-managing work teams (SMWTs) and their leadership in the section on leadership.

In the management process, inputs from the external environment and internal systems, procedures and strategies are taken into account during the planning and decision-making process. After the planning has been completed, the work is organised to effect smooth implementation of the plan. Implementation is the step where the plans are brought into reality through actual activities by the organisation and not by management alone. The evidence of the implemented plans is noticeable in the organisational outcomes, such as profits realised, the changes accomplished and the services rendered to satisfied customers. Figure 2.1 gives an overview of the management process.

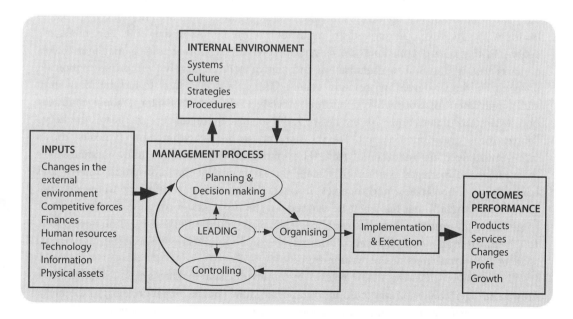

Figure 2.1: Overview of the management process

Responses from the outcomes are brought back into the managerial process through a system of outcomes or performance reviews. The results of such reviews are the inputs to the controlling function, which has the aim of correcting or improving outcomes by revising plans or formulating new ones.

Note that the managerial process also acts upon the internal environment, when necessary, while internal environmental factors are taken into account during the planning process. The managerial process may have only limited influence on the external environment, depending on the target of influence and the ability of the organisation to influence that target.

2.1.2 The roles managers play

In addition to the functions of management, there are other roles a manager has to play due to his/her position in an organisational unit. Mintzberg (1975) classifies managerial work into three broad categories with ten different roles in those categories. Table 2.1 summarises the different roles managers play.

Table 2.1 Managerial roles

Interpersonal roles	Informational roles	Decisional roles
Figurehead	Monitor	Entrepreneur (see chap. 20)
Leader	Disseminator	Disturbance handler
Liaisor	Spokesperson	Resource allocator
		Negotiator

By virtue of his/her position as head of an organisational unit, every manager must perform some duties of a ceremonial nature. These duties involve little serious communication and no important decision making, but are important to the organisation's image. This is the role of the *figurehead*. Because a manager is in charge of an organisational unit, he/she is responsible for the work, the development and the motivation of the people of that unit. The manager's actions in this regard constitute the *leadership* role. In the *liaison* role, a manager makes contacts beyond the unit or organisation and cultivates these largely to find information, in order to build an external information system. This role is necessary to aid the motivation and achievement of worthwhile economic objectives. By virtue of interpersonal contracts, both with subordinates and with the network of contacts, the manager emerges as the nerve centre of the organisational unit. He/She may not know everything, but typically knows more than any other member of staff. The manager performs three informational roles. As *monitor,* the manager continuously scans the environment for information, interrogates contacts and subordinates, and receives unsolicited information, mostly due to the network of personal contacts that has been developed. In the *disseminator* role, the manager passes some of the information directly to subordinates, who would otherwise have no access to it. In the *spokesperson role,* he/she sends information to people outside the unit, through formal speeches, informal meetings or written reports.

The manager plays the major role in the unit's decision-making system. Traditionally, only a manager could commit the unit to important new courses of action, and, as the nerve centre, only a manager has full information to make the decisions that determine the unit's strategy. Four roles describe a manager as a decision maker. As an *entrepreneur,* a manager seeks to improve the unit and adapt it to changing conditions in the environment. In this role, a manager is always on the lookout for new ideas. When a good one appears, a development project may be initiated that results in new or improved business. The manager is the initiator of change. In the role of *disturbance handler,* a manager has to respond to changes and pressures beyond his/her control. Action must be taken because the pressures of the situation are too severe to be ignored, like strikes, natural disasters and accidents. The manager has to decide on the division and allocation of different resources to different people, projects and activities within the unit, in the role of *resource allocator.* Managers spend a lot of time *negotiating* for new resources, information or changes to be implemented.

2.2 THE PLANNING FUNCTION

Planning is the managerial function through which the organisation determines how it will achieve its objectives. During the planning process, the 'what?', 'when?', 'where?', 'who?', 'how?' and 'why?' of certain things that need to be done to achieve the organisation's objectives are determined. The planning process is illustrated in figure 2.2.

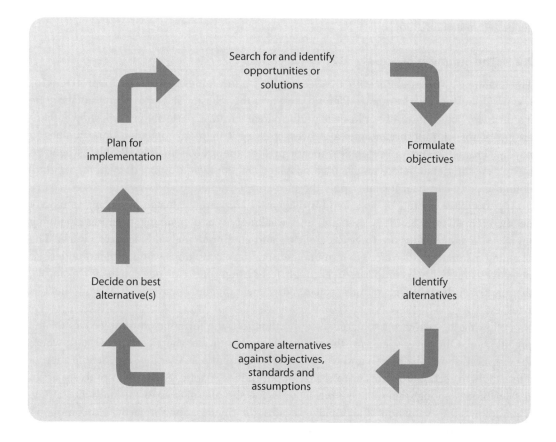

Figure 2.2 The planning process

Planning is done in advance of acting (implementation and execution). Decision making and planning are interrelated. A decision is the choice among alternatives. A decision is not a plan, because it need neither involve an action nor the future. Decisions are, however, necessary at every stage of the planning process and are therefore an integral part of planning.

Planning focuses on objectives and objectives are the incentives for planning. Plans formulate the direction and purpose essential to the long-term effectiveness and survival of an organisation. They assist in motivating participants towards goal accomplishment. Objectives provide the standards against which an organisation measures its performance. Plans spell out the ways and means of accomplishing goals. Planning facilitates more explicit attention to the establishment of objectives and the accomplishment of goals, and leads to more effective performance.

Planning alone does not lead to action. Leadership is necessary to implement what has been planned through people development and motivation, organising, and controlling actual performance against planned objectives. Performance measures involve effectiveness and efficiency measurements. Effectiveness measurements reflect the degree to which goals are accomplished, and efficiency measurements reflect the ratio of resources used to the outputs generated in attaining goals.

2.2.1 The planning process

The planning process starts with the search for opportunities, responses to problems or reviews of business performance. Planning requires the realistic diagnosis of the situation, be it a problem, an opportunity or historic performance. Historic performance will show if there are problems with the organisational process itself or if there are potential opportunities to pursue. Problems and opportunities do not present themselves readily. Problems need to be analysed in detail before an accurate problem definition is formulated. With an accurate definition, one can plan for and apply the best corrective actions.

Opportunities should be searched for diligently. They are evaluated according to market needs, competition, business strengths and weaknesses, and planning premises. Planning premises or assumptions are forecasts, policies, and environmental and legal constraints. The planning assumptions therefore determine the boundary conditions within which the desired objectives are to be accomplished. Through forecasting, one can determine, for example, the markets to be in, sales volumes, prices, costs, technologies, wage rates, taxes and interest rates, etc.

Following the search for solutions or opportunities is the formulation of objectives for the organisation. Objectives specifying the expected results indicate what needs to be done, where the emphasis should be placed and what is to be accomplished. After this, the identification of alternatives to rectify problems or to realise opportunities is done. A common problem arises not in finding enough alternatives, but in reducing the alternatives to the most promising. Then follows the evaluation of alternatives against the organisation's objectives, assumptions, strengths and weaknesses. The best alternative may sometimes bear the highest risk, while the less profitable may have a better risk profile over time.

After evaluating all the alternatives, the best course of action is selected. This is the point at which a plan is adopted, where the real decisions are made.

Sometimes, a decision is made to follow two or more courses of action to see which one actually performs best in the future. At a later stage, a decision can be made to abandon the option with the poorer performance.

Once the course of action has been decided on, the amount of subsequent planning can increase considerably. The last phase involves the formulation of implementation plans, human resources and budgets to bring the plan to fruition. This may involve several subsidiary plans, like plans to buy a new plant, hire more employees, train for new technologies, acquire additional finance or market new products. Coupled to these plans are the budgets, which indicate the subsidiary plans in terms of financial numbers. Budgets become one of the ways in which planning progress and implementation results are measured and controlled. It is at this point, when plans are put into numbers, that one should assess plans against available resources. Often, it is only realised at this point that, with the available resources and given constraints, the selected course will not be achievable. The planning process then has to start afresh.

The planning process is iterative and continuous. The more the planning process commits resources to the future, the more important it is for managers to periodically check assumptions and expectations, and redraw plans as necessary to maintain the course towards the desired objectives.

2.3 THE ORGANISING FUNCTION

Organising is the management function of creating structures or work units for the organisation or organisational unit that will enable people to work together effectively towards achieving its objectives through the best utilisation of resources.

2.3.1 Organisation structure

The function of organising is to divide the work an organisation has to perform into tasks and to group related tasks into divisions, functions or departments. This is the basis of the functional organisation. Another way is to group the work into projects and to form team structures with the necessary skills to execute the tasks or projects. This is the basis of the team-based and projects-based organisation.

The organisation structure is the primary means of division of work. For example, financially oriented work can be grouped in a finance or administration department. Work related to sales, the supply of services, forecasting market variables and market research is often grouped in a marketing department. Activities relating to one particular product – e.g. the production, financing, marketing and selling of that product – can be grouped in an independent business unit.

The organisation structure is an established pattern of relationships among the different components or units of the organisation. In an organisation, the structure is formed by the design of the sub-systems of the business system and by the formation of relationships among these sub-systems.

An organisation structure can be defined in terms of:
- its pattern of formal relationships, evidenced by organisation charts, job descriptions, positions and lines of authority. This pattern is referred to as the *hierarchy*;
- the manner in which activities and duties are assigned to different units or people. This is referred to as *differentiation of work*;

- the way activities are co-ordinated (or *co-ordination* of work);
- the relationships among power, status and positions within the organisation. This is referred to as the distribution of *formal authority*; and
- the policies, procedures, guidelines and controls that regulate activities and relationships in the organisation. This is referred to as the *systems* of a business.

The basic forms of organisation structure managers can choose from when designing an organisation are as follows.

- **The functional organisation:** The organisation is grouped into different functions such as finance, personnel, engineering, production and marketing.
- **The product or project organisation:** This organisation deals in several products, projects or classes of products simultaneously, and is grouped around each product, project or group of products. Each product group or project may have its own production, engineering and marketing functions, and may or may not share other functions with other product groups.
- **The customer organisation:** This organisation follows the same divisional principles as the product organisation. The only difference is that the groupings are around specific customer groups, not products. For example, banks may typically have industrial, personal, corporate and offshore clientele and will set up their structures to cater for the different clients accordingly.
- **The matrix organisation:** One typically encounters this organisation in engineering, construction, and research and development environments. The structure looks like a project structure, but is horizontally superimposed on a functional structure to form a grid pattern. Inside the matrix, the organisation members have both functional and project responsibilities. The main advantages of this structure are its flexibility and its ability to utilise human and other resources effectively. However, there are often some co-ordination problems.
- **The team-based structure:** Teams are very much in vogue. Recently, the self-managing work team (SMWT) or high-performance work team (HPWT) has become popular in leading-edge organisations. Organisations who break away from the traditional hierarchical organisation format are employing team structures. To function, they need highly empowered team members to fulfil many of the traditional managerial roles and functions discussed so far. The members have other roles too, and the role of the team leader or manager changes, often dramatically.

2.3.2 Authority, responsibility and accountability

Authority is the right to demand compliance by subordinates based on formal position and control over rewards and sanctions. Authority is impersonal and conforms to the position rather than the individual. Authority is the means for integrating the activities of subordinates with objectives and provides the basis for direction and control. Responsibility is the assigned obligation of a subordinate to carry out a delegated task or activity. Although authority can be delegated to a subordinate, responsibility cannot. Whereas responsibility can be assigned to a subordinate, the superior is still ultimately responsible for ensuring that the job is done properly. Since the superior retains ultimate responsibility, the delegation of authority always entails the creation of accountability. Thus, subordinates automatically become accountable to their superior for the performance of the tasks assigned to them. Authority and responsibility flow in direct lines, vertically, from the highest to the lowest level in the organisation. The vertical lines through the hierarchy are the lines of authority, and establish superior–subordinate relationships.

2.3.3 Delegation

Delegation is the assignment of tasks and the authority to carry out those tasks to a subordinate who assumes accountability for performing them. Delegation requires effective communication. A manager has duties that must be assigned to subordinates. To perform these tasks properly, subordinates need to understand exactly what they involve. Delegation also involves motivation, influence and leadership. As in all communication and influence processes, both parties are crucial to the success of the outcome.

2.3.4 Span of management

The span of management refers to the number of subordinates that a superior can manage effectively. There is a limit to the number of subordinates that can be managed effectively. Wide spans result in flat organisation structures, whereas short spans result in tall, multi-layered organisation structures. The span is determined by several factors, including:

- the similarity of subordinate functions (the more similar the functions, the wider the span can be);
- the geographic closeness of subordinates (if all subordinates are in the same location, the wider the span and the better the control can be);
- the complexity of subordinate functions;
- the extent and nature of the direction required by subordinates;
- the time needed to perform the tasks;
- the skills levels and maturity of subordinates; and
- the subordinates' ability to work as cohesive teams and to support each other. The better they can function as teams, the less management direction is needed.

2.3.5 Line, staff and teams

The traditional way of distinguishing between departments is to view them as either line or staff departments. Line functions are those that have direct responsibility for accomplishing the objectives of the organisation, e.g. production and marketing departments. Staff functions assist and advise line managers in accomplishing objectives. They do not necessarily have less important roles than line managers, and many essential activities, including personnel, quality control and purchasing, fall into the staff category.

There are three basic types of staff. Advisory staff normally comprise specialists who advise line management in their areas of expertise. Examples are lawyers, technologists, and training and counselling specialists. Service staff provide essential services to business departments, such as personnel, market research, corporate planning and public relations. Personal staff include secretaries and personal assistants who provide direct assistance to their managers.

High-performance work teams are assembled using skilled people from the traditional line and staff classification of work division. HPWTs may be permanent or temporary, depending on the tasks, the type of organisation or the projects to be completed. The main point is that HPWTs should comprise the right people with the right skills to perform the task in the most effective way, with minimum interference from organisational management and red tape.

2.3.6 Co-ordination

Co-ordination is the process of achieving unity of effort among various business units or people to accomplish the organisation's objectives. The greater the differentiation of activities and specialisation of labour, the more co-ordination is required. Co-ordination is essential whenever two or more interdependent individuals, groups or departments seek to achieve a mutual goal. Co-ordination in organisations can be achieved through various means.

- Informal communication can co-ordinate the relevant groups or people assigned to achieve a specific objective.
- Rules and procedures standardise work processes and facilitate co-ordination. Work is done in a predictable way and can be planned on a routine basis. Rules and procedures are applicable to continuously recurring activities, such as purchasing stock or paying wages.
- The design of the organisation structure also facilitates co-ordination. Depending on the organisation type, it may be better to use a product-type structure rather than a functional structure or a geographical divisional structure.
- The use of staff departments to assist with research, liaison, special knowledge, planning or problem solving can enhance co-ordination.
- The use of interdepartmental committees, task forces or project teams can achieve co-ordination for special problems or opportunities.
- The standardisation of skills and knowledge through training and on-the-job experience can greatly enhance work co-ordination. Workers will know what to do, how to do it and why it needs to be done in a certain way.
- An independent project leader is not attached to any of the departments or parties he/she co-ordinates and may therefore be perceived to be impartial by staff. This makes for more effective co-ordination.

2.3.7 Human resource capacity planning

Capacity planning is part of organising. It is defined as the filling of positions in the organisation structure by identifying workforce requirements, recruitment, selection, placement, promotion, orientation, appraisal, rewarding, training and development of the necessary employees. Capacity planning is an important responsibility of line and team managers. It is not the responsibility of the human resources or personnel function.

The personnel department assists the line or team management in the capacity-planning process as part of the staff service function. Managers have to fill the positions in their organisational units with the necessary qualified people.

To select employees requires a clear understanding of the nature and purpose of the positions to be filled. An objective analysis of the position requirements must be made and the jobs designed to meet organisational and individual career needs. In addition, positions must be evaluated and compared so that the incumbents are treated fairly and equitably. Among the factors to consider are the skills required, e.g. technical, interpersonal and conceptual skills. The need for these skills varies with the level of the position in the organisation hierarchy and the personality characteristics needed for the position. See chapter 9 for information on production capacity planning.

2.4 THE CONTROLLING FUNCTION

The process of organisational controlling is shown in figure 2.3. The managerial function measures actual performance, compares it against pre-set standards or goals, and, where there are deviations, these are communicated as areas for review and correction. The causes for the deviations are analysed to define the most appropriate corrective steps. The corrective actions are planned for the relevant causes and implemented to rectify negative performance deviations. Alternatively, one can build on positive deviations from goals or standards.

Management control is the function in the managerial process concerned with maintaining activities within the allowable limits, measured against expectations. The planning function provides the benchmarks, standards and performance objectives against which the controlling process is executed. Responses from the controlling process often identify the need for new plans or the revision of existing plans.

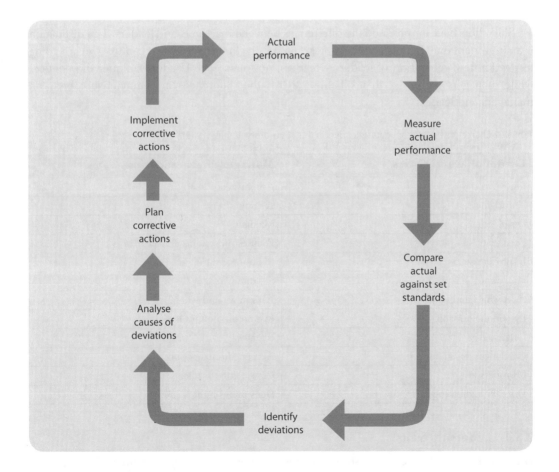

Figure 2.3 The controlling process

Control standards can be expressed in money values, time, quantity or quality, or a combination of these. Some standards are quantitative. Controlling is accomplished through various mechanisms, including monthly progress reports, budgets, sales reports, profit and loss statements, balance sheets, Pareto analysis, score cards (Kaplan & Norton, 1996), project reports, deviation analysis reports, quality control reports, information management systems, management and board meetings, and informal interpersonal conversations.

2.5 LEADERSHIP

Leadership is the ability to inspire people, a group, a team or an organisation to work together towards achieving worthwhile goals. It involves influencing, motivating and directing individuals or teams in such a way that they willingly pursue the goals and objectives of the group, team or organisation.

Over the decades, leadership theorists have attempted to define leadership in many ways and have ascribed leadership abilities to the personal characteristics of the leader – so-called *leadership trait theories*. Others have established theories on the situational factors that leaders encounter in daily work situations as determining aspects of leadership roles, behaviour and styles. These are the so-called *contingency* or *situational theories* of leadership.

Still others have emphasised the differences between managers and leaders. The distinction is an important one. Bennis (1989:7) notes: 'To survive in the twenty-first century, we are going to need a new generation of leaders – leaders, not managers. Leaders conquer the context, while managers surrender to it.' Table 2.2 lists the differences in characteristics between managers and leaders.

Table 2.2 Characteristics of managers vs leaders in the 21st century (Bennis, 1989)

Leader characteristics	Manager characteristics
Innovates	Administers
An original	A copy
Develops	Maintains
Focuses on people	Focuses on systems and structures
Inspires trust	Relies on control
Long-range perspective	Short-range view
Asks what? and why?	Asks how? and when?
Eye on the horizon	Eye on the bottom line
Originates	Imitates
Challenges the status quo	Accepts the status quo
Own person	Classic good soldier
Does the right thing	Does things right

2.5.1 Leadership traits

Dunn (1995) views the leadership personality in terms of the outcomes flowing from the leadership process. Leadership outcomes are twofold: they give direction and create followership. Certain essential characteristics promote these two important leadership outcomes and characterise the personality traits of most effective leaders.

- **Vision:** All successful leaders have this in common: the ability to create and pursue a vision or picture of the future. Leaders know where they want to go; they know what they want to achieve. Often they will have a clear picture of a product before knowing how it can be realised.
- **Focus:** Leaders have an ability to concentrate or focus on issues that are key to attaining desired objectives and, subsequently, to channel time and effort, with the necessary resources, in ways that support the achievement of these objectives.
- **Energy:** This refers to the capacity to act vigorously and decisively, and the drive to achieve. Leaders think in terms of accomplishment.
- **Creativity:** This is the ability to generate ideas, to find creative solutions to problems and generally to promote innovative approaches. Leaders are 'possibility thinkers', adept at finding new ways of making progress.
- **Empathy:** This is a unique kind of insight that allows the leader to see others in total perspective. Those lacking empathy often fail to read the behavioural clues around them – they look but do not see, listen but do not hear, socialise but don't identify, associate but don't belong. Empathy is of real importance in leaders, allowing them to relate to others effectively. Leaders with empathy know when to talk and when to listen; they are adept at making suggestions because they understand the other person.
- **Influence:** This is the ability to communicate and persuade.
- **Endurance:** This is the capacity to sustain effort over long periods and fix one's attention on essential issues without becoming distracted.
- **Stability:** This is the capacity for emotional and mental stability and resilience. This keeps the leader steady under pressure, objective under criticism and consistent in behaviour. It reflects a person's capacity for coping with stress, pressure, criticism and adversity.
- **Faith:** This is the capacity to believe. Leaders are not always in a position to 'play it safe'. At times, they have to be willing to face the unknown. They need the capacity to accept change and adapt to it, and often have to be prepared to take risky decisions when confronted with future uncertainty. This boldness to act comes from a special personal characteristic called faith. It leads to the attainment of goals that might not otherwise have been attempted, far less achieved.

Not all leaders possess all of the above traits to the same degree; the composition will vary from person to person. Experience indicates that they can be regarded as typical in the personality make-up of most leaders.

2.5.2 Situational or contingency theories

The behaviour pattern of the leader, as perceived by others, is defined as the leadership style. It is one of the most important elements of leadership. Leaders develop their style over time, based on experience, education and training. It is not how managers see themselves that is important, but rather how they come across to the people they are trying to influence. This is often difficult for leaders to understand. For example, if a leader's followers think that he/she is a hard, task-oriented leader, this is very valuable information. It makes little difference whether the leader thinks he/she is a relationship-oriented, democratic leader, because the followers will behave according to their perceptions. Leaders must learn how others perceive them.

Path-goal leadership theory

One of the popular approaches to leadership is the path-goal theory developed by Robert House (House & Mitchell, 1974). In essence, the path-goal theory attempts to explain the impact leader behaviour has on subordinate motivation, satisfaction and performance. The theory incorporates four major types, or styles, of leadership. Briefly, these are as follows:

- **Directive leadership:** This style can be described as authoritarian leadership. Subordinates know exactly what is expected of them and the leader gives specific directions. There is no participation by subordinates.
- **Supportive leadership:** The leader is approachable and shows a genuine concern for subordinates.
- **Participative leadership:** The leader asks for and uses suggestions from subordinates, but still makes the decisions.
- **Achievement-oriented leadership:** The leader sets challenging goals for subordinates and shows confidence that they will attain these goals.

The path-goal theory suggests that these styles can be and are used by the same leader in different situations. Two of the situational factors that have been identified are the personal characteristics of subordinates and the environmental pressures and demands facing subordinates.

Using one of the four styles, depending on the situational factors, the leader attempts to influence subordinates' perceptions and motivate them. In turn, this leads to their role clarity, goal expectancies, satisfaction and performance. The leader accomplishes this by:

- recognising subordinates' needs for outcomes over which the leader has some control;
- increasing personal rewards to subordinates for work goal attainment;
- making the path to those payoffs easier through coaching and direction;
- helping subordinates to clarify expectancies;
- reducing frustrating barriers; and
- increasing the opportunities for personal satisfaction linked to effective performance.

By doing the above, the leader attempts to make the path to subordinates' goals as smooth as possible. However, to accomplish path-goal facilitation, the leader must use the appropriate style, taking into consideration the situational variables present.

The essence of the theory is that it is the leader's job to assist his/her followers in attaining their goals; in addition, it is to provide the necessary direction and support to ensure that their goals are compatible with the objectives of the group or organisation. The term 'path-goal' is derived from the belief that effective leaders clarify the path to help their followers get from where they are to the achievement of their work goals. Furthermore, the journey along the path is made easier by reducing or removing obstacles.

Likert's four systems of leadership

Likert (1967) proposed four systems of organisational leadership. Table 2.3 on the following page summarises these four styles, called systems of leadership. The manager who operates under a system 1 approach is very authoritarian and actually tries to exploit group members. The system 2 manager is also authoritarian, but in a paternalistic way. The benevolent autocrat keeps strict control and never delegates authority to workgroup members, but 'pats them on the back' and 'does it in their best interests'. The system 3 manager uses a consultative style. This manager asks for and receives participative input from workgroup members, but maintains the right to make the final decision. The system 4 manager uses a democratic style. This manager gives some direction to workgroup members, but provides for total participation and decision by consensus and majority.

Table 2.3 Likert's systems of leadership theory

Leadership variable	System 1: Exploitative autocratic	System 2: Benevolent autocratic	System 3: Participative	System 4: Democratic
Confidence and trust in subordinates	Manager has no confidence or trust in subordinates	Manager has condescending confidence and trust, such as a master has in a servant	Manager has substantial but not complete confidence and trust; still wishes to keep control of decisions	Manager has complete confidence and trust in subordinates in all matters
Subordinates' feeling of freedom	Subordinates do not feel at all free to discuss things about the job with their superior	Subordinates do not feel very free to discuss things about the job with their superior	Subordinates feel rather free to discuss things about the job with their superior	Subordinates feel completely free to discuss things about the job with their superior
Superiors seeking involvement with subordinates	Manager seldom gets ideas and opinions of subordinates in solving job problems	Manager sometimes gets ideas and opinions of subordinates in solving job problems	Manager usually gets ideas and opinions and usually tries to make constructive use of them	Manager always asks subordinates for opinions and always tries to make constructive use of them

Hersey and Blanchard's situational theory

Hersey and Blanchard's (1982) situational theory is one that focuses on followers. Successful leadership is achieved by selecting the right leadership style, which Hersey and Blanchard argue depends on the level of the followers' readiness or maturity. The emphasis on followers in leadership effectiveness reflects the reality that it is they who accept or reject the leader. Regardless of what the leader does, effectiveness depends on the actions of his/her followers. This important dimension has been underemphasised in most leadership theories.

- **Readiness** is the *ability* and *willingness* of people to take responsibility for directing their own behaviour.
- **Ability** is the *knowledge, experience* and *skill* an individual or group brings to a particular task or activity.
- **Knowledge** is how much *understanding* a follower demonstrates to be able to accomplish a specific task.
- **Experience** is how often a follower has already performed a specific task.
- **Skill** is how well a follower performs a specific task.
- **Willingness** has to do with *confidence, commitment* and *motivation* to accomplish a specific task or activity.
- **Confidence** is a follower's demonstrated self-assurance in accomplishing a specific task.
- **Commitment** is a follower's demonstrated dedication to accomplishing a specific task.
- **Motivation** is a follower's demonstrated desire to accomplish a specific task.

Situational leadership uses the two leadership dimensions *task* and *relationship behaviours*. However, Hersey and Blanchard consider each dimension as either high or low and then combine them into four specific leadership styles: *telling, selling, participating* and *delegating*.

Style 1 (Sl): Telling *(high task – low relationship)*
The leader defines the roles and tells people what to do and how, when and where to do various tasks. It is directive behaviour.
Style 2 (S2): Selling *(high task – high relationship)*
The leader provides both direction and support.
Style 3: (S3): Participating *(low task – high relationship)*
The leader and the follower share in decision making. The main role of the leader is that of a facilitator and communicator.
Style 4 (S4): Delegating *(low task – low relationship)*
The leader provides little direction or support. The subordinate is ready to direct activities and monitor results.

The four readiness (R) stages of the follower are as follows.

R1: People are both unable and unwilling to take responsibility for doing something. They are neither competent nor confident.
R2: People are unable but willing to do the necessary tasks. They are motivated but currently lack the necessary skills.
R3: People are able but unwilling to do what the leader wants.
R4: People are both able and willing to do what is asked of them.

Figure 2.4 on the next page integrates the various components of the situational leadership model. As followers reach high levels of readiness, the leader responds by not only decreasing control over activities, but also decreasing relationship behaviour.

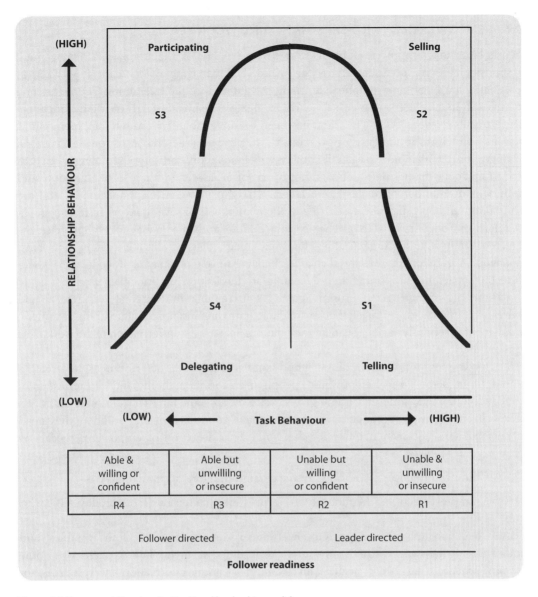

Figure 2.4 Hersey and Blanchard's situational leadership model

If workers are both unable and unwilling to take responsibility for doing something, they are at readiness stage 1. The supervisor or manager should provide them with clear and specific directions (applying the 'Telling' or S1 leadership style). When workers are at readiness stage 2, both high-task and high-relationship behaviour are needed. The supervisor or manager should apply the 'Selling' leadership style (S2). Supervisors of workers who are unwilling or unsecure (R3 readiness stage) should use a supportive, non-directive, participative style, called the 'Participating' leadership style (S3). Leaders of willing and able (R4) workers do not have much to do and could adopt the 'Delegating' leadership style.

2.6   EMPOWERMENT AND SELF-MANAGING WORK TEAMS

Workplaces that empower employees to assume many of the traditional management responsibilities are spreading across the world. Self-managing work teams (SMWTs), in particular, are beginning to appear in many private and public operations. People assume numerous management tasks and work in flexible teams instead of rigid functional departments (Fisher, 1993). The trend has been to move away from traditional, hierarchical pyramid structures to flatter, more responsive and leaner structures. Most of the structural changes aim at obtaining greater flexibility and innovation, and employ changes such as cross-functional teams, delegating authority to lower levels within the organisation and reducing the number of management layers. The gap between top management and the shop-floor employee has decreased due to the elimination of several layers of management (Clifford & Sohal, 1998:77–84). The existing middle and line managers within these de-layered manufacturing organisations find that their roles within the organisation have changed. A traditionally dictatorial style is changing to one of situational leadership, where directing, coaching, supporting and delegating are the leadership principles used.

These big changes often prompt managers to re-think fundamental hierarchical and bureaucratic practices and beliefs that are fundamental to organisational culture. Empowerment, implemented seriously, is not just an incremental change: it is a transformation of the workplace (Fisher, 1993).

Empowerment is a function of four important variables: authority, resources, information and accountability. To feel empowered, people need formal authority and all the resources (budget, equipment, time and training) necessary to do something with the new authority. They also need timely, accurate information to make good decisions. In addition, they need a personal sense of accountability for the work. This definition of empowerment can be expressed as follows:

Empowerment = f (authority, resources, information, accountability);

Empowerment = 0 (if authority, resources, information or accountability = 0)

In this equation, we can see that empowerment is a function of the four variables, and that if any of the variables rates zero, there is no empowerment. This explains why some empowerment initiatives are unsuccessful. Authority without information and resources, for example, is only permission. Telling team members that they should go ahead and make decisions or solve problems without providing access to accurate business information, skills training and budgets, and the time to accomplish the task is a prescription for failure. Not sharing accountability is paternalistic and condescending. It sends the message that the empowerment is not real. Only when all four elements are present will people feel responsible and act responsibly. SMWTs are the most advanced form of empowerment. Whether they are called employee involvement, a high-performance system, a partnership, semi-autonomous work teams or any of the myriad names referring to organisations based on SMWT concepts, many organisations have been using them aggressively (Fisher, 1993).

A self-managing work team is a group of employees who have day-to-day responsibility for managing themselves and the work they do, with a minimum of direct supervision. Members of self-managing teams typically handle job assignments, plan and schedule work, make production- and/or service-related decisions, and take action on problems.

Whereas traditional workgroups are typically organised into separate, specialised jobs with rather narrow responsibilities, these teams comprise members who are jointly responsible for whole-work processes, with each individual performing multiple tasks. Whereas a traditional organisation may be divided into groups of functional specialists, SMWTs are usually responsible for the delivery of an entire service or product, or responsible for a geographic or customer base. This is done to create, wherever possible, small, self-sustaining businesses that can be jointly managed by the organisational membership. For other essential differences between self-directed work teams and traditional organisations, see table 2.4.

Table 2.4 SMWTs vs traditional organisations

Self-managing work teams	Traditional organisations
Customer driven	Management driven
Multi-skilled workforce	Workforce of isolated specialists
Few job descriptions	Many job descriptions
Information shared widely	Information limited
Few levels of management	Many levels of management
Whole-business focus	Function/department focus
Shared goals	Segregated goals
Seemingly chaotic	Seemingly organised
Purpose-achievement emphasis	Problem-solving emphasis
High worker commitment	High management commitment
Continuous improvements	Incremental improvements
Self-controlled	Management controlled
Values- and principles-based	Policy/procedure-based

2.6.1 Seven competencies of SMWT leaders

The overall role of an SMWT leader is to be a boundary manager. This includes being an organisation designer, infrastructure builder and cross-organisation collaborator, and many other responsibilities. For example, boundary managers often play the role of translator, as they help team members comprehend the reality of the outside world. They also block certain disruptions from entering the team, shielding it from inappropriate distractions and unnecessary confusion. Nevertheless, as diverse as these team leader responsibilities appear, all of them seem to require a common set of personal competencies (Fisher, 1993).

The role of team leader is particularly important to understand, for it requires a variety of competencies that have not necessarily been associated with managers and supervisors in the past (Rayner, 1992). The role of team leader requires a different set of abilities from those of the traditional manager. The classic roles of planning, controlling and directing must be replaced by leading, empowering, facilitating, mentoring and coaching.

The SMWT leader role has seven core attributes:
1. leader;
2. living example;
3. coach;
4. business analyser;
5. barrier buster;

6. facilitator; and
7. customer advocate.

- The *leader* unleashes energy and enthusiasm by creating a vision that others find inspiring and motivating.
- The *living example* serves as a role model for others by 'walking the talk' and demonstrating the desired behaviours of team members and leaders.
- The *coach* teaches others and helps them develop to their potential, maintains an appropriate authority balance and ensures accountability in others.
- The *business analyser* understands the big picture and is able to translate changes in the business environment into opportunities for the organisation.
- The role of the *barrier buster* is to challenge the status quo and break down artificial barriers to the team's performance.
- The *facilitator* brings together the necessary tools, information and resources for the team to get the job done, and facilitates group efforts.
- The *customer advocate* develops and maintains close customer ties, articulates customer needs and keeps priorities in focus with the desires and expectations of the customers.

You will find more information on teamwork and self-managing teams in chapter 5 ('Managing People and Teams').

2.7 APPLICATION OF GENERAL MANAGEMENT PRINCIPLES WHEN MANAGING PROJECTS

Functional managers should apply the general management principle of planning, leading, organising and controlling in each of their functional areas (e.g. maintenance department, quality assurance department, marketing department, human resource department and production/operations department) during the operational stage of an organisation. General management principles are, however, also applied in the management of projects. In chapter 13 ('Principles of Project Management') the concepts of project management are explained in detail and will not be repeated here. The object of this section is to describe how the principles of general management are applicable to project management and how they are applied differently during the operational stage of an organisation.

As described in chapter 1, a facility's (e.g. a plant, mine or factory) life cycle can be described by two life stages, the project development stage and the operational stage. More generally, the first stage is when the concept of the project is developed into a fully functional deliverable, e.g. a product, system, plant or organisation. Stage two is where the fully functional deliverable is taken over by an operational function and operated in a stable fashion.

The best way to describe the differences between project and operational management is by way of an example. Let us consider a new mine or plant. In each case the project stage would first progress through its project life cycle. It will be developed through its concept, definition, implementation and, finally, close-out phases. After the close-out phase, the project, plant, mine or system is ready to move into the second organisational stage, which is when it becomes operational. At the end of the project close-out phase, the deliverable of the project, whether it is a new mine, new plant, newly developed product or system, is handed over to the operational function of the organisation. If an operational function does not exist, it is usually created to take over the project and operate it in a steady fashion. Let us consider the project management stage first.

2.7.1 The project management stage of the organisation life cycle

The general management process functions are executed during the project management stage of the organisation as well, although they may look different to the management functions normally observed in operational organisations. The functions of planning, decision making, organising, leading and controlling are executed through a variety of project management tools, procedures and systems. The management processes in the project management stage are focused on delivering the project outcome within pre-defined scope, budget, time frame and quality parameters.

The planning and decision-making function during a project is executed through different project development plans for each phase of the project. As the project progresses through its different phases, the plans become more detailed. Other tools used for the project-planning function are project milestones, schedules, phase budgets, baselines, work breakdown structures (WBSs) and work packages.

The project control function is executed through utilising the same tools as mentioned above. Large and complex projects make use of additional tools such as project cost control budgets and formal change control procedures to ensure that such projects are delivered in time, within the budget and the agreed deliverable specifications.

The project organisation function is usually organised through a matrix-type project structure as illustrated in figure 2.5. For small and less complex types of projects that are not spanning across multiple functions in an organisation, a structure such as the team-based structure or functional structure would suffice. Many small projects are done through the establishment of temporary teams or self-directed work teams with an assigned team leader. The composition of such teams would ideally include all the necessary skilled people with the requisite competencies to ensure the successful completion of the project deliverable. After the project is delivered at its close-out phase to become fully operational, these teams are usually disbanded and the people resume their original jobs in the functions from where they were seconded.

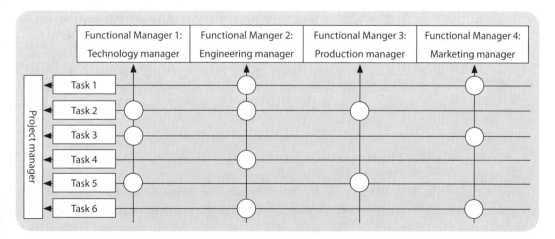

Figure 2.5 Example of a project matrix structure

The project leadership function is executed by the project manager. It is important to note that the project manager is responsible for all integrative management of the project without having direct line authority over those people who are actually doing the work. The matrix structure in figure 2.5 illustrates this point. The project manager in the project organisation has ultimate responsibility for delivering the project; however, he/she has to deliver it through people who are reporting to their respective functional managers on their respective functional inputs to the project. The functional managers' roles are to ensure that the work packages their people deliver to the project are within the correct functional standards, procedures and quality parameters (i.e. technical standards such as design specifications for high-pressure vessels). The people in the project organisation also have to report to the project manager on the specific work packages they have to deliver within the agreed work package schedules, budgets and agreed work package scopes, in order for the project to be completed as a whole on schedule, within budget and within the project's agreed scope. Thus each project worker has a dual reporting function within a matrix organisation structure. This type of structure makes it particularly challenging to manage due to its dual reporting requirements. The project manager overcomes this challenge through a variety of integrative management tools at his/her disposal. The integrative management functions of the project manager are explained in more detail in chapter 13.

2.7.2 The operational stage of the organisation life cycle

An organisation evolves into the operational stage of its life cycle when the project stage delivers its final fully operational deliverable, e.g. a new plant, product or system.

The management processes in the operational stage are now turned towards the effective and efficient delivery of the products and/or services from the new system, plant or mine.

The organisation structure used to manage this stage of the life cycle is usually designed differently to the organisation structure used to manage the project through its different development phases. The structure as it existed for the project stage is usually disbanded and another more permanent structure is put in place. Typical structures that could be used are the functional, divisional, customer or product organisation. Sometimes the operational structure is designed and implemented before the project is finally completed in order to ensure a smooth transition from project stage to the operational stage. For example, when a new manufacturing plant or a mine is put into operation the organisational structure that will manage the operations will already be in place by the time the project is delivered for commissioning. Most often a project will be delivered into an already existing organisational structure that will continue with the operation of the new project. An example is a new information management system that has been developed to improve the efficiency of an older version. The organisation will merely take over the new system and put it into operation while the old system is taken off-line.

Planning and decision-making systems used during the operations stage are focused on the effective day-to-day functions of the organisation. Thus the planning and control systems are designed to handle activities that are usually more repetitive in nature when compared to the planning and control systems used for more distinctive activities during the project stage. Typical planning systems one would employ during this stage are personnel recruitment and payroll, production planning and forecasting, marketing and sales planning, and budgets for different operational periods. The control function would normally rely on measurements against these plans and budgets and any deviations would be used to improve plans and budgets. Policies, procedures, guidelines and controls will be designed in sufficient detail to enable operational personnel to do every task effectively in order to ensure that the product or service is delivered at the desired cost and quality levels without circumventing safety and environmental precautions.

The leadership and management functions ensure that the operations make effective use of all the different inputs into the organisation, i.e. finance, human resources, technology, information and other assets. They must also ensure that these inputs are applied in such a way that the products or services generate the required profits and growth opportunities for the organisation.

Table 2.5 Summary: Differences between operational and project management processes

Management processes	Project management	Operational management
Planning & decision making	Project scope, project milestones, schedules, phase budgets, baselines, work breakdown structures (WBSs) and work packages	Operational budgets Production plans Marketing & sales plans Recruitment plans Technology development plan Functional strategic plan
Controlling	Control project scope, schedule, cost and quality parameters	Control product and service costs, quality, and functional budgets
Organising	Matrix structure or project teams	Functional structure: product, divisional or customer organisational structures
Leading and management	Integrative management, cross-functional, multi-disciplinary leadership Focus is effective delivery of pre-defined project deliverable within scope, cost, quality and time boundaries	Focus is profitable outcome of the organisation and future growth Functional management Strategic leadership of people, skills, resources, functions, products, technology and services

Self-assessment

2.1 List the functional areas or departments in your company (or a company that you are familiar with) and describe their interdependence.

2.2 Describe the role of management at a plant, factory, mine or any other company that you are familiar with.

2.3 Give a few reasons why management is considered indispensable in any organisation.

2.4 List the four most important elements in the general management process and describe them briefly.

2.5 True or false?

a) Planning involves goal setting for the organisation.

b) Organising involves the design of the organisation structure.

    c)   Controlling includes recording actual results.

    d)   Leading involves deciding how to use human resources.

2.6   Explain why planning is considered the starting point of the management process.

2.7   Describe the close relationship between planning and the other elements of the management process.

2.8   What factors must be considered when alternative plans are developed?

2.9   Define organising.

2.10 Define authority.

2.11 Differentiate between authority and responsibility.

2.12 Differentiate between line and staff authority. Give an example of each.

2.13 Briefly describe the following forms of organisational structure and give examples of such organograms:

- functional organisation
- product or project organisation
- customer organisation
- matrix organisation
- team-based structures.

2.14 Define co-ordination.

2.15 List the mechanisms in your company that promote co-ordination or integration among departments.

2.16 What are the dangers of having little or no co-ordination among departments?

2.17 Describe the concept 'span of management'.

2.18 Would you describe your company's structure (or span of management) as broad or narrow? Give reasons for your answer.

2.19 What are the dangers of having a very high structure?

2.20 List five factors that have influenced the design of your company's organisation structure.

2.21 Define management leadership.

2.22 Define delegation.

2.23 Define control.

2.24 Describe the control process in detail.

2.25 Describe the interrelation of the management functions of planning and controlling.

2.26 Choose any person in a company that you are familiar with and describe this person's task of general management by referring specifically to the following functions:

- planning
- organising and co-ordination
- leadership
- controlling.

2.27 Briefly describe the four fundamental management tasks of planning, organising, leading and controlling.

2.28 Describe the extent to which the above functions may influence one another and whether they should be executed in a specific order.

2.29 Explain the role of controlling in achieving objectives and production targets. Describe how effective control should be exercised.

2.30 List the objectives of your employer and explain how they should provide direction for all employees' activities.

2.31 Briefly describe the roles that managers play according to Mintzberg.

2.32 Bennis says that a new generation of leaders is required in order for organisations to survive in the 21st century. Briefly explain how he differentiates between leaders and managers.

2.33 Briefly describe the following leadership traits: vision, focus, energy, creativity, empathy, influence, endurance, stability and faith.

2.34 Describe the path-goal leadership theory.

2.35 Describe Likert's four systems of leadership.

2.36 Briefly describe Hersey and Blanchard's situational leadership model.

2.37 Describe how employees should be empowered.

2.38 Differentiate between self-managing work teams and traditional organisations.

References

Bennis, W.G. 1989. 'Managing the dream: Leadership in the 21st century.' *Journal of Organizational Change Management*, 2 (1), p. 7.

Clifford, G.P. & Sohal, A.S. 1998. 'Developing self-directed work teams.' *Management Decision, 36 (2), pp. 77–84.*

Dunn, J. 1995. The Effective Leader. Eastbourne: Kingsway.

Fisher, K. 1993. Leading Self-Directed Work Teams: A Guide to Developing New Team Leadership Skills. New York: McGraw-Hill.

Hersey, P. & Blanchard, K. 1982. *Management of Organizational Behavior: Utilizing Human Resources.* 4th edition. Englewood Cliffs: Prentice-Hall.

House, R.J. & Mitchell, T.R. 1974. 'Path-goal theory of leadership.' *Journal of Contemporary Business, Autumn, pp. 81–97.*

Kaplan, R.S. & Norton, D.P. 1996. The Balanced Scorecard. Boston: Harvard Business School Press.

Likert, R. 1967. *The Human Organization.* New York: McGraw-Hill.

Mintzberg, H. 1975. 'The manager's job: Folklore and fact.' *Harvard Business Review*, July/ August, pp. 299–323.

Rayner, S.R. 1992. *Recreating the Workplace.* Essex Junction: Oliver Wight.

Wageman, R. 1997. 'Critical success factors for creating superb self-managing teams.' *Organisational Dynamics, 26 (1), pp. 49–61.*

3   Human Resource Management

Marius Meyer

Study objectives

After studying this chapter, you should be able to:
- explain the human resource planning process
- describe the recruitment process
- compile a job advertisement
- describe the selection and interviewing process
- conduct a selection interview
- explain the phases of the training process
- implement a performance management system
- identify and prevent rating errors when evaluating employee performance

The aim of this chapter is to gain an understanding of the human resource management process.

3.1   INTRODUCTION

Human resource management is a relatively new field in the South African business environment. During the 1980s, many companies employed managers to head up personnel departments to deal with people issues in their organisations. Today, these managers are called human resource managers, which means that they have an important role in ensuring that companies utilise human resources effectively and efficiently in order to achieve their goals.

If we have qualified human resource managers, why do engineers need to be knowledgeable about human resource management? While human resource professionals facilitate the human resource management process in an organisation, the line managers are responsible for the implementation of human resource practices. The human resource function is a staff or support function to the rest of the organisation. Engineers and other line managers are responsible for the day-to-day operation of the company and they contribute directly to the bottom line of the business. The challenge for engineers is to create a work environment where human resource processes can be integrated into business processes, to ensure people contribute optimally to organisational effectiveness. Engineering managers must be able to implement human resource activities for their specific group of employees. Every engineer's human resource actions can have major consequences for his/her company.

Human resource management can be defined as all the processes, methods, systems and procedures employed to attract, acquire, develop and manage human resources in order to achieve the goals of an organisation. The human resource process starts with human resource planning and an analysis of the skills needed to achieve company objectives. Employees are recruited, selected, orientated and trained to occupy specific jobs. In addition, through the process of performance management, the performance of employees is directed towards delivering the required outputs.

3.2 HUMAN RESOURCE PLANNING

Many companies invest large amounts of financial and physical resources in planning their production processes with meticulous care and dedication. But this is often not the case with their human resources, even though people are a company's most valuable asset. Without proper human resource planning, the best-intended engineering designs might not be able to achieve the company's objectives. The competitive business environment necessitates well-developed human resource planning to contribute to business plans and strategies. *Human resource planning* can be defined as the process of systematically reviewing the current human resource profile and planning proactive actions to forecast future requirements in terms of human resources and skills needed, by taking cognisance of the external and internal environments. The internal environment refers to the factors within the organisation that have an impact on human resource planning, e.g. skills availability to meet future challenges. The external environment refers to all the external influences outside the company that have an impact on human resource planning, e.g. labour legislation such as the Employment Equity Act and the Skills Development Act, and changes in technology.

3.2.1 The human resource planning process

Like any other aspect of business planning, human resource planning consists of a management process to ensure that the right people are available, at the right time and at the right place (see figure 3.1). In most organisations, the human resource manager drives the human resource planning process. However, without the input of line managers, such as production managers, the process will be ineffective. For instance, the best person to decide how many engineers, technicians or labourers will be required in an operations department and the skills that they should have is the operations manager. It is therefore essential that engineering managers have the necessary knowledge to assist with the human resource planning process. Below are given the essential steps in the human resource planning process.

Step 1: Review the strategic business plan

This entails reviewing the strategic business plan according to the external business environment. For example, an overview of the strategic business plan may indicate the need to focus on increased globalisation, high-technology products and a more diverse customer profile in terms of race and gender.

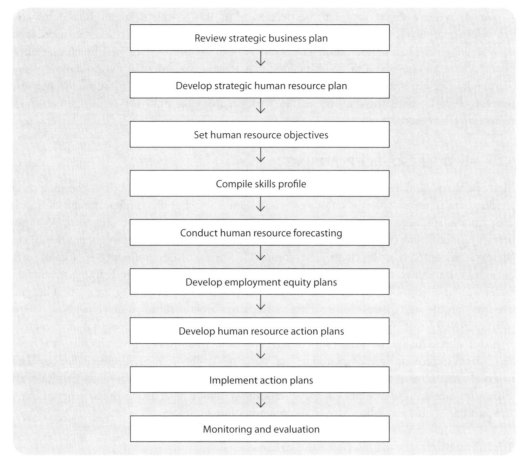

Figure 3.1 Human resource planning process

Step 2: Develop a strategic human resource plan

A strategic human resource plan must be developed to ensure that the company acquires and develops the necessary human resources to achieve its strategic plan. If we take the example in step 1 again, the strategic human resource plan will focus on acquiring staff who will be able to contribute to the strategic objectives, e.g. employing more blacks and females, more product designers and engineers, or consultants and managers who are experts in international business trends and developments.

Step 3: Set human resource objectives

According to Czanik (1996), human resource planning objectives vary according to aspects such as the type of environment in which an organisation operates, its strategic and operational plans, the current design of jobs and employee work behaviour. For example, in a highly technological company in which new technology is introduced into the production process, highly skilled employees will be needed. Employee training, development and retraining to use new technology could be important human resource objectives.

Step 4: Compile a skills profile

Once the human resource objectives have been set, a skills profile of the current workforce should be compiled. The skills profile will help the company to compare the skills level of the workforce with those personnel required in the future. This will give an indication of the number of internal employees who can be utilised to achieve human resource objectives, and the number that should be recruited from external sources. Unfortunately, skills shortages exacerbate the problem in South Africa. A recent study showed that South Africa produces 15 times fewer graduate engineers per million of the population than Japan, eight times fewer than the USA, and six times fewer than Australia (Boyd, 1997).

Step 5: Conduct human resource forecasting

There are two types of human resource forecasting: supply forecasting and demand forecasting. Supply forecasting answers the question: How many present employees will be available at some future time to meet the human resource staffing needs? Internal supply forecasting makes use of variables such as age distribution and geographical location. External supply forecasting is used to analyse the potential pool of employees in the external market.

Demand forecasting is done to determine the total number of employees required. By ascertaining the expected workload and skills requirements according to the number and complexity of engineering and other projects, one can forecast the demand for human resources. This also requires detailed job analysis and productivity measures.

Step 6: Develop employment equity plans

The next step is to develop employment equity plans in line with the Employment Equity Act. These plans should include the setting of numerical targets for future employment and development of employees from designated groups. These groups are blacks (African, Indian and Coloured), females and people with disabilities.

Step 7: Develop human resource action plans

Once human resource forecasting has been done and employment equity plans have been developed, human resource action plans can be developed to ensure that future business goals are achieved. For example, if the human resource forecast indicates that a company will need more engineers in the future and there is a shortage in the external market, it could introduce the following action plans.

- Identify staff such as work planners and technicians who could be educated and trained as engineers.
- Launch a bursary programme for engineering students.
- Develop more competitive remuneration packages to attract more engineers.

Step 8: Implement human resource action plans

Now the action plans must be implemented. Examples of implementation activities follow.

- Compile a memorandum to internal staff inviting applications from staff members to be trained as engineers.
- Screen internal applicants and select the most appropriate candidates.
- Nominate technicians for learnerships so that they can improve their skills levels.

- Enrol these employees at a university.
- Design a newspaper advertisement that offers bursaries to engineering students.
- Select the most suitable student candidates.
- Allocate financial resources to pay for student fees.

Step 9: Monitoring and evaluation

The purpose of monitoring and evaluation is to determine whether the planning process achieves its objectives. For example, when specific employment equity targets are set to increase the number of blacks and females in managerial positions, these must be monitored to determine whether the company is reaching its targets. The plan can be altered to make provision for new developments, or corrective action can be taken.

3.3   RECRUITMENT

Once the process of human resource planning is in place and the need for more staff has been identified, suitable candidates can be recruited. Smit (1996) defines recruitment as the process of attracting applicants who may comply with the criteria of a position to be filled in a company. The purpose of recruitment is therefore to invite, in a relatively cost-effective manner, a pool of job applicants who are potentially qualified to do a particular job.

Effective recruitment can assist the selection process by reducing the number of candidates that are unsuitable for a position. If the job specifications are clear and realistic, it is easier to match the skills and competencies of the applicants with the specifications. Moreover, recruitment can play a role in increasing individual and organisational performance.

If the right candidate is recruited and selected, this will result in a positive match between the company and the incumbent, and ensure a satisfied and productive employee and a productive organisation. Recruitment can be done from either inside or outside the organisation.

3.3.1 Internal recruitment

Internal recruitment occurs when internal applicants within an organisation are invited to apply for a position. Internal recruitment has several advantages for a company, including the opportunity for internal staff to be promoted and developed. In addition, internal applicants are familiar with the organisation's policies, procedures and systems, which will shorten the orientation and training time of employees in their new positions.

A disadvantage of internal recruitment is that internal applicants sometimes tend to lack the creativity and new ideas required to make the company grow and develop.

3.3.2 External recruitment

External recruitment refers to the recruitment of candidates outside an organisation by placing job advertisements in newspapers, on company websites or in other advertising media. Some companies make use of the services of employment agencies to do the recruitment for them. The major advantage of external recruitment is that new people bring new ideas and skills into the organisation that are needed to make the company grow and develop.

However, external recruitment has several drawbacks. For example, new employees may be resented if current employees also applied for the position and were unsuccessful. Most external appointees do not become immediately productive, because they have to settle, learn and adapt to the new organisation. In addition, external recruitment is more expensive than internal recruitment.

3.3.3 Advertising

Advertising is a popular internal and external recruitment method. In smaller companies, engineering managers draw up their own advertisements, but in large organisations human resource managers or recruitment specialists fulfil this function. However, engineers often need to provide input in terms of job requirements and specifications when technical jobs are advertised.

A good job advertisement should meet the following requirements.
- The title of the position should be very clear and specific, e.g. production manager, project manager, etc.
- The layout of the advertisement should be attractive and interesting.
- The type of organisation and location should be indicated (e.g. whether it is a mine, factory, plant or municipality). Some employers indicate the name of their companies to make it clear to applicants who the employer is.
- A brief summary of the job and the main functions or responsibilities should be provided, to make it clear to candidates what they will be expected to do.
- The particular job requirements or specifications must be indicated, to make it clear to readers whether they qualify for the position or not.
- There should not be any grammatical, spelling or technical errors.
- Contact information of the company that will receive the applications should be provided (telephone number, fax number, e-mail and postal address).
- The closing date for applications should be realistic and allow enough time for candidates to apply.
- Information should be provided on the organisational culture, e.g. whether innovation, teamwork and creativity are encouraged. Other cultural components could be that it is a highly technological environment, or that there are many deadlines to meet in stressful situations.

3.4   SELECTION

Once a suitable range of candidates has been recruited, the selection process can start. There are different types of selection methods, including tests and assessment centres, but the most popular method is the selection interview.

A selection interview is a discussion between an interviewer (or interviewers) and interviewee(s) about a particular position. The interviewer wants to determine whether the candidate is suitable to fill the position. The interviewee, however, is making use of the opportunity to show the interviewer that he/she is capable of occupying a particular position. Interviewees also see the interview as an opportunity to obtain more information about the position and the company, to find out whether their career and personal expectations will be met.

3.4.1 Types of interviews

Most employers either make use of individual interviews or panel interviews.

- **Individual interviews:** Most small companies prefer individual interviews to make selection decisions, mainly because only one person needs to stop work to conduct the interview. The major disadvantage of individual interviews is that a single interviewer can make a wrong decision that could be very counterproductive for the company. It is also easier for an individual to allow bias and personal feelings to influence the selection decision. This form of subjectivity could compromise the quality of the selection decision.

- **Panel interviews:** Large companies tend to make use of panel interviews. A panel interview entails several interviewers interviewing applicants. For example, for the position of mechanical engineering assistant, the interview panel could consist of a manager such as the maintenance manager, an engineer, the human resource manager and a trade union representative. Each panel member has a different role to fulfil.

  For instance, the human resource manager, who is qualified to conduct interviews and assess the people skills of candidates, will ask questions related to human relations, teamwork and career expectations. The human resource manager is also responsible for all the planning, administration and follow-up activities related to the interview. The maintenance manager and engineer (in whose department the successful candidate will work) will ask questions related to the technical and functional skills required for the job. The role of the trade union representative is twofold: firstly, to ascertain whether the candidate has the ability to foster sound labour relations in the company and, secondly, to assess other members of the panel in terms of the fairness of the questions being asked and the overall selection process. After the interview, the panel members will discuss the reasons for their ratings and decide on the most suitable candidate.

  The main advantage of panel interviews is that there is more than one opinion. Different interviewers may observe and assess different aspects of the candidate. They can also compare their interpretations and observations. Panel interviews are therefore more objective and fair. The quality of the selection decision is also higher if more than one person contributes to the decision-making process.

3.4.2 Types of questions

Questions form the basis of interviews and often determine their success. These days it is essential that the panel members prepare a standard set of questions so that all candidates are asked the same questions in the interest of fairness and equity. Questions must be phrased in an appropriate way to obtain the right information from the interviewees. There are two main types of questions: closed questions and open-ended questions.

Closed questions

These questions are formulated in such a way that they require one specific answer. Examples of closed questions include the following.

- How long did you work at Avtex Engineering?
- Have you completed any occupational safety courses?
- At which university did you study?
- Which computer packages are you familiar with?
- What salary do you require?

Open-ended questions

These questions force the applicant to give a complete and more comprehensive answer, rather than a simple 'yes' or 'no' response. Such questions are very useful because the interviewee provides a lot of information the interviewer(s) can use to assess the candidate. These questions usually start with the words 'what', 'where', 'when', 'who', 'why' and 'how'. Examples of open-ended questions include the following.

- Why did you apply for this position?
- What do you know about our company?
- What is the most difficult aspect of your current job and how do you handle it?
- If you had to conduct a safety investigation, how would you go about planning it?
- Can you give me an example of a conflict situation you have been involved in over the last year and how was it resolved?
- What do you think is the most important challenge in the field of engineering?

3.4.3 The interviewing process

Regardless of whether an individual or a panel interview is used, it is essential to ensure that it is conducted in a professional and effective manner. Effective interviewing encompasses the careful planning of activities before, during and after the interview.

Before the interview

- Ensure that effective shortlisting has been done and that only the best candidates who meet the selection criteria are invited to be interviewed. However, be flexible in making sure that there is a sufficient pool of candidates to select from. Sometimes the best candidate declines the job offer and there should be viable alternatives to select from.
- When scheduling interviews, allow enough time for each interview and time between interviews to review and summarise observations and assessments.
- Peruse the candidates' CVs or application forms the day before the interview and on the day the interviews will take place. Make notes and underline key aspects about the candidates beforehand.
- Make arrangements for a private venue that will be free from interruptions (no phone calls, personal visits, noise, etc.).
- Ensure the applicants are correctly informed of the date, time and venue of the interview.
- Inform security guards, reception staff and secretaries whom to expect and when, and ensure that the waiting area is comfortable and neat.

During the interview

- Welcome the job applicant, introduce yourself and the other panel members, and ask the candidate to sit down. Give a brief indication of what the candidate can expect in terms of the sequence and process of the interview.
- Start with easier and more comfortable questions. For instance, ask the candidate to describe his/her current duties, as this is familiar ground that will put the candidate at ease. More complex questions should be posed during the latter part of the interview, when the candidate is more relaxed.
- Be friendly and courteous towards candidates, and show interest in what they are saying by maintaining eye contact throughout the interview.
- Let the candidate do most of the talking by posing more open-ended questions to obtain the information you need.
- Be a good listener and make notes of the impressions you get of the candidate.
- Pose questions in a very clear and unambiguous way.
- The interviewer should refrain from interrupting the candidate.
- Do not ask questions that lead the candidate to an answer that he/she thinks the interviewer wants to hear. For example, an interviewer may ask the following leading question: 'At Mechpro, the working environment is very demanding in terms of workload and deadlines. Employees are often asked to work later than usual. Would you have a problem with that?' This question will not help the interviewer obtain new information, as most candidates will answer something like this: 'No problem, I don't mind working long hours.'
- Give the candidate an opportunity to ask questions.
- End the interview on a positive note, thanking the candidates for their time and interest in the company. Also give an indication of when the candidates can expect an answer from you.

After the interview

- Once the candidate has left, the interviewer(s) should assess the outcome of the interview. Based on the selection criteria and the performance of the candidates during the interviews, a preliminary decision can be made on who the most suitable candidates were.
- It is useful to do a reference check, subject to the candidates' permission during the interviews. The purpose of the reference check is to obtain information from the candidates' previous employers about their abilities and employment record. Many applicants provide the names of colleagues, friends or pastors. This information is very subjective and therefore of little relevance. Try to obtain reference information from managers, supervisors or lecturers, who are in a better position to comment on the applicants' skills, knowledge, abilities, potential and performance.
- After carefully considering the most suitable candidates, a final selection decision must be made. Remember that preference must be given to blacks, women and people with disabilities, according to the Employment Equity Act.

Keep detailed records of all documents, including written reasons why certain candidates were selected and others not. This information may be requested by an inspector from the Department of Labour, or used as evidence should grievances and lawsuits arise.

Inform the successful and unsuccessful candidates in writing of the outcome of the interview and the selection decision. Prepare a written offer of employment for the successful candidate or negotiate a salary. Also draw up a contract of employment.

3.5　TRAINING AND DEVELOPMENT

Once employees have been recruited and selected, they have to be trained to acquire the knowledge and skills they need to perform. There is continuous and rapid change in the South African business environment that requires continuous training and development of all employees and managers. Not only do South African managers experience change in the external environment in the form of globalisation, increased competition, technological advancement, and new and amended legislation, they also face dramatic internal changes such as new systems, policies, procedures, methods and techniques.

Line management and staff managers such as human resource managers, training managers and human resource development managers share responsibility for training and development. Large companies such as Sasol, Eskom and Goldfields all have technical training departments. Effective training requires a partnership between line and staff managers. They must work together on all phases of the training process and recognise their shared responsibility for planning and implementing training interventions.

3.5.1　An overview of training legislation

Education and training is regulated in South Africa by skills development legislation. The two main acts are the South African Qualifications Authority Act and the Skills Development Act. These two pieces of legislation have a major impact on the way in which training is conducted in the workplace.

South African Qualifications Authority Act

This act provides for the development and implementation of the National Qualifications Framework (NQF). The NQF is a key strategy for human resource development that facilitates access to quality learning opportunities for all learners. The NQF seeks to remove unnecessary constraints to entry and progression within the learning system, and creates processes and bodies for quality assurance in education and training. The South African Qualifications Authority (SAQA) is the statutory body responsible for overseeing the implementation of the NQF.

National Standards Bodies (NSBs) are responsible for regulating national standards in 12 fields, of which Engineering, Science and Technology is one. However, the standards are generated by Standards Generating Bodies (SGBs) in various sub-fields. In addition to the setting of national standards, a system of quality assurance is instituted to assess the quality of education and training provided. The bodies responsible for this function are called Education and Training Quality Assurers (ETQAs).

Skills Development Act

The Skills Development Act forms an integral part of the government's commitment to overall human resource development in the workplace. The act makes provision for a minimum level of training investment (1 per cent of payroll). The objectives of the Skills Development Act are to provide for a skills development strategy that is flexible, accessible, decentralised, demand-led and based on partnerships between the public and private sectors. It also attempts to improve the competency level of the workforce and to achieve higher levels of productivity and competitiveness. A key component of the act is the introduction of learnerships. A learnership includes both structured experience in the workplace and instructional learning with a training provider or inside a company. Learnerships are registered with the particular Sector Education and Training Authority (SETA) according to the economic sector that you find yourself in. For example, if you work in a production facility, you will be registered with the Manufacturing, Engineering and Related Services SETA (MERSETA) (see the website at the end of the chapter). The SETAs are responsible for promoting skills development in different industries.

Organisations should implement the following activities to comply with the Skills Development Act:

- appoint a skills development facilitator for the company;
- form a training committee representative of all stakeholders;
- pay the necessary skills development levies to the South African Revenue Services;
- align their skills development strategies with the overall goals of the company;
- conduct skills audits to determine skills gaps to be addressed;
- develop a workplace skills plan to address the gaps;
- liaise with the relevant SETA to identify and address sector skills needs in the industry at large;
- design and present outcomes-based learning programmes to promote skills development;
- develop quality management systems for skills development;
- submit their workplace skills plans and annual training reports to the SETA;
- align training programmes to unit standards;
- implement learner support mechanisms to assist learners with skills development;
- participate in learnerships to enhance the skills levels of employees and the unemployed; and
- keep abreast of all SETA, NQF and ETQA developments.

MERSETA is considered to be one of the best-performing SETAs in South Africa. It focuses in particular on learnerships and apprenticeships that even address the unemployed, in order to help them to increase their skills and employability. Some companies in the sector like Highveld Steel, Volkswagen and BMW have embraced these challenges by implementing learnerships and skills programmes to accelerate skills development (MERSETA, 2004).

3.5.2 Training process

Like many other professional management practices, training consists of a process of sequential phases that must be followed to ensure the effective acquisition and transfer of knowledge and skills. The phases of the training process are outlined in figure 3.2.

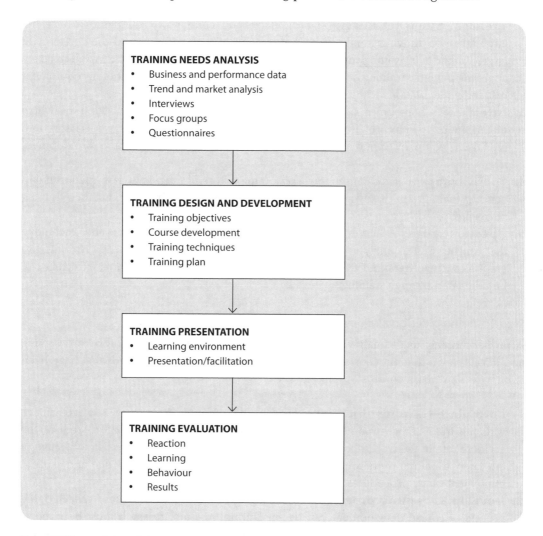

TRAINING NEEDS ANALYSIS
• Business and performance data
• Trend and market analysis
• Interviews
• Focus groups
• Questionnaires

TRAINING DESIGN AND DEVELOPMENT
• Training objectives
• Course development
• Training techniques
• Training plan

TRAINING PRESENTATION
• Learning environment
• Presentation/facilitation

TRAINING EVALUATION
• Reaction
• Learning
• Behaviour
• Results

Figure 3.2 Phases of the training process

Step 1: Training needs analysis

The first step in the training process is to conduct a training needs analysis or a skills audit. This requires an assessment of the current performance of employees, a department or section of the company, or the organisation as a whole, compared to the ideal level of performance. In other words, the analysis should assess whether there is a gap between current performance and ideal performance. Identifying this gap is done through training needs analysis. The following methods can be used to analyse training needs.

1. **Business and performance data:** By studying business reports or the results of employee and departmental performance, one can determine a particular training need. For example, a high accident rate in a workshop may be the result of a lack of knowledge about safety procedures and regulations. Other performance indicators are customer complaints and product defects.

2. **Trend and market analysis:** Progressive and competitive companies realise the importance of studying trends and developments in their industry and the economy at large. Rather than waiting for a problem to occur, a proactive approach is adopted by analysing potential training needs, e.g. new computer systems, high-technology equipment or pending legislation.

3. **Interviews:** Sometimes performance problems exist in an organisation of which management is unaware. By conducting interviews with employees, supervisors and managers, company input can be obtained on particular training and development needs.

4. **Focus groups:** Because of the time-consuming nature of interviews, it is often useful to conduct focus-group discussions with groups of employees to find out what their training needs are.

5. **Questionnaires:** A survey instrument such as a questionnaire is another technique that can be used to analyse training needs. These questionnaires can be distributed to employees, managers and even customers to find out whether problems exist that can be addressed through training.

Step 2: Training design and development

A proper training needs analysis leads to the next phase of the training process: designing and developing a training course or programme that will address the needs that have been identified. This phase consists of four steps.

1. **Formulate training objectives:** First ask the question 'What do we want to achieve with the training programme?' A manager, for example, may decide to reduce the accident rate in the workshop. A training objective can be formulated: 'The objective of the safety training programme is to equip participants with the knowledge and skills to implement safety procedures in the workshop.'

2. **Develop a training programme:** If training objectives are clear and specified, the process of course development can begin. During this step, research must be done to obtain the required information. To develop a training programme on workshop safety, various sources must be consulted. Some of these sources are the Occupational Health and Safety Act, the company's safety regulations, safety inspectors, safety representatives and workshop supervisors. Once the information has been gathered and analysed, decisions must be made regarding the content of the programme. Then the learning manual can be developed.

3. **Decide on training techniques:** The programme compilers must decide which training techniques will be the most appropriate to achieve the programme objectives. Most companies use a combination of the following techniques: lectures, discussions, case studies, role-plays, electronic learning, DVDs or on-the-job training. The nature and scope of the training content will determine which technique to use. For example, a lecture is

probably the most suitable technique to present information on the Occupational Health and Safety Act, using PowerPoint slides or flipcharts.

4. **Compile a training plan:** A comprehensive training plan should be compiled to ensure the presentation of effective training. A training plan should indicate:

- the duration of the programme and its various sections;
- the objectives and content of the programme;
- the resources required, e.g. flip charts, DVDs, manuals, stationery, computers, lunch, etc.
- a communication strategy to reach delegates and their supervisors, e.g. memorandums, e-mail, posters, pamphlets, Intranet, etc.;
- who will present the training (internal or external trainers);
- who will provide support services, such as administration and planning; and
- how the training will be transferred to the workplace.

Step 3: Presentation of training

After all the preparation work has been completed, it is time to present the training programme. It is important to create a learning environment in which participants feel comfortable and open to learning. The environment does not have to look like a traditional classroom or be a classroom. Participants should be encouraged to work in groups and be actively involved in the learning process.

Sometimes engineers are expected to participate in training programmes and transfer their technical knowledge to trainees in the form of a training presentation. It is therefore essential that engineers develop effective presentation skills. The following guidelines can be used when delivering a training course or any other form of presentation.

- Make sure that the content of your presentation is well prepared.
- Start with a forceful introduction in which you reveal some of the interesting facts or knowledge that you will share with the trainees. Remember that the purpose of your introduction is to stimulate interest so that your audience will be positively influenced to listen to the rest of your presentation. Give an outline of the objectives and main topics you will discuss.
- Display a positive attitude and enthusiasm towards your topic and the trainees. A sense of humour and a friendly disposition can contribute to the creation of a learner-friendly environment.
- Use training aids such as handouts, colourful pictures, flip charts or slides to make the content interesting and facilitate the learning process.
- Vary your tone of voice to prevent monotony.
- Provide practical examples when dealing with complicated technical information.
- Show an interest in the needs of the trainees by inviting them to ask questions and make contributions during the discussion.
- Maintain eye contact with all the trainees.
- Do not criticise a trainee for his/her remarks. Rather facilitate the discussion by providing additional information and steering the trainee in the right direction.
- Ask questions to establish whether the trainees understand the content of the presentation.

▪ Conclude the presentation by summarising the key points and asking the trainees whether the objectives have been achieved. Once again, emphasise the importance of the session and the follow-up action required.

Step 4: Training evaluation

The last phase in the training process is to evaluate training. The purpose of evaluation is to determine whether the objectives of training have been met. It also provides a measurement tool to determine whether the investment in training yielded the expected results. In other words, it indicates whether individual and organisational performance improved as a result of the training.

Kirkpatrick (1998) developed a widely used framework for the evaluation of training programmes. His four levels of evaluation are as follows.

Level 1: Reaction evaluation

According to Van Dyk *et al.* (1997), reaction evaluation indicates how well the trainees liked a particular programme by measuring the participants' feelings and perceptions. Reaction evaluation forms or questionnaires are used for this.

Level 2: Learning evaluation

This form of evaluation seeks to determine whether learning took place. Pre- and post-learning exercises can be used, including projects, case studies and tests, to determine whether trainees can apply the acquired knowledge and skills.

Level 3: Behaviour evaluation

Van Dyk *et al.* (1997) refer to behaviour evaluation as assessing a change of behaviour in the work situation. For example, supervisors may observe the behaviour of employees after the safety course to determine whether they are more safety conscious.

Level 4: Results evaluation

Here the results achieved in terms of organisational effectiveness need to be determined. Measures must be quantifiable, e.g. fewer accidents, reduced costs or improved productivity. These measures are then converted to a financial value in order to determine the rand value return on investment (ROI) of the intervention. For example, a large chemical company calculated a ROI of more than 200 per cent on technical training provided to staff.

3.6   PERFORMANCE MANAGEMENT

After employees receive orientation and training, their performance must be managed on a daily basis to ensure that they deliver the required outputs. Performance management is not merely about appraising the performance of employees – it is much broader than that. It also encompasses creating an environment in which employee performance can be optimised to achieve organisational objectives. Of all the human resource processes, performance management is the most important one for engineers. Not only do engineers manage their own performance, they also contribute to their team's performance. The success of an engineering project depends to a large extent on the quality of the performance management system.

What is performance management? Watt (1998) defines it as

... a continuous process, supported by an effective and streamlined system, which is in the first place developmentally-focused and implemented as a managerial accountability. It is manifested in both formal and informal interactions between manager and subordinate, and is the basis for a line management-human resource management interface, serving as the basis for other human resource management decisions.

From this definition it is evident that the aim of performance management is not to identify employees' shortcomings, but rather to create an environment in which improved performance can be developed. Moreover, employee performance is dependent on management support. Co-operation and sound relationships are of paramount importance in implementing a performance management system.

3.6.1 Performance planning

Performance planning entails developing a system in which performance management can be implemented. Managers and employees should be trained to acquire the skills to implement performance management. Some of these skills are target setting, communication, listening, feedback, and giving and receiving criticism.

The objectives of performance management should also be clarified. The benefits of performance management become very clear if one considers its objectives, which are:
- to create a supportive environment in which performance improvement becomes part of the organisation's culture;
- to establish clear standards and criteria for performance;
- to ensure the implementation of business plans and strategies;
- to provide feedback to employees on their performance;
- to promote the development of people by identifying training, counselling and coaching needs and opportunities;
- to improve career development by discussing plans for career advancement and promotion;
- to improve communication and relationships in the workplace by establishing mutual goals; and
- to establish a framework for linking pay to performance.

The performance plan should indicate the following.
1. **Purpose of the job:** For instance, the purpose of the job of mechanical engineering assistant is to provide an efficient and accurate mechanical drafting service.
2. **Critical performance areas (CPAs):** These are the main areas of responsibility of a job. There are usually between two and five CPAs. For example, the CPAs for a mechanical engineering assistant may be:
   - preparing drawings;
   - investigating sites;

- ☐ co-ordinating administration; and
- ☐ conducting safety inspections.

3. **Tasks:** Each CPA should be broken down into two or more specific tasks, e.g. plan site investigation; draw up investigation control sheets; write reports.

4. **Performance standards:** Performance standards are written statements describing how well a job should be performed to meet or exceed expectations. In other words, performance standards provide a benchmark against which work performance can be evaluated. Both the manager and the employee will know what the expectations are for the performance of essential functions and related tasks when performance standards are in place. This common understanding provides the basis for ongoing feedback and performance counselling between evaluations, and for the formal performance evaluation process.

A participative approach should be used in which employees work with their supervisors to develop the performance standards for their positions. Both the manager and the employee bring valuable information to the process, and the end result is more likely to be supported by everyone involved. Mutual agreement, understanding and recognition of the standards are very important. The following guidelines can be used to write performance standards.

- Describe the behaviours and results that would constitute the minimum acceptable performance for the task or function. Performance that satisfies those standards will receive the rating 'meets standards'.
- Standards should be written in clear language, describing the specific behaviours and actions required for work performance to meet, exceed or fail to meet expectations. Strive for a clear definition at all times. Standards that are clearly defined minimise doubt and ambiguity.
- Use specific terms to describe measurable or verifiable features of performance. These terms should describe specific performance criteria such as quality, quantity, cost and time.
- Be careful not to set the standards too high or too low. The standards must be realistic, reasonable and appropriate, no matter who is performing the job.
- Quantifiable measures may not apply to all functions. In such cases, describe in clear and specific terms the characteristics of performance in terms of quality that are verifiable and would meet expectations.
- Accomplishment of organisational objectives should be included where appropriate, such as cost control, improved efficiency, productivity, project completion or public service. Many organisations use the balanced scorecard approach to performance management with the purpose of integrating financial, process, customer and people dimensions of performance. Not only does this approach highlight quantifiable measures, it also ensures that performance targets filter through to all levels of the organisation.

The design of a performance appraisal form constitutes an important component of performance planning. Figure 3.3 is an example of a performance appraisal form. Note how specific the performance standards are.

Score	Performance level	Performance description
4	Outstanding Performance	Exceeds all standards in all instances
3	Superior Performance	Exceeds standards in some instances
2	Standard Performance	Meets standards
1	Underperformance	Fails to meet standards in some instances
0	Unacceptable Performance	Fails to meet standards in all instances

Figure 3.3 Example of performance appraisal form

3.6.2 Performance support

Having a performance plan with well-defined standards does not automatically mean that employees will perform optimally. The manager must provide the opportunity and support for employee performance. The following guidelines can be used to ensure that effective performance support takes place.

- Provide the necessary resources for employees to perform tasks.
- Be available for guidance, support and coaching.
- Have regular meetings and discussions to keep communication lines open.
- Give continuous feedback on performance.
- Provide opportunities for job enrichment and challenges.
- Give sufficient authority to employees to carry out tasks and make decisions.
- Make specific notes of successes and problems experienced.
- Encourage innovative thinking and creativity.
- Provide the necessary training and development opportunities.
- Reward performance.
- Take corrective action in the event of deviation from performance standards.

CPA	TASKS	STANDARDS	SCORES
Drawing preparation	Prepare drawings from sketches	Sketches are according to acceptable computer-aided drafting standards and completed within the specified time schedule.	0 1 2 3 4
	Prepare drawings for tender	Drawings meet tender specifications and client requirements. Additional notes are provided in an understandable and logical format.	0 1 2 3 4

Site investigations	Identify measurements to be taken	The correct measurement is selected for the particular investigation using the correct measurement equipement.	0 1 2 3 4
	Calculate quantities	Calculations are accurate.	0 1 2 3 4
	Record data in investigation reports	Data is recorded accurately.	0 1 2 3 4
Administration	Compile documents that accompany tenders	Accurate record keeping of drawings and made accessible to relevant parties.	0 1 2 3 4
	Supervise typists and supports staff	Tender documents are checked to ensure correct typing and editing thereof.	0 1 2 3 4
Safety inspections	Record inspections	Inspection records are updated once a month and meet the provisions of the Occupational Health and Safety Act.	0 1 2 3 4
	Consolidate inspection reports	Reports indicate the adherence to safety regulations and are disseminated to safety representatives and workshop engineer.	0 1 2 3 4
Final score			
SUPERVISOR'S COMMENT:			
EMPLOYEE'S COMMENT:			

Figure 3.4 Example of a performance plan

3.6.3 Performance evaluation and labour productivity

Performance evaluation entails the assessment of employee performance according to specific standards. Many companies allocate numerical values to specific criteria. This means that a performance evaluation form, like the one in figure 3.3, is used to assess performance against the set performance standards.

Performance evaluation has an important role to play in measuring performance and identifying action plans for performance improvement. It is essential that managers have information available to conduct the performance evaluation. In other words, the evaluator must have knowledge of employees' outputs.

According to Liebenberg (1996), managers should avoid the following rating errors when evaluating employee performance.

■ **Bias, prejudice and stereotyping:** This is where the appraiser makes a decision about an employee based on a personal belief or view about a personal characteristic (e.g. race, gender, age, religion, education, family background) and not on objective performance information.

- **Trait assessment:** This is where too much emphasis is placed on characteristics that have nothing to do with the job and are difficult to measure. Examples include characteristics such as flexibility, sincerity and friendliness.
- **Halo effect:** The appraiser makes the mistake of focusing on only one or two of the employee's good or weak attributes and allowing these attributes to affect all the ratings. For example, all the ratings could be positively influenced by the employee's strong creativity or all the ratings could be negatively influenced by the employee's impatience.
- **Leniency:** The appraiser regards people as good performers on principle and allocates high marks. This often happens when the appraisers want to 'buy' good relationships with employees.
- **Strictness:** The appraiser believes, irrespective of the facts, that all employee performance is below standard, and therefore allocates low scores. Some managers adopt this approach in the mistaken belief that it will encourage future performance.
- **Central tendency:** All employees are appraised as average performers. This often happens when the appraiser has insufficient performance information and allocates a safe mark that will not 'rock the boat'.
- **Errors of logic:** The appraiser makes this type of error by incorrectly grouping tasks that only appear to be related, and allocating the same mark for each task. For example, if the appraiser awards a high mark for good drawing, he/she may automatically award a high mark for administration. The two performance areas are not necessarily related.
- **Similarity:** The appraiser uses the same criteria he/she would use to assess him-/herself when appraising employees. For example, a technician with the same analytical skills as an appraising engineer may be awarded high performance scores.
- **Contrast:** Low scores are given to employees if they differ in attitude, attributes and approach from the appraiser.
- **Political considerations:** This error is made more consciously when the appraiser deliberately adjusts his/her appraisal for specific reasons. If the appraiser wants to promote an employee, high scores are given.

The problem with the performance problems above is that many companies do not have real output-driven performance management systems in place. The more you focus on real outputs (e.g. number of units, decrease in defects, etc.) the less subjective your performance ratings will be. The real challenge therefore is to measure your labour productivity. Very often remarkable improvements in performance and productivity can be achieved, e.g. if you consider the production process in a drawing office and attempt to measure productivity. The National Productivity Institute (NPI) has productivity standards called Drawing Office Time Standards for the purpose of performance monitoring. Afrox is one company that used this approach. Its drawing office has managed to determine the time required to complete a drawing far more accurately. A spreadsheet assigns time to planned or actual drawings on the basis of the codes determined for them. The results were excellent. The lead time for work to go through the drawing office has halved. Furthermore, weak and slow contractors are quickly identified and either helped to achieve the required standards or replaced. Rework has been reduced and the need for contractors has diminished so that 30 per cent fewer are now employed. Not only has the standards been effectively upgraded, but performance as measured against standards has improved as well. Productivity has more than doubled (NPI).

3.6.4 Performance discussion

A performance discussion should take place after the manager has completed the performance evaluation form. The manager and employee should meet and discuss the performance period under review.

During the appraisal discussion the employee should be encouraged to discuss those aspects of the job that have given him/her most satisfaction, and those areas he/she wishes to develop further or in which he/she feels he/she needs support. If appropriate, the appraiser should attempt to advise the employee on how his/her strengths and talents might be used within the wider business context, and what support could be given to develop his/her performance. The following guidelines can be used to ensure effective performance discussions.

1. State the purpose of the discussion

This helps to set the tone for the meeting by helping the employee to focus on the nature of the discussion that is to follow, e.g. 'The purpose of our meeting today is to discuss your performance over the last three months and to decide jointly on action plans for the next year.'

2. Get the employee talking

This can be achieved by asking open-ended questions. The main goal is to find out why employees perform the way they do, e.g. 'How do you feel about your performance over the past 12 months?' or 'Tell me about the successes and failures in your job over the past year.'

3. Focus on specific issues

Listen carefully to the employee's answers and focus on specific issues that require further discussion. These are usually your agenda items, e.g.: 'So you feel that you have been very successful in developing the technicians. What exactly have you done?' or 'You say that the last engineering project was a failure due to poor communication. What do you think contributed to the break-down of communication?'

4. Give feedback on the employee's performance

Having given the employee the opportunity to do most of the talking, this stage allows the appraiser to give his/her interpretation of the employee's performance. At this stage, the employee may be more receptive to ideas and input. During this stage, raise any items on the agenda that have not yet been discussed. It would also be the right time to show and discuss any written evaluations or rating forms that have been completed, e.g. 'I like the time and effort you put into training and developing the technicians, especially the coaching sessions' or 'We need to develop a plan to improve communication between you and your team.'

5. Set mutually agreed, specific goals

These goals will normally be based on the last two steps and must be mutually agreed upon. It is a good idea to record any goals that are agreed upon, e.g. 'To improve communication with the project team by having weekly meetings and e-mail updates.'

6. Close the discussion

End on a positive note by summarising the discussion. A date and time should be set for a follow-up meeting to assess whether the goals set will be achieved and are appropriate. If not, this is the opportunity to revise goals with the employee.

3.6.5 Follow-up after the discussion

If the manager and the employee have committed themselves to discussing or reviewing performance in a few months' time, they must ensure that the follow-up discussion materialises. They should jointly decide on the support the appraiser can provide in achieving the performance goals, including resources required, or training and development opportunities.

3.7 CONCLUSION

In this chapter you were introduced to the main human resource management processes. The process starts with proper human resource planning to determine the human resources needed to do the company's work. We also explored the recruitment and selection of human resources with specific reference to interviews. We then described the training process to ensure that employees have the necessary knowledge and skills to perform their jobs effectively. The chapter concluded with a discussion on the importance and methodology of performance management in ensuring that employees perform according to predetermined performance standards.

Human resource management forms an integral part of engineering management in a complex and changing environment. By managing human resources effectively, engineers will be able to utilise, develop and manage an organisation's human resources in order to contribute to organisational goals and objectives.

Self-assessment

3.1 Explain why engineers need knowledge of human resource management.

3.2 Identify the steps in the human resource planning process and indicate how these steps can be applied in practice.

3.3 Distinguish between the advantages and disadvantages of internal and external recruitment.

3.4 Draw up a job advertisement for the position of an artisan you want to employ. Make sure you meet the requirements of a good advertisement.

3.5 Develop a checklist of questions you can ask the candidates for the position of artisan when you are conducting the interviews. Indicate the guidelines you would follow to ensure that the interview is successful.

3.6 Describe the two main pieces of training legislation in South Africa.

3.7 Explain how engineers can implement the phases of the training process.

3.8 Suppose you have to present a training course for technicians on the design requirements of the ISO 9000 quality system. Indicate how you would apply the guidelines for delivering the presentation.

3.9 Develop a performance appraisal form you can use to evaluate the performance of an artisan.

3.10 Identify the rating errors you should guard against when evaluating employee performance.

3.11 Explain how you would conduct a performance discussion.

3.12 Provide examples of labour productivity measurement at your organisation.

3.13 Visit any one of the four SETA websites provided at the end of this chapter. Identify skills priorities in these industries.

References

Boyd, L. 1997. 'Engineers: A critical shortage looming in South Africa.' *South African Mechanical Engineer*, 47, June, pp. 11–13.

Cascio, W.F. 1998. *Managing Human Resources: Productivity, Quality of Work Life, Profits*. Boston: McGraw-Hill.

Carrell, M.R., Elbert, N.F., Hatfield, R.D., Grobler, P.A., Marx, M. & Van der Schyf, S. 1998. *Human Resource Management in South Africa*. Upper Saddle River: Prentice-Hall.

Coughlin, P.M. 1993. 'Manpower planning and recruitment.' In Lock, D. (ed.). *Handbook of Engineering Management*. 2nd edition. Oxford: Butterworth-Heinemann.

Czanik, M. 1996. 'Human resource planning.' In Pieters, M. (ed.). *Textbook for Human Resource Practitioners*. Pretoria: Kagiso.

Erasmus, B.J. & Van Dyk, P.S. 1999. *Training Management in South Africa*. 2nd edition. Halfway House: International Thomson.

Human, L., Bluen, S. & Davies, R. 1999. *Baking a New Cake: How to Succeed at Employment Equity*. Randburg: Knowledge Resources.

Kirkpatrick, D. 1998. 'Evaluating training programs: The four levels.' Paper presented at the ASTD International Conference, San Francisco, 31 May.

Liebenberg, J.J. 1996. 'Performance evaluation.' In Pieters, M. (ed.). *Textbook for Human Resource Practitioners*. Pretoria: Kagiso.

MERSETA. 2004. 'Learnerships take centre stage.' *Merseta News*. Johannesburg: MERSETA.

Meyer, M. & Kirsten, M. 2005. *Introduction to Human Resource Management*. Cape Town: New Africa.

National Productivity Institute (NPI). 1997. *Productivity Improvement in the Drawing Office at Afrox Limited*. Consulting Update Case Study 4. Pretoria: NPI.

Meyer, T. 1996. *Creating Competitiveness through Competencies: Currency for the 21st Century*. Randburg: Knowledge Resources.

Smit, E. 1996. 'Recruitment.' In Pieters, M. (ed.). *Textbook for Human Resource Practitioners*. Pretoria: Kagiso.

South Africa. 1998. *Employment Equity Act*. Pretoria: Government Printer.

Spangenberg, H. 1994. *Understanding and Implementing Performance Management.* Cape Town: Juta.

Swanepoel, B.J. (ed.). 1998. *South African Human Resource Management: Theory and Practice.* Cape Town: Juta.

Thomas, A. & Robertshaw, D. 1999. *Achieving Employment Equity: An Implementation Guide.* Randburg: Knowledge Resources.

Van Dyk, P.S., Nel, P.S., Loedolff, P. van Z. & Haasbroek, G.D. 1997. *Training Management: A Multidisciplinary Approach to Human Resource Development in Southern Africa.* 2nd edition. Halfway House: International Thomson.

Watt, D. 1998. 'Competency-based strategic performance management.' Paper presented at the SA Forum of ASTD Conference, Roodepoort, 7 August.

Websites

American Society of Training and Development: www.astd.org.

Association of Personnel Service Organisations: www.apso.co.za.

ASTD Global Network South Africa: www.astd.co.za.

Chartered Institute of Personnel and Development: www.cipd.co.uk.

Chemical Industries Education and Training Authority: www.chieta.org.za.

Construction Education and Training Authority: www.ceta.org.za.

Department of Labour: www.labour.gov.za.

Equity Skills Web: www.equityskillsweb.com.

Employment Equity Act, Skills Development Act, SAQA Act: www.gov.za.

Engineering News: www.engineeringnews.co.za.

HR Future: www.hrfuture.net.

HR Highway: www.hrhighway.co.za.

Institute of People Management: www.ipm.co.za.

International Society for Performance Improvement: www.ispi.org.

Investors in People: www.iipuk.co.uk.

Manufacturing, Engineering and Related Services SETA: www.merseta.org.za.

Mining Qualifications Authority: www.mqa.org.za.

Mining Weekly: www.miningweekly.co.za.

National Productivity Institute: www.npi.co.za.

Skills Portal: www.skillsportal.co.za.

Society for Human Resource Management: www.shrm.org.

South African Board of Personnel Practice: www.sabpp.co.za.

South African Qualifications Authority: www.saqa.org.za.

Work Information: www.workinfo.com.

4 The Impact of Employment Relations and Labour Legislation on an Organisation

Marinda Conradie

Study objectives

After studying this chapter, you should be able to:

- list the advantages of practising sound employment relations in an organisation
- identify all the relevant stakeholders in the employment relationship and explain what their respective roles are
- compile a service contract according to the requirements set out in the BCEA
- distinguish between an employee and an independent contractor
- deal with employee grievances
- conduct a fair disciplinary hearing
- distinguish among dismissals, unfair labour practices and automatically unfair dismissals
- formulate procedures to follow in cases of misconduct, incapacity and operational requirements (e.g. retrenchments)
- assist in the preparation of CCMA cases
- comply with legal requirements to ensure protected strikes and lock-outs
- consult all the relevant labour legislation

The aim of this chapter is to gain an understanding of the employment relations legislation and practises.

4.1 INTRODUCTION

Technical people are often employed as supervisors and managers. They are also self-employed as consultants, contractors and business owners. It is therefore important that such people acquire a basic understanding and knowledge of employment relations in the workplace.

Employment relations deal with everything that emanates from or impacts on the employment relationship (Nel *et al.*, 2005:6). The employment relationship is that relationship that exists between the employer and the employee in the working environment. It is crucial that this relationship be effective and successful. If a sound employment relationship is maintained in an organisation, it will result in motivated employees, which will in turn have a positive influence on productivity. However, ineffective and poor employment relations in an organisation can have a detrimental effect on the long-term employer–employee relationship. Furthermore, it can be extremely costly for an organisation because if people are ignorant of the legal requirements and rights of employees and employers, they may end up at the Labour Court or the employees may embark on a strike.

The question may then be raised: What is the difference between industrial relations, labour relations and employment relations? 'Industrial relations' was the term initially used, since the study focus was primarily on employment relationships in the manufacturing sector. Today the term 'labour relations' (many organisations prefer the term 'employment relations') is used, since more and more employees fall under the ambit of labour legislation, both in the public and private sectors. These employees are not only employed in the manufacturing sector, but in a multitude of other sectors (Finnemore, 1997:1).

It is evident, therefore, that it is important to practise sound employment relations, but what are the advantages of sound employment relations?

4.1.1 Advantages of sound employment relations

The advantages of sound employment relations are:
- better performance;
- low labour turnover;
- improvement in quality;
- increase in productivity;
- reduction in throughput time;
- rapid innovation due to more decision making by employees;
- fewer dismissals; and
- reduced absenteeism.

The most important role-players active in employment relations are briefly discussed in the next section.

4.2 ROLE-PLAYERS IN EMPLOYMENT RELATIONS

The three parties most involved in employment relations are: the employer (management), the employee (trade unions) and the state. This is called a tripartite employment relationship (see figure 4.1).

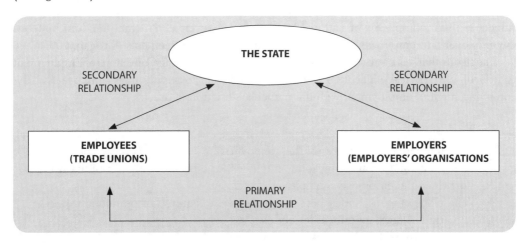

Figure 4.1 The tripartite relationship

The role of each party can be explained as follows (Slabbert & Swanepoel, 2002:10).

▨ **The employee:** An employment relationship (primary relationship) between the employer and employee is created as soon as one person employs another. This employment relationship is an economic relationship because one party (the employee) is prepared to do work in exchange for compensation. Trade unions' main functions are to negotiate on behalf of their members for better wages and working conditions and to defend and advance workers' rights.

▨ **The employer:** The employer must ensure that his/her organisation's goals are met through effective planning, organising, leading and controlling. The employer's emphasis will be on output and the delivery of quality service and products. Furthermore, the employer will clarify the roles and responsibilities of the employees, including job descriptions for all the employees. He/she must also ensure that all the relevant labour legislative requirements are met.

▨ **The state:** The employment relationship between employers and employees is complex due to the fact that they have shared as well as conflicting interests. The state thus enters as a third party in order to provide a legal framework within which relations between the two primary parties can be managed. The framework will give guidance in respect of aspects such as collective bargaining, the settling of disputes, the handling of grievances, how to ensure fair dismissal and strikes. The state is therefore the watchdog in employment relations, as it ensures that employees and employers conduct their relationship according to the law.

Various factors have a direct influence on the nature of the employment relationship, such as the contract of employment. The contract of employment is an agreement that exists between the employer and the employee that details the respective duties and obligations of each of the parties in the employment relationship.

4.3 THE CONTRACT OF EMPLOYMENT AND THE BASIC CONDITIONS OF EMPLOYMENT ACT

The contract of employment can be utilised as an instrument to manage the employment relationship. The parties are free to regulate their respective rights and duties in the contract, subject to the requirements of national legislation. However, contractual freedom between the employer and employee has been limited by collective agreements (Grogan, 2005:67).

The statute that sets basic conditions for all employees is the Basic Conditions of Employment Act 75 of 1997 (BCEA). The BCEA requires employers to supply their employees at the commencement of employment with the particulars listed in figure 4.2.

▨ The full name and address of the employer
▨ The name and occupation of the employee, or a brief description of the work for which the employee is employed
▨ The place of work, and, where the employee is required or permitted to work at various places, an indication of this
▨ The date on which the employment began
▨ The employee's ordinary hours of work and days of work

- The employee's wage or the rate and method of calculating wages
- The rate of pay for overtime work
- Any other cash payments that the employee is entitled to
- Any payment in kind that the employee is entitled to and the value of the payment in kind
- How frequently remuneration will be paid
- Any deductions to be made from the employee's remuneration
- The leave to which the employee is entitled
- The period of notice required to terminate employment, or if employment is for a specified period, the date when employment is to terminate
- A description of any council or sectoral determination that covers the employer's business
- Any period of employment with a previous employer that counts towards the employee's period of employment
- A list of documents that form part of the contract of employment, indicating a place that is reasonably accessible to the employee where a copy of each can be obtained

Figure 4.2 Information that employers should supply to employees in terms of the BCEA

The BCEA stipulates all the minimum basic conditions of employment. A brief summary of some of the detail of the Act is listed in table 4.1.

Table 4.1 Summary of the BCEA

EMPLOYMENT ISSUE	CONDITION OF EMPLOYMENT
Application of the Act	The Act applies to all employees and employers, except to members of the South African National Defence Force, the National Intelligence Agency, the South African Secret Service and unpaid charity workers.
Ordinary hours of work	The Act does not apply to senior managerial employees, employees engaged as sales staff who travel extensively and employees who work less than 24 hours a month. Ordinary weekly working hours may not exceed 45 hours. Employees working 5 days a week will work a maximum of 9 hours per day and employees working 6 days a week will work a maximum of 8 hours per day.
Overtime	An employee may not be permitted to work more than 12 hours per day, normal as well as overtime hours included. An employee may not work more than 10 hours overtime per week. This may be increased by a collective agreement to 15 hours per week for 2 months in a 12-month period. Overtime must be compensated by paying the employee 1,5 times the normal wage or granting the employee time off equivalent to the value of the overtime pay.
Meal intervals	An employer must give the employee who works continuously for more than 5 hours a meal interval of one continuous hour. An agreement in writing may reduce this to not less than 30 minutes.

Daily and weekly rest periods	An employee must have a daily rest period of 12 consecutive hours and a weekly rest period of 36 consecutive hours, which, unless otherwise agreed, must include Sunday.
Pay for work on Sundays	An employer must pay an employee who works on a Sunday double the employee's wage for each hour worked, unless the employee ordinarily works on a Sunday, in which case the employer must pay the employee 1,5 times the employee's wage for each hour worked.
Night work	This is worked performed after 18:00 and before 06:00 the next day. An employer may require an employee to perform night work, if so agreed, and if the employee is compensated by the payment of an allowance and if transportation is available to the employee at the commencement and end of the shift.
Public holidays	An employee who does not work on a public holiday must be paid the wage that the employee would ordinarily have received for work on that day. An employee who works on a public holiday must receive double the ordinary wage for that day.
Annual leave	An employee is entitled to 21 consecutive days' annual leave per leave cycle.
Sick leave	During the leave cycle of 36 months an employee is entitled to an amount of paid sick leave equal to the number of days the employee would normally work during a period of 6 weeks. If you work 5 days a week, you will have 30 days' paid sick leave over 3 years. The employer is not required to pay for sick leave where the employee has been absent for more than 2 consecutive days or on more than two occasions during an 8-week period, unless the employee produces a medical certificate.
Maternity leave	An employee is entitled to at least 4 consecutive months of maternity leave. The employer is not required to pay an employee on maternity leave, but the employee may claim benefits from the Unemployment Insurance Fund.
Family responsibility leave	An employee is entitled to 3 days' paid family responsibility leave. This only applies to employees who work 4 or more days in a week.
Termination of employment	The notice period that must be given by an employee is as follows: ▪ one week if the employee has been employed for 6 months or less; ▪ two weeks if the employee has been employed for more than 6 months but less than one year; and ▪ four weeks if the employee has been employed for one year or more.
Severance pay	When an employee is dismissed due to the employer's operational requirements, the employer must pay the employee at least one weeks' salary for each completed year of continuous service.
Prohibition of employment of children	It is a criminal offence to employ a child under 15 years of age.

Variation of basic conditions of employment	A collective agreement concluded by a bargaining council may replace or exclude any basic condition of employment except the following: ■ the duty to arrange working time with regard to the health and safety, and family responsibility of employees; and ■ the prohibition of child and forced labour. Such an agreement may not: ■ reduce the protection afforded to employees who perform night work; ■ reduce annual leave to less than 2 weeks; ■ reduce entitlement of maternity leave; or ■ reduce entitlement to sick leave to less than the extent permitted.

Various industries make use of independent contractors. It is important to correctly distinguish between employees and independent contractors. Some employers, in an attempt to avoid constraints placed on them by labour legislation, have drawn up contracts with their employees that portray them as 'independent contractors' although they are indeed employees.

4.3.1 Employee versus independent contractor

It is imperative for engineers to distinguish between an employee and an independent contractor, because independent contractors are not covered by the Labour Relations Act and the Basic Conditions of Employment Act. Thus, they will not be entitled to aspects such as annual leave, sick leave or to claim unfair dismissal.

In terms of labour legislation, an employee is defined as:

'(a) Any person, excluding an independent contractor, who works for another person or for the State and who receives, or is entitled to receive any remuneration; and

(b) Any other person who in any manner assists in carrying on or conducting the business of an employer.'

According to the Labour Relations Act and the Basic Conditions of Employment Act, a person is presumed to be an 'employee' if certain circumstances exist. These are rebuttable presumptions and are basically a codification of the common law. A person who works for or renders services to another is presumed, until the contrary is proved, to be an employee, regardless of the form of the employment contract, if any one or more of the following factors are present:

■ the manner in which the person works is subject to the control or direction of another person;

■ the person's hours of work are subject to the control or direction of another person;

■ in the case of a person who works for an organisation, the person is a part of that organisation;

■ the person has worked for that other person for an average of at least 40 hours per month over the last three months;

- the person is economically dependent on the other person, for whom that person works or renders services;
- the person is provided with tools of trade or work equipment by the other person; or
- the person only works for or renders services to one person.

The employee vs contractor issue is further discussed below.

'Labour Relations Act fuzzy on contracts'

Ivan Israelstam www.iol.co.za

An important aspect that affects employment is whether people are hired as employees or as independent contractors. The difference between them is extremely important because it is only the employees that are covered by our labour legislation and not the independent contractor.

The Labour Relations Act defines an employee as anyone who does work for a company/organisation except for independent contractors and members of the armed forces. However, the LRA doesn't define an independent contractor. Case law indicates that the courts see an independent contractor as a person who is not obliged to render his services personally and is allowed to delegate this obligation, does not have to keep fixed hours and is not paid a regular wage or salary, is not subject to the alleged employer's disciplinary code, is not entitled to benefits, such as membership of a pension fund or medical aid scheme and is not subject to a degree of control by the employer.

An example of an independent contractor is the external plumber who is called out when a tap is leaking. The plumber is under no legal obligation to repair the tap on the date and at the time the client says he must. The plumber comes when he decides it is convenient for him, does the work without the supervision of the client, and is paid for fixing the tap, not for being available.

There are workers employed under contracts labelled as independent contractor agreements who are indeed employees and they are not sure what their legal rights are.

Employers have, up to now, been able to avoid a multitude of labour laws by hiring workers as independent contractors. For example, they can in this way avoid providing benefits such as annual leave, they do not have to comply with affirmative action requirements and they do not need to follow disciplinary procedures.

We have found that a large proportion of workers hired on the basis of independent contracts are, in fact, employees according to labour law. Where such employees are treated as independent contractors, this is illegal and could result in serious ramifications for the employer.

If the CCMA, at the arbitration hearing for unfair dismissal, reclassifies the workers as employees, the employer is likely to be found to be in the wrong because employees have the right to fair labour practice!

The above illustrates that an organisation's classification of employees and independent contractors should be precisely defined.

Because employees and employers have shared ideas as well as conflicting interests, it is important that organisations have procedures in place such as grievance and disciplinary procedures in order to resolve potential conflict. This is discussed in the next section.

4.4   THE GRIEVANCE PROCEDURE

Most employees, in a 40–45 year career, spend 80 000 hours in the workplace and spend over half their waking hours at work. Most of an employed person's waking hours are spent with co-workers rather than with anybody else. Little wonder then that conflict is common in the work environment (www.jetson.net.au).

The grievance procedure is a form of upward communication from employees to employers, by which employees may raise their working grievances to higher levels. A written grievance procedure that is known to employees can be helpful in creating a positive working environment. A grievance procedure informs employees on how to manage and resolve grievances and it makes management aware that problems may exist in an organisation.

It is very important that employees' grievances be dealt with as soon as possible, because unresolved grievances can lead to unproductiveness and even industrial action (e.g. strikes). In figure 4.3 on the next page you will find the steps that may be followed in a grievance procedure. Please note that grievance procedures will vary from organisation to organisation depending on the nature and size of the organisation.

4.5   DISCIPLINE MANAGEMENT

The other vital side of employment relations is discipline. An effective grievance procedure that employees can turn to and an effective disciplinary procedure employers may use will ensure that a fair relationship exists between the parties.

It is the prerogative of the employer to ensure that discipline is maintained in an organisation. These disciplinary procedures must be fair, just and equitable for the entire workforce, irrespective of race, sex and religion or job category. The main purpose of the disciplinary procedure is to correct unacceptable behaviour and adopt a progressive approach in the workplace. This also creates certainty and consistency in the application of discipline. A disciplinary enquiry must be both procedurally and substantively fair and a code of good practice is contained in the Labour Relations Act, which provides guidelines when dealing with discipline. The act also distinguishes between misconduct, incapacity (poor performance/ill health) and operational requirement (i.e. retrenchments) dismissals. Knowing these guidelines and your rights plays an important role in preparing for a disciplinary enquiry.

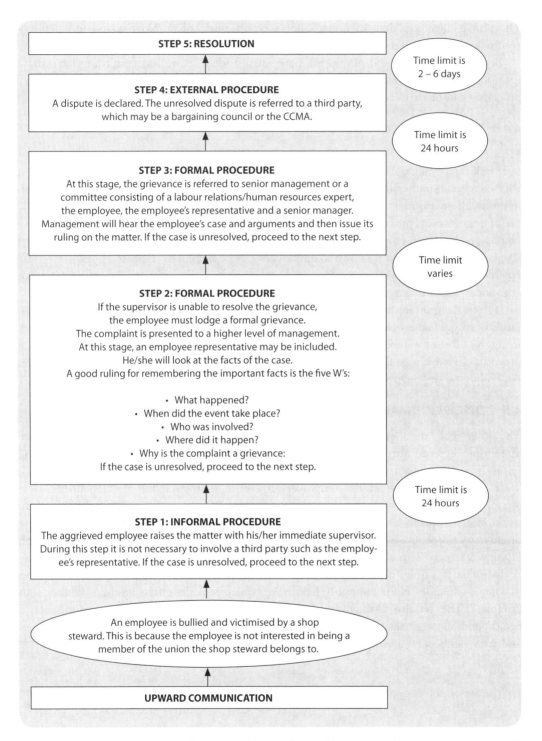

STEP 5: RESOLUTION

Time limit is
2 – 6 days

STEP 4: EXTERNAL PROCEDURE
A dispute is declared. The unresolved dispute is referred to a third party, which may be a bargaining council or the CCMA.

Time limit is
24 hours

STEP 3: FORMAL PROCEDURE
At this stage, the grievance is referred to senior management or a committee consisting of a labour relations/human resources expert, the employee, the employee's representative and a senior manager. Management will hear the employee's case and arguments and then issue its ruling on the matter. If the case is unresolved, proceed to the next step.

Time limit
varies

STEP 2: FORMAL PROCEDURE
If the supervisor is unable to resolve the grievance, the employee must lodge a formal grievance.
The complaint is presented to a higher level of management.
At this stage, an employee representative may be inicluded.
He/she will look at the facts of the case.
A good ruling for remembering the important facts is the five W's:

- What happened?
- When did the event take place?
- Who was involved?
- Where did it happen?
- Why is the complaint a grievance:
If the case is unresolved, proceed to the next step.

Time limit is
24 hours

STEP 1: INFORMAL PROCEDURE
The aggrieved employee raises the matter with his/her immediate supervisor. During this step it is not necessary to involve a third party such as the employee's representative. If the case is unresolved, proceed to the next step.

An employee is bullied and victimised by a shop steward. This is because the employee is not interested in being a member of the union the shop steward belongs to.

UPWARD COMMUNICATION

Figure 4.3 Steps in the grievance procedure (adapted from Grobler *et al.* (2002:524) and Swanepoel *et al.* (2003:676))

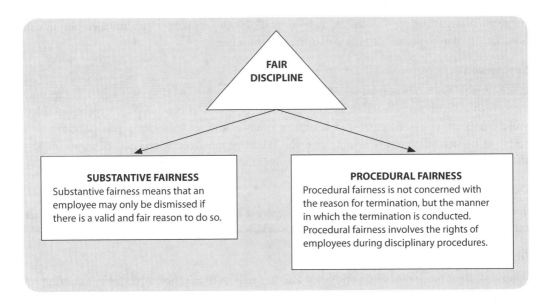

Figure 4.4 Fair discipline

Disciplinary action may take the following forms, depending on the seriousness of the matter. According to Landis and Grossett (2005:211) and Nel et al. (2005:253), various acceptable disciplinary sanctions apply with the progressive application of discipline. These are discussed below.

4.5.1 The disciplinary sanctions

Verbal warnings

Verbal warnings are given in cases of less serious transgressions and shall constitute an informal disciplinary procedure. Written documentation of a verbal warning is placed in the employee's personal file. Example: late-coming.

Written warnings

These warnings are reserved for more serious transgressions or where the verbal warning did not have the desired effect. The manager will investigate the matter by interviewing the employee and, if relevant, witnesses. Example: absenteeism.

Final written warnings

This is in cases of very serious transgressions or where the written warning had no effect. All warnings must be directed to the employee in the presence of his/her representative, if any. He/she must be informed if a similar offence is committed within a certain period of time. He/she may then be dismissed. Example: alcohol or drug abuse.

Transfer

This sanction will only be applied if management is convinced that it will solve the particular problem and have a rehabilitating effect. For example, if a person does not fit into the department/organisation (i.e. he/she is incompatible) the employer can transfer him/her to another division/department rather than dismiss him/her.

Suspension

It is advisable not to suspend a person without pay due to the fact that the Labour Court may interpret this action as an unfair labour practice. Suspension on full pay is recommended when an employee is suspected of a serious offence and an investigation is pending as a result of his/her offence. For example, if the financial manager is accused of fraud; he/she should be suspended with pay pending the outcome of the hearing.

Demotion

This is an alternative to the sanction of dismissal; an employee is demoted because of his/her incompetence or inability to meet required working standards.

Dismissal

After investigations have been conducted and it is believed that the employee has committed a serious offence, he/she will be furnished with a 'Notice to Attend a Disciplinary Hearing'. Action to dismiss an employee should only be taken in cases where the trust relationship has been irreparably damaged and/or after the employer has adopted all reasonable measures to rectify the employee's behaviour, but with no success. Example: fraud.

Table 4.2 An example of a disciplinary code (adapted from Grosset, 1999:252)

OFFENCES	FIRST OFFENCE	SECOND OFFENCE	THIRD OFFENCE	FOURTH OFFENCE
MINOR OFFENCES Arriving late for work Loafing	Verbal warning	First written warning	Final written warning	Dismissal
SERIOUS OFFENCES Absenteeism Leaving company premises without permission Sleeping on duty Abusing sick leave benefits	Written warning	Final written warning	Dismissal	
VERY SERIOUS OFFENCES Refusal to carry out lawful instructions Being under the influence of alcohol/drugs in the workplace Forging time cards	Final written warning	Dismissal		
DISMISSABLE OFFENCES Theft Fraud Violation of safety regulations Assault at the workplace Deliberately damaging company property	Dismissal			

4.5.2 Employee rights during a disciplinary hearing

The employee is entitled to the following rights regarding a disciplinary inquiry:

- the right to be informed of the nature of the complaint;
- the right to a timeous hearing;
- the right to have an interpreter;
- the right to be represented;
- the right to plead to the complaint;
- the right to state his/her case (*audi alterem partem* rule);
- the right to ask questions to witnesses of the initiator;
- the right to call his/her own witnesses and to challenge the evidence brought against him/her;
- the right to address the tribunal prior to a decision on guilt;
- the right to be informed by the chairperson of the decisions reached;
- the right to address the tribunal on mitigating circumstances prior to a decision on punitive action;
- the right to consideration of his/her record of service and other factors that may influence the punitive action taken against him/her;
- the right to be informed by the chairperson of the penalty to be imposed; and
- the right to be informed by the chairperson of his/her right to appeal.

'Case law provides detail for LRA'

Ivan lsraelstam

The Star, 2 August 2004

Various labour legislation deals largely with wide-ranging legal principles but is frequently deficient when it comes to detail. Included in the Labour Relations Act is schedule 8 that gives guidelines as to whether a dismissal is fair. However, these guidelines cannot indicate what is fair in each individual situation. The employer's own procedures, policies and terms and conditions of employment significantly contribute to determining what discipline is fair. To explain this – it is very difficult to discipline an employee if he/she was unaware of a certain rule applicable in the organisation. In a certain case the employee was dismissed for reasons including the fact that he swore. However, the arbitrator found that no rule existed in the organisation that prohibits swearing. This led the arbitrator to find the dismissal unfair and to award the employee 12 months' remuneration as compensation.

The Labour Relations Act states clearly that employers should inform their employees of their rules. Thus, the onus of proving that the rule exists and that the employee knew of the rule falls on the employer. Due to the fact that labour legislation is constantly changing, employers' rules and policies should encapsulate the latest labour laws so that, when management applies the policies, they are in line with the law.

4.5.3 Steps in the disciplinary procedure

NOTIFICATION
Notification should be in writing. Adequate notice is imperative. Notification letter should include the time, venue, date, allegations and rights of the employee.

OPENING LETTER OF PROCEEDINGS
Confirm that everyone is present and introduce them.

LEVEL OF DISCIPLINARY ACTION
Confirm that the employee is aware of the level of the disciplinary action initiated and that the formal disciplinary action is being taken i.t.o. the organisation's disciplinary policy.

CONFIRM THE ALLEGATIONS
Confirm that the employee is aware of the allegations. Read the allegations.

PROCESS OF THE PROCEEDINGS
1. Each party gives a brief overview of the event.
2. Initiator calls witnesses to substantiate the facts.
3. Employee responds to the allegations and summarises his/her version of the incidents.
4. Empoyee and/or representative may cross-examine the initiator and witnesses.
5. Employee/representative puts his/her side of the story.
6. Witnesses may be called to substantiate facts.
7. Initiator may cross-examine these witnesses.

CHAIRPERSON TO REACH A VERDICT
Chairperson will state and motivate the decision as to whether the employee is guilty or not guilty with regard to the allegations.

MITIGATION
Elements to consider are: situational factors, personal problems, employee's record and personal circumstances.

AGGRAVATION
Initiator may address the chairperson on the subject of a suitable penalty and why.

DETERMINE THE PENALTY AND COMMUNICATE THE DECISION
The chairperson will state and motivate the decision with regard to a suitable penalty.

RIGHT TO APPEAL
Chairperson must advise the employee of his/her right to appeal.

Figure 4.5 Downward communication

It is of paramount importance that management see to it that the internal procedures, in this case the grievance and disciplinary procedures, are employee-friendly and that employees have access to these internal procedures. The Labour Relations Act prohibits any form of unfair treatment in the workplace. Let's look at what activities are classified as dismissal, unfair labour practices and automatically unfair dismissal.

4.5.4 Dismissal, unfair labour practices and automatically unfair dismissals

According to the Labour Relations Act, an employee has the right not to be unfairly dismissed. Sections 186 and 187, given below, stipulate a list of reasons that are not acceptable reasons for dismissals.

186. Meaning of dismissal and unfair labour practice

1) "Dismissal" means that –

   a) an employer has terminated a contract of employment with or without notice;

   b) an employee reasonably expected the employer to renew a fixed term contract of employment on the same or similar terms but the employer offered to renew it on less favourable terms, or did not renew it;

   c) an employer refused to allow an employee to resume work after she took maternity leave in terms of any law, collective agreement or her contract of employment

   d) an employer who dismissed a number of employees for the same or similar reasons has offered to re-employ one or more of them but has refused to re-employ another; or

   e) an employee terminated a contract of employment with or without notice because the employer made continued employment intolerable for the employee;

   f) an employee terminated a contract of employment with or without notice because the new employer, after a transfer in terms of section 197 or section 197A, provided the employee with conditions or circumstances at work that are substantially less favourable to the employee than those provided by the old employer.

2) "Unfair labour practice" means any unfair act or omission that arises between an employer and an employee involving –

   a) unfair conduct by the employer relating to the promotion, demotion, probation (excluding disputes about dismissals for a reason relating to probation) or training of an employee or relating to the provision of benefits to an employee;

   b) the unfair suspension of an employee or any other unfair disciplinary action short of dismissal in respect of an employee;

   c) a failure or refusal by an employer to reinstate or re-employ a former employee in terms of any agreement; and

   d) an occupational detriment, other than dismissal, in contravention of the Protected Disclosures Act, 2000 (Act No. 26 of 2000), on account of the employee having made a protected disclosure defined in that Act.'

187. Automatically unfair dismissals

1) A dismissal is automatically unfair if the employer, in dismissing the employee, acts contrary to section 5 or if the reason for the dismissal is-

   a) that the employee participated in or supported, or indicated an intention to participate in or support, a strike or protest action that complies with the provisions of Chapter IV;

   b) that the employee refused, or indicated an intention to refuse, to do any work normally done by an employee who at the time was taking part in a strike that complies with the provisions of Chapter IV or was locked out, unless that work is necessary to prevent an actual danger to life, personal safety or health;

   c) to compel the employee to accept a demand in respect of any matter of mutual interest between the employer and employee;

   d) that the employee took action, or indicated an intention to take action against the employer by –

      i) exercising any right conferred by this Act; or

      ii) participating in any proceedings in terms of this Act;

   e) the employee's pregnancy, intended pregnancy, or any reason related to her pregnancy;

   f) that the employer unfairly discriminated against an employee, directly or indirectly, on any arbitrary ground, including, but not limited to race, gender, sex, ethnic or social origin, colour, sexual orientation, age, disability, religion, conscience, belief, political opinion, culture, language, marital status or family responsibility;

   g) a transfer, or a reason related to a transfer, contemplated in section 197 or 197A; or

   h) a contravention of the Protected Disclosures Act, 2000, by the employer, on account of an employee having made a protected disclosure defined in that Act.

2) Despite subsection (1)(f) –

   a) a dismissal may be fair if the reason for dismissal is based on an inherent requirement of the particular job;

   b) a dismissal based on age is fair if the employee has reached the normal or agreed retirement age for persons employed in that capacity.

According to the Labour Relations Act, there are only three grounds that justify dismissal. These are the following:

- **misconduct:** if an employee intentionally or negligently breaks a rule at the workplace, e.g. steals company goods;
- **incapacity:** if an employee cannot perform duties properly due to illness, ill health or inability; and
- **operational requirements:** if a company has to dismiss employees for reasons that are related to purely business needs (i.e. retrenchments) and not because of some failing on the part of the employee.

Each case for dismissal is unique. Therefore, as already explained, prior to the dismissal of an employee, it is vital that substantive and procedural fairness be considered. The fairness elements are listed in detail in tables 4.3, 4.4, 4.5 and 4.6, as adapted from Meyer and Kirsten (2005:194–8).

Table 4.3 Fairness elements for misconduct

SUBSTANTIVE FAIRNESS	PROCEDURAL FAIRNESS
Consider the following guidelines: - Did the employee contravene a workplace rule? - Was the rule reasonable and valid? - Was the employee aware or could the employee reasonably be expected to have been aware of the rule? - Has the employer consistently applied the rule? - Is dismissal an appropriate sanction for the contravention of the rule? *Consider also the following:* - A reason is fair if a continued employment relationship is impossible. Dismissal must be considered as a last resort. - There is no fixed rule about the number of warnings that must precede a dismissal. - The seriousness of the misconduct. - The nature of the misconduct. - Always be consistent and treat similar cases alike. - The employee's previous disciplinary record. - The nature of the post and of the workplace. - The employee's personal circumstances.	- Notify the employee of the charge against him/her and what his/her rights are at the hearing. - Hold the disciplinary hearing as soon as possible after the incident, but the employee must be granted sufficient time to prepare his/her case. - The employee must state his/her case, may call witnesses and has the right to cross-examine witnesses called by management. - The employee also has the right to an interpreter and to be represented or assisted by a co-employee. - The chairperson must be unbiased. - The employee must be informed of the final decisions, the sanction(s) and the reasons for them in writing. - The employee must be informed of his/her right of appeal - The chairperson at the appeal must be a person other than the person who chaired the disciplinary hearing. - The employee must be informed of the outcome of the appeal.

Table 4.4 Fairness elements for incapacity – poor work performance

SUBSTANTIVE FAIRNESS	PROCEDURAL FAIRNESS
Consider the following guidelines: Whether or not – ▪ the employee failed to meet a performance standard; ▪ the employee was aware, or could reasonably be expected to have been aware, of the required performance standard; ▪ the employee was given a fair opportunity to meet the required performance standard; and ▪ dismissal was an appropriate sanction for not meeting the required performance standard.	▪ Dismissal during the probation period should be preceded by an opportunity for the employee to state his/her case and to be assisted by a trade union representative or a fellow employee. ▪ Employers making a decision about the fairness of a dismissal during the probationary period ought to accept reasons for dismissal that may be less compelling than would be the case in dismissals effected after the completion of the probationary period. ▪ After probation, the employee should not be dismissed unless the employer has given the employee appropriate evaluation, instruction, training, guidance or counselling and a reasonable amount of time to improve. ▪ Investigate to establish the reasons for the unsatisfactory performance and consideration of other ways short of dismissal to remedy the matter.

Table 4.5 Fairness elements for incapacity – ill-health

SUBSTANTIVE FAIRNESS	PROCEDURAL FAIRNESS

▪ Determine whether or not the employee is capable of performing the work.
▪ If the employee is not capable, determine:
  ❒ the extent to which the employee is able to perform the work;
  ❒ the extent to which the employee's work circumstances might be adapted to accommodate him/her;
  ❒ whether it is possible to adapt the employee's duties; and
  ❒ the availability of any suitable alternative work.
	▪ Conduct an investigation to establish the extent of the incapacity or injury.
	▪ Investigate alternatives to dismissal if the employee is unable to perform or will be absent for an unreasonably long period.
	▪ The employee has the right to be heard and represented.
	▪ Consider the degree of incapacity and the cause of incapacity.
	▪ If the employee is temporarily unable to work, consider the following relevant factors:
	  ❒ the nature of the job;
	  ❒ the period of absence;
	  ❒ the seriousness of the illness or injury; and
	  ❒ the possibility of securing a temporary replacement.
	▪ If the incapacity is permanent, ascertain the possibility of securing alternative employment, or adapting the duties or work circumstances of the employee to accommodate his/her disability.

Table 4.6 Fairness elements for retrenchments due to operational requirements

SUBSTANTIVE FAIRNESS	PROCEDURAL FAIRNESS
The employer must have a valid economic reason for the retrenchment.	*Consultation* must take place as soon as the employer contemplates retrenchment. An employer has to *disclose the following information:* ▦ reasons for retrenchment; ▦ alternatives considered and the reason if they were rejected; ▦ the number of employees likely to be affected and their job categories; ▦ the proposed method of selection; ▦ the time when the dismissals are likely to take effect; ▦ the proposed severance pay; ▦ assistance that the employer will be offering (e.g. time off for employees to attend interviews, early release should a new job be found, issuing letters of reference and psychological counselling); ▦ the possibility of future re-employment; ▦ the number of employees employed by the employer; and ▦ the number of employees that the employer has retrenched in the preceding 12 months. The parties in a consultation process must attempt to reach consensus. The employer must give the consulting party an opportunity to make presentations, to consider them and if he/she does not agree, to state his/her reasons for disagreement. *Alternatives to retrenchment* that should be considered by employers, such as: ▦ stopping the recruitment of new employees; ▦ requesting older employees to take early retirement; ▦ transferring employees; ▦ retraining employees; ▦ spreading termination of employment over a period of time; ▦ working short time; ▦ stopping the using of outside contractors; and ▦ work sharing. When it is necessary to retrench only part of the workforce, then criteria for selecting those to be retrenched must be fair and objective. Examples of *selection criteria* are: LIFO (last in first out), conduct, early retirement and asking for volunteers. The following *payments* must be made to employees who are retrenched. ▦ Severance pay: Employees should be paid at least one week's remuneration for each completed and continued (pro rata) year of service (e.g. at least 4 weeks of pay should be received for $3\frac{1}{2}$ years of service). ▦ Outstanding leave must be paid out. ▦ Notice pay: If employed for six months or less – one week's notice pay is required. If the employee has been employed for more than six months but not for more than one year, two weeks' notice pay is required, and if the employee has been employed for one year or more, four weeks' notice pay is required. ▦ Other: Depending on the employment contract, the following may be relevant: pro rata payment of bonus, pension and provident fund. *The undertaking to re-employ* Various unions demand that retrenched employees be given priority should vacancies arise in the future.

As mentioned earlier, the Labour Relations Act prohibits unfair treatment. Therefore, guidelines have been included in the 'Code of Good Practice: Dismissal' (Schedule 8 of the LRA) to ensure that dismissals occur in a substantive and procedurally fair manner. If employees feel that they have been unfairly treated or that the penalty of dismissal imposed on them was unfair, they may use external mechanisms such as the Commission for Conciliation Mediation and Arbitration to resolve the matter.

4.6 THE COMMISSION FOR CONCILIATION MEDIATION AND ARBITRATION (CCMA), LABOUR COURT AND LABOUR APPEAL COURT

The CCMA is an independent dispute resolution body. It was established by the Labour Relations Act 66 of 1995. The main objective of the CCMA is to promote a sound working relationship, to prevent labour disputes from arising and to settle disputes that do arise by means of conciliation, arbitration, conciliation arbitration (con-arb) and pre-dismissal arbitration. A brief discussion of these mechanisms will follow (See also table 4.7 on the next page).

4.6.1 Pre-dismissal arbitration

Pre-dismissal arbitration is arbitration that takes the place of a disciplinary hearing. The CCMA may at the request of an employer and with the employee's consent send a commissioner to conduct a pre-dismissal arbitration to establish whether the employee is guilty of alleged misconduct or incapacity and, if so, the sanction to be applied (Grogan, 2005:443).

4.6.2 Conciliation

The LRA requires a commissioner to resolve a dispute through conciliation. The commissioner at the conciliation proceedings will make use of mediation and fact-finding exercises and will then make recommendations to the parties. The meeting is conducted in an informal way. The commissioner has no power to make a decision, but will assist the parties in attempting to reach an agreement. The commissioner must attempt to resolve the dispute within 30 days of the date that the CCMA received the referral. The commissioner must issue a certificate to state as to whether the dispute has been resolved or not (Van Niekerk, 2002:96). It is important that parties exhaust internal procedures (grievance and disciplinary procedures) prior to referring a dispute to the CCMA (www.ccma.org.za).

4.6.3 Conciliation arbitration (con-arb)

Con-arb means that as soon as the CCMA has issued a certificate for the non-resolution of a dispute during the conciliation, arbitration must commence (Du Plessis et al., 2002:329).

4.6.4 Arbitration

If a dispute cannot be resolved by conciliation, the parties to the dispute may, within 90 days from the date of the certificate, refer the case to arbitration or the Labour Court. The act specifies which dispute goes to which forum. If a dispute remains unresolved after the conciliation process and a certificate has been issued, then any party may refer the dispute to arbitration. At an arbitration hearing a commissioner of the CCMA gives the parties the opportunity to state their cases. The commissioner then makes a decision on the issue in dispute. This is an arbitration award and is legally binding on both parties (www.ccma.org.za).

4.6.5 Labour Court

The Labour Court has the same powers as a provincial division of the Supreme Court. The Labour Court has jurisdiction in respect of all matters that in terms of the act are to be determined by it. However, the Labour Court does not have jurisdiction to adjudicate an unresolved dispute if the act requires the dispute to be resolved through arbitration (Du Plessis *et al.*, 2002:330).

4.6.6 Labour Appeal Court

A party subject to a decision of the Labour Court may apply for leave to appeal against such decision. The Labour Appeal Court has the same status as the Appellate Division of the Supreme Court. Its primary role is to hear appeals from the Labour Court both with regard to its judgements in respect of disputes over which it has jurisdiction and in respect of reviews by the Labour Court of arbitration awards (Van Niekerk, 2002:101).

Table 4.7 Dispute handling forums (Landis & Grossett, 2005:402)

DISPUTE	CONCILIATION	ARBITRATION	LABOUR COURT
Automatically unfair dismissals	Refer within 30 days to the CCMA for conciliation.		Refer within 90 days to the Labour Court for adjudication.
Dismissal for misconduct	Refer within 30 days to the CCMA for conciliation.	Refer within 90 days to the CCMA for arbitration.	
Dismissal for incapacity/poor work performance	Refer within 30 days to the CCMA for conciliation.	Refer within 90 days to the CCMA for arbitration.	
Constructive dismissal	Refer within 30 days to the CCMA for conciliation.	Refer within 90 days to the CCMA for arbitration.	
Dismissal because the employee made protected disclosure i.t.o. the Protected Disclosure Act	Refer within 30 days to the CCMA for conciliation.		Refer within 90 days to the Labour Court for adjudication.
Unfair employer conduct relating to promotion, demotion, training or provision of benefits	Refer within 90 days to the CCMA for conciliation.	Refer within 90 days to the CCMA for arbitration.	
Unfair suspension or other disciplinary action short of dismissal	Refer within 90 days to the CCMA for conciliation.	Refer within 90 days to the CCMA for arbitration.	
Disputes over exercising organisational rights	Refer to the CCMA for conciliation (no time limit).	Refer within 90 days to the CCMA for arbitration.	
Unprotected strike/lock-out			Labour Court may be approached for an interdict.

Disputes concerning unilateral changes to conditions of employment	Refer to the CCMA for conciliation (no time limit). Referral to require employer to restore status quo. If not, give notice (48 hours or 7 days) and then strike or lock-out commences.		Labour Court may be approached for interdict.

4.6.7 Referring a dispute to the CCMA

The following list of steps to be taken will assist employees when referring a case to the CCMA (www.ccma.org.za).

Step 1

In the case of unfair dismissal disputes, the employee has 30 days from the date on which the dispute arose to open a case. With discrimination cases, employees have six months.

Step 2

The employee needs to complete a CCMA case referral form, also known as an LRA form 7.11.

Step 3

Once the form is completed, it must be ensured that a copy is delivered to the other party. The employee must be able to prove that the other party received a copy.

Step 4

It is not necessary to bring the referral form to the CCMA in person. You may also fax the form or post it. Make sure a copy of the proof that the form has been served on the other party is also enclosed.

Step 5

The CCMA will inform the parties as to the date, time and venue of the first hearing.

Step 6

The first meeting is called a conciliation hearing. Only the parties, trade union or employer organisation representatives and the CCMA commissioner will attend. The purpose is to reach an agreement acceptable to both parties. Legal representation is not allowed.

Step 7

If no agreement is reached, the commissioner will issue a certificate to that effect. Depending on the nature of the case, it may be referred to the CCMA for arbitration or the Labour Court.

Step 8

For an arbitration hearing, an LRA form 7.13 must be completed. The employee must complete and lodge this CCMA form within 90 days of the date on which the certificate is issued. A copy must be served on the other party.

Step 9

Arbitration is a more formal process, and evidence (including witnesses and documents) may be necessary to prove your case. Parties may cross-examine each other. Legal representation may be allowed. The commissioner will make a final and binding decision called an arbitration award within 14 days.

Step 10

If a party does not comply with the arbitration award, it may be made an order of the Labour Court.

Unresolved grievances and feelings of dissatisfaction may lead to strikes. However, it is of importance to follow the correct procedure in order to ensure a protected strike, so that no employees may face dismissal.

4.7   STRIKES AND LOCK-OUTS

A strike occurs when a group of workers refuse to perform their duties partially or fully. The reason for striking is to force the employer(s) to address their grievance or to resolve a dispute that they are having with their employer. Strikes are used by employees as a tool to force management to accept their demands. A lock-out is the employer's counterpart of a strike. Employers confronted with an unprotected strike are to lock the employees out until the employees comply with the employer's proposal. The Labour Relations Act gives employees the right to strike. This is described in sections 64 and 65 of the LRA as follows.

> ### 64. Right to strike and recourse to lock-out
>
> 1) Every employee has the right to strike and every employer has recourse to lock-out if –
>
>   a) the issue in dispute has been referred to a council or to the Commission as required by this Act, and –
>
>   i) a certificate stating that the dispute remains unresolved has been issued; or
>
>   ii) a period of 30 days, or any extension of that period agreed to between the parties to the dispute, has elapsed since the referral was received by the council or the Commission; and after that –
>
>   b) in the case of a proposed strike, at least 48 hours' notice of the commencement of the strike, in writing, has been given to the employer, unless –

i) the issue in dispute relates to a collective agreement to be concluded in a Council, in which case, notice must have been given to that council; or

ii) the employer is a member of an employers' organisation that is a party to the dispute, in which case, notice must have been given to that employers' organization; or

c) in the case of a proposed lock-out, at least 48 hours' notice of the commencement of the lock-out, in writing, has been given to any trade union that is a party to the dispute, or, if there is no such trade union, to the employees, unless the issue in dispute relates to a collective agreement to be concluded in a council, in which case, notice must have been given to that council; or

d) in the case of a proposed strike or lock-out where the State is the employer, at least seven days' notice of the commencement of the strike or lock-out has been given to the parties contemplated in paragraphs (b) and (c).

65. Limitations on right to strike or resource to lock-out

1) No person may take part in a strike or a lock-out or in any conduct in contemplation or furtherance of a strike or a lock-out if –

a) that person is bound by a collective agreement that prohibits a strike or lock-out in respect of the issue in dispute;

b) that person is bound by an agreement that requires the issue in dispute to be referred to arbitration;

c) the issue in dispute is one that a party has the right to refer to arbitration or to the Labour Court in terms of this Act;

d) that person is engaged in –

i) an essential service; or

ii) a maintenance service.

4.7.1 The consequences of a protected strike and lock-out

According to Landis and Grossett (2005:366), the following are consequences of a protected strike and lock-out.

- Workers may not be dismissed for participating in a protected strike action. Such a dismissal will constitute an automatically unfair dismissal.
- Involvement in a protected strike does not constitute a breach of contract. Civil legal proceedings may not be instituted against a striking person.
- Employers are not obliged to remunerate striking workers except if remuneration includes payment in kind.
- Workers may be dismissed due to their conduct during a strike after procedural and substantive requirements have been followed.

- Workers may be dismissed due to the economic implications the strike had on the organisation after the procedural and substantive requirements have been followed.
- Some limitations apply in respect of the employer's ability to employ replacement workers.

It is imperative that the correct procedure is followed as set out in the Labour Relations Act to ensure a protected strike, the reason being that employee's services may be terminated when participating in unlawful strikes.

4.8 OTHER IMPORTANT ASPECTS OF THE LABOUR RELATIONS ACT

4.8.1 Freedom of association

Every employee has the right to participate in forming a trade union and to join a trade union. Furthermore, members of trade unions may participate in their lawful activities; participate in the election of any of their office-bearers, officials or trade union representatives; and stand for election as office bearers or officials and/or trade union representatives. Employees may not be discriminated against for exercising any right in relation to freedom of association. Employers have the same rights as employees regarding employers' organisations.

4.8.2 Collective bargaining

The initial step in collective bargaining occurs when an employer recognises a trade union as the bargaining agent for its employees in a particular bargaining unit. This relationship is formalised in a document known as a recognition agreement. The Labour Relations Act accords the following organisational rights to registered, representative unions in an organisation.

Table 4.8 Organisational rights of unions

SUFFICIENTLY REPRESENTATIVE UNIONS	MAJORITY UNIONS
Trade union access permitted to the workplace to recruit and communicate with members. Also entitled to hold meetings with employees outside working hours.	Trade union representatives may be elected according to a set formula at a workplace where a union enjoys majority support.
Deduction of trade union subscriptions or levies permitted. Any employee who is a member of a representative trade union may authorise the employer in writing to deduct subscriptions or levies payable to that trade union from an employee's wages.	An employer must disclose all relevant information to the majority trade union that will allow it to perform its functions effectively.
An employee who is an office-bearer is entitled to take reasonable leave during working hours to perform his/her functions.	

4.8.3 Bargaining councils

Bargaining councils conduct collective bargaining between the respective interested parties. Furthermore, they also resolve disputes and promote training and education schemes.

Statutory councils are smaller versions of bargaining councils that cater for more mundane matters such as benefits, pensions, provident funds, etc.

4.8.4 Workplace forums

Workplace forums represent the employees in the workplace and are established to allow employees to consult with management. Workplace forums and employers must meet often to discuss matters that are of mutual interest to both parties. They can only be formed if there are more than 100 employers employed in a workplace. A union that represents the majority of employees can apply to form a workplace forum.

4.9 OTHER IMPORTANT LABOUR LEGISLATION THAT GOVERNS AND PROTECTS THE EMPLOYMENT RELATIONSHIP

4.9.1 Employment Equity Act (EEA) 55 of 1998

The purpose of the EEA is to achieve equity in the workplace by promoting equal opportunities and fair treatment by eliminating unfair discrimination and by implementing affirmative action measures to redress the disadvantages in employment experienced by designated groups. No person may unfairly discriminate, directly or indirectly, against an employee in any employment policy or practice on one or more grounds, including race, gender, pregnancy, marital status, family responsibility, ethnic or social origin, colour, sexual orientation, age, disability, religion, HIV status, conscience, belief, political opinion, culture, language and birth. In terms of this act, designated groups are classified as black people, women and people with disabilities.

A designated employer is an employer who employs 50 or more employees, and employers who employ fewer than 50 employees but have a total annual turnover that is equal to or above the applicable annual turnover of a small business, a municipality or an organ of the state. A designated employer must implement affirmative action measures for designated groups to achieve employment equity. In order to implement affirmative action measures, a designated employer must consult with employees, conduct an analysis, prepare an employment equity plan and report to the Director-General on progress made in the implementation of the plan.

Purpose of the employment equity plan

An employment equity plan is the essence of an employment equity programme. This document represents the parties' commitment on how the organisation will progress from the current structure to the new diverse structure. The plan will include aspects such as a workforce analysis determined by occupational groups where there was under-representation of one or more designated groups, and an employment systems review, which identified policies and procedures that could constitute unfair employment barriers to designated group members. Furthermore, it will set out goals, including a timetable for implementation. Benchmarks will also be included that will permit assessments of success and clearly indicate who is accountable and responsible for the implementation of the various components of the plan.

4.9.2 Skills Development Act 97 of 1998

This piece of legislation was designed to develop the skills of the South African workforce and to increase investment by employers in the education and training of their workforces. It is achieved by way of the Skills Development Levies Act, which imposes on employers a levy equal to 1 per cent of their total wage bill, payable to the South African Revenue Services (SARS). Portions of this levy can be recovered by the employer as a grant on meeting certain requirements set out in the Skills Development Act and its regulations. This act puts great emphasis on learnerships.

What is a learnership?

Sector Education and Training Authorities (SETAs) must approve every learnership programme before it can be forwarded to the Department of Labour. Their approval will depend on aspects such as whether the learnership has properly prepared learning material and whether it includes practical work experience. The SETAs have been given this role because they are well placed to assess whether an intended learnership programme will meet a need and whether it is addressing an occupation where there are likely to be jobs or self-employment opportunities.

Learnerships are new paraprofessional and vocational education and training progammes. They combine theory and practice and culminate in a qualification that is registered within the National Qualifications Framework (NQF). A person who successfully completes a learnership will have a qualification that signals occupational competence and that is recognised throughout the country, almost like the apprenticeship system that was used previously.

Learnerships are important because they emphasise outcomes. The learner will be able to practically use the skills that he/she has been taught. Furthermore, it combines both theory and practice. Also refer to the chapter entitled 'Human Resource Management' (chapter 3).

4.9.3 Unemployment Insurance Act 63 of 2001

The Unemployment Insurance Fund (UIF) provides benefits and security for unemployed people and for the dependents of deceased contributors. Illness and maternity benefits are also paid from this fund. Revenues are obtained from the contributions of employees, their employers and government.

Employers must register themselves and their workers with the UIF and pay their contributions every month. Employers must deduct 1 per cent of their workers' pay for UIF. Employers must pay the 1 per cent they deduct from workers, together with 1 per cent from themselves to the UIF or SARS before the seventh day of every month.

The Unemployment Insurance Act and Unemployment Insurance Contributions Act apply to all employers and workers, but not to:

- workers working less than 24 hours a month for an employer;
- learners;
- public servants;
- foreigners working on contract;
- workers who get a monthly state (old age) pension; or
- workers who only earn commission.

4.9.4 Occupational Health and Safety (OSH) Act 85 of 1993

This act places an obligation on employers to reduce risks to health and safety in the workplace and provides for the regulation and monitoring of workplaces in order to protect the health and safety of employees and other persons in the workplace and in the use of plant and machinery.

Duties of employers

All employers must:
- provide and maintain a safe, healthy working environment;
- ensure workers' health and safety by providing information, instructions, training and supervision;
- inform health and safety representatives of incidents, inspections, investigations and inquiries; and
- report to an inspector incidents in which people are killed, injured or become ill; dangerous substances are released; or machinery fails or runs out of control.

Duties of employees

Workers must report health and safety incidents to their employer, a health and safety representative, or a health and safety inspector.

Duties of safety committees

One or more health and safety committees must be formed when employers have appointed two or more health and safety representatives. Health and safety committees make and keep records of recommendations to employers and inspectors, and discuss reports and keep records of incidents in which someone is killed, injured or becomes ill.

Duties of safety representatives

Employers who employ 20 or more workers must appoint representatives to monitor health and safety conditions. Engineering industries must have at least one representative for every 50 workers or part thereof. However, an inspector may order an employer to appoint more. Representatives must be full-time workers who are familiar with the workplace. They must be trained during working hours. Representatives monitor, investigate and report on health and safety matters; accompany inspectors during inspections; and attend health and safety committee meetings.

4.9.5 Compensation for Occupational Injuries and Diseases Act 130 of 1993

The Compensation for Occupational Injuries and Diseases Act applies to all employers and casual and full-time employees who as a result of a workplace accident or work-related disease are injured, disabled or killed or become ill. Compensation claims for occupational injuries and diseases are calculated according to the seriousness of the injury or disease. Injuries or diseases caused by the negligence of a worker's employer or another worker may result in increased compensation. Workers are entitled to compensation if they are injured while working or contract any work-related disease. The types of compensation paid to workers for injuries or diseases are medical aid, temporary disablement, permanent disablement and fatalities.

4.10 CONCLUSION

It is essential to practise successful employment relations. To be educated and up to date with labour legislation applicable to your organisation is vital. One of the major objectives of organisations and the relevant stakeholders is to promote sound employment relations and to maintain fair labour practices and equality in the organisation. Successfully applying the above will result in a highly motivated and a more productive labour force, which is indispensable for world-class organisations. You may find table 4.9 useful in achieving sound employment relations.

Table 4.9 Checklist to ensure sound employment relations

EMPLOYMENT RELATIONS ASPECTS	YES	NO
■ Do you involve the relevant role-players in the effective management of your organisation?		
■ Do you have employment contracts that distinguish between employees and independent contractors?		
■ Do you comply with the legal requirements as set out in the labour legislation pertaining to the clauses included in the employment contracts?		
■ Do you have a well-defined and user-friendly grievance procedure?		
■ Do you have a disciplinary code and procedure that is legally compliant and protects the needs of your organisation?		
■ Do you have guidelines on how to distinguish among dismissals, unfair labour practices and automatically unfair dismissals?		
■ Do you have a policy and guidelines on how to deal with misconduct and incapacity cases?		
■ Do you have a well-defined and legally compliant retrenchment policy and procedure?		
■ Do you have a set of guidelines on how to refer a case to the CCMA?		
■ Do you have a set of guidelines on the legal requirements to ensure a protected strike and lock-out?		
■ Do you have a clear understanding pertaining to the following labour legislation:		
⬚ the Labour Relations Act		
⬚ the Employment Equity Act		
⬚ the Skills Development Act		
⬚ the Unemployment Insurance Act		
⬚ the Occupational Health and Safety Act		
⬚ the Compensation for Occupational Injuries and Diseases Act		

Self-assessment

1.1 Identify the role-players active in employment relations and describe briefly the role allocated to each one.

1.2 List the various aspects according to the BCEA that must be included in the contract of employment.

4.3 Explain the provisions set out in the BCEA with regard to the following:
a) sick leave
b) termination of employment.

4.4 According to the BCEA, certain circumstances must exist for a person to be presumed an employee. List five of these circumstances.

4.5 If an employee is unsatisfied with the outcome of a disciplinary hearing, may he/she make use of the grievance procedure to resolve it?

4.6 Compile the following documents for the organisation you are currently working for or any other organisation of your choice:
a) a disciplinary code
b) disciplinary procedures.

4.7 Give a brief definition of each of the following concepts:
a) dismissal
b) automatically unfair dismissal
c) unfair labour practice.

4.8 Identify three grounds that will justify dismissal and write a paragraph on the procedure that needs to be followed in each case.

4.9 Distinguish between the conciliation and arbitration processes.

4.10 Name the consequences of a protected strike.

4.11 List five different pieces of legislation applicable to employment relations.

References

Du Plessis, J.V., Fouche, M.A. & Van Wyk, M.W. 2002. *A Practical Guide to Labour Law.* 5th edition. Durban: Butterworths.

Finnemore, M. 1997. *Introduction to Labour Relations in South Africa.* Durban: Butterworths.

Finnemore, M. & Van Rensburg, R. 2002. *Contemporary Labour Relations.* 2nd edition. Durban: Butterworths.

Grobler, P.A., Warnich, S., Carrell, M.R., Elbert, N.F. & Hatfield, R.D. 2002. *Human Resource Management in South Africa.* 2nd edition. London: Thomson Learning.

Grogan, J. 2005. *Workplace Law.* 8th edition. Cape Town: Juta Law.

Grossett, M. 1999. *Discipline and Dismissal: A Practical Guide for South African Managers.* 2nd edition. Johannesburg: Thomson.

Israelstam, I. 2001. 'Labour relations act fuzzy on contracts.' Available at http://www.iol.co.za, accessed 28 November 2001.

Israelstam, I. 2004. 'Case law provides detail for LRA.' *The Star, 2 August.*

Landis, H. & Grossett, L. 2005. *Employment and the Law: A Practical Guide for the Workplace.* 2nd edition. Cape Town: Juta.

Meyer, M. & Kirsten, M. 2005. *Introduction to Human Resource Management*. Claremont: New Africa Books.

Nel, P.S., Swanepoel, B.J., Kirsten, M., Erasmus, B.J. & Tsabadi, M.J. 2005. *South African Employment Relations: Theory and Practice*. 5th edition. Pretoria: Van Schaik.

Slabbert, J.A. & Swanepoel, B.J. 2002. *Introduction to Employment Relations Management: A Global Perspective*. 2nd edition. Durban: Butterworths.

South Africa

Compensation for Occupational Injuries and Diseases Act 130 of 1993. Pretoria: Government Printer.

Occupational Health and Safety Act 85 of 1993. Pretoria: Government Printer.

Labour Relations Act 66 of 1995. Pretoria: Government Printer.

Basic Conditions of Employment Act 75 of 1997. Pretoria: Government Printer.

Employment Equity Act 55 of 1998. Pretoria: Government Printer.

Skills Development Act 97 of 1998. Pretoria: Government Printer.

Unemployment Insurance Act 63 of 2001. Pretoria: Government Printer.

Swanepoel, B., Erasmus, B., Van Wyk, M. & Schenk, H. 2003. *South African Human Resource Management: Theory and Practice*. 3rd edition. Cape Town: Juta.

Van Niekerk, A. 2002. *What You Must Know about Unfair Dismissal*. Cape Town: Siber Inc.

Websites

Commission for Conciliation, Mediation and Arbitration: www.ccma.org.za.

Cosatu: www.cosatu.org.za.

Department of Labour: www.labour.org.za.

Independent online: www.iol.co.za.

IRASA: www.irasa.org.za.

Nedlac: www.nedlac.org.za.

Jetson: www.jetson.net.au.

5 Managing People and Teams

Marius Meyer

Study objectives

After studying this chapter, you should be able to:
- indicate the importance of human relations skills for engineers
- provide guidelines for maintaining sound human relations in the workplace
- identify causes of conflict
- distinguish among different types of conflict
- explain the strategies for dealing with conflict
- describe different conflict management styles
- provide guidelines for conflict resolution
- name the advantages of teamwork
- discuss the stages in team development
- identify the attributes of effective teams
- define diversity management
- indicate the benefits of diversity management
- explain the concept of *ubuntu*
- provide guidelines for managing a diverse workforce

The aim of this chapter is to equip engineers with people management skills.

5.1 INTRODUCTION

The ability to manage people is an essential skill for all engineers in the modern business environment. Engineers must be able to establish and maintain sound human relations in the workplace in order to achieve project and organisational goals. They should have the necessary knowledge and skills to handle conflict and the ability to enhance effective teamwork, since most work is done in teams today. Moreover, the reality of a diverse South African workforce, where employees from different backgrounds and cultures must function effectively, requires engineers to show an understanding of these differences to manage effectively.

This chapter will focus on the important skills needed to manage people effectively. These skills are summarised in table 5.1 on the next page.

Table 5.1 People skills for managers

HUMAN RELATIONS	CONFLICT MANAGEMENT	TEAMWORK	DIVERSITY MANAGEMENT
▪ Communication ▪ Trust building ▪ Positive attitude ▪ Open mindedness ▪ Appreciation ▪ Networking	▪ Problem definition ▪ Collect information ▪ Problem focus ▪ Communication ▪ Understanding ▪ Consider options ▪ Agree on solutions ▪ Implement solutions ▪ Follow-up action	▪ Clear objectives ▪ Openness ▪ Trust & Support ▪ Co-operation ▪ Information sharing ▪ Consensus ▪ Flexible leadership ▪ Review teamwork ▪ Interpersonal relations ▪ Individual development ▪ Mutual help ▪ Needs of individual, task, team integrated	▪ Open communication ▪ Multi-cultural respect ▪ Non-discriminatory behaviour ▪ Representativeness ▪ Celebrate diversity ▪ Sensitivity ▪ Understanding & Trust

5.2 HUMAN RELATIONS

The term 'human relations' refers to all types of interactions among people. According to Reece and Brandt (1993), the study of human relations is the study of why our beliefs, attitudes and behaviours sometimes cause problems in our relations with others. The study of human relations emphasises the analysis of human behaviour, prevention strategies and the resolution of behavioural problems. Well-developed human relations skills are essential for engineers and technicians for the following reasons.

1. Engineers often need to manage people in the workplace. They have to mobilise employees to achieve engineering project goals. If engineers do not know how to get the most out of people and solve day-to-day problems, engineering work will not be as effective as it could be.

2. The South African Constitution, with its emphasis on human rights, forms the basis of human relations in South Africa. Labour legislation, such as the Skills Development Act, Labour Relations Act and Employment Equity Act, focuses to a large extent on the human side of business. Engineers or managers who lack human relations skills will find it difficult to implement these laws, which could cause problems for their companies in the form of labour unrest, law suits and low productivity.

3. The modern business trend towards focusing on quality-management and customer-satisfaction strategies requires engineers to liaise more directly with customers and to design products to meet customers' needs.

4. Engineers also need to establish and maintain sound human relationships with suppliers, in order to obtain the right goods and equipment for engineering design.
5. The increasing imperative of globalisation requires engineers to liaise with business partners world-wide. Establishing contact and building partnerships and co-operation agreements require human relations skills in order for engineers to network effectively.

5.3 GUIDELINES FOR SOUND HUMAN RELATIONS

The following guidelines can be used to build sound human relations in the workplace.

■ **Communication:** The importance of communication as the basis of human relations cannot be overemphasised. Communication is the process whereby we come to an understanding of ourselves and of other people. It is therefore essential to clearly communicate your thoughts, feelings and ideas to other people. Moreover, your ability to listen to other people and to understand their needs and ideas is also essential.

■ **Trust:** Sound human relations, whether among colleagues or between managers and employees, are based on trust. When a lack of trust exists, relationships suffer and productivity decreases. Employees communicate less information to management and are reluctant to express their true feelings and opinions. Conversely, when a climate of trust is present, open discussion of problems and issues, and a free exchange of ideas and information are encouraged. The end result is a productive work environment based on trust, co-operation, commitment and openness.

■ **Positive attitude:** People who have negative thoughts and attitudes tend to exhibit negative behaviours that can make people around them negative. When you screen your thoughts to accentuate the positive, you will be a more pleasant person and people will enjoy working with you. A positive attitude, coupled with the ability to inspire and encourage others, even in difficult circumstances, is a powerful human relations skill. No matter how good your engineering design or your product, if the manager or people who execute the work have a negative attitude, the effectiveness of your product or service may be adversely affected.

■ **Open mindedness:** Bad managers often make unilateral decisions and are reluctant to consider other points of view that might lead them to question their ideas. Effective managers have the skill to integrate employees' contributions by being open minded to new ideas, concepts and approaches. The modern fast-changing business environment needs managers who are dynamic, modern and future-orientated in the way they do business.

■ **Appreciation:** All people like to be valued and appreciated. Failure to express appreciation can damage human relations in the workplace. A technician who works very hard to complete an urgent task for an engineer during a lunch hour or after hours is likely to feel upset if that extra effort is ignored. The need for recognition and appreciation is a basic human need that all people require to help them stay focused and motivated in what they have to do.

■ **Networking:** Modern companies cannot function as isolated business units. Different companies need to be connected as part of a value chain to deliver products and services to meet the needs of increasingly demanding customers. Engineers should create opportunities for continuous networking with people across departments, both inside and outside the organisation. They should share ideas about successful and unsuccessful projects and offer help where it is needed. It is a very good idea to try to remember names and keep records of people you have met. The better your relationships, the more effective you will be as a manager or engineer.

5.4 CONFLICT

Conflict occurs when individuals or groups are simultaneously confronted with incompatible and opposing needs, goals or behavioural activities. This process often results in hostility, anxiety and a breakdown in relationships. Conflict should not, however, be viewed from a negative perspective only. In fact, a degree of conflict is essential to resolve problems and encourage creativity and innovation. The key is to manage conflict constructively in order to optimise organisational performance. In addition, by being proactive, it is possible to prevent unnecessary and destructive conflict from occurring.

5.4.1 Causes of conflict

Conflict is caused by a range of factors, depending on a particular situation.

- A breakdown in communication occurs when an individual communicates a message to another individual, but the receiver misunderstands the message or the sender communicates it unclearly, e.g. the wrong equipment is purchased by a storekeeper due to an incorrect specification by an engineer.
- Personality clashes can arise, e.g. a technician who needs a great deal of reassurance and social interaction may have problems relating to an engineer who makes quick decisions and expects immediate results.
- Conflict can also ensue when people experience inconsistent and unfair managerial and employee behaviour such as favouritism or discrimination.
- People resent an autocratic management style in the modern business world and the potential for conflict is very high if a manager adopts such a style.
- A lack of resources may cause employees and departments to compete for the same resources. The end result is that some departments 'win' and others 'lose', which leads to conflict. This typically happens when certain departments or sections are considered to be 'more important' than others.
- People from different cultures who do not understand, recognise and appreciate cultural differences may become involved in conflict situations.

5.4.2 Types of conflict

Different types of conflict can be experienced, depending on the particular situation and the people involved.

- *Intrapersonal conflict* occurs within an individual when two equally attractive and irreconcilable options are presented. The individual finds it difficult to make a decision and experiences a high level of stress because one of the options needs to be relinquished. For example, Jabulani applies for the position of engineer with two companies that both offer the same remuneration package. He receives job offers from both employers and experiences intense anxiety and intrapersonal conflict when making a decision.

- *Interpersonal conflict* develops between two individuals as a result of different needs and goals. For example, two engineers may approach a project from different perspectives and cannot reach agreement on certain issues.
- *Intergroup conflict* develops from the expression of incompatible ideas by two or more groups of people. For example, conflict ensues when a trade union confronts an employer organisation on salary increases for its members.

5.4.3 Conflict management strategies

There are several ways of dealing with conflict. These can be grouped into three basic conflict management strategies (Reece & Brandt 1993:355).

- **Win-lose strategy**
  This approach eliminates the conflict by having one party achieve its goals at the expense of the other. Although the conflict may be solved over the short term, it usually does not address the cause of the problem. The 'loser' is likely to resent the solution and conflict could develop again.

  In the work situation, this strategy can be applied in two ways: the manager rules or the majority rules. Following the first option, the engineer decides on the solution to the conflict and states that it is final. Employees do not have a say and will probably resent the lack of involvement. Following the second option, a vote can be taken. This means that one party will be on the losing side and will not favour the solution. The win-lose strategy can be applied when the two parties cannot agree to a solution.

- **Lose-lose strategy**
  All parties lose when the lose-lose strategy is used. For example, if both parties are asked to compromise, each one must 'give in' in some way to the other. When the sacrifices are too great, both parties may feel that too much has been given. Another example is the conflict that develops when an employee's request for flexible working hours is rejected. The employee becomes negative and does not perform, and the company loses productivity. Both parties are therefore losers.

- **Win-win strategy**
  The purpose of this strategy is to resolve the conflict in such a way that both parties end up as winners. Its success depends on a clearly defined problem, open communication and listening, and the trust and commitment of the parties to find a workable solution. The end result of the win-win strategy is a solution to the problem that caused the conflict, and one that both parties will accept and commit to implement jointly.

5.4.4 Conflict management styles

Depending on their personalities and experience in dealing with conflict, individuals and members of groups develop their own conflict management styles. People can learn how and when to adapt their conflict management style to effectively deal with conflict situations. Maddux (in Reece & Brandt, 1993) suggests that five different conflict management styles may be used, depending on the situation.

1. Avoidance

Some people tend to avoid conflict when the potential for it begins to arise. They also avoid the other party, hoping that the conflict will go away. This style is appropriate when the conflict is too minor or too great to resolve. In general, however, avoidance does not address the issue, and very often the problem remains unresolved and recurs at a later stage.

2. Competing

A person who uses this style stands firm and rejects the views of the other party. Power, authority and aggression are used in an attempt to resolve the conflict. This style often causes resentment and hostility. However, there may be situations in which you have no alternative. For instance, if you are expected to do business based on unethical principles it may be necessary for you to state that certain principles are 'non-negotiable'.

3. Compromising

The parties negotiate and compromise to find an acceptable solution for both. This is a matter of give and take to reach an agreement.

4. Accommodating

An individual or group neglects its interests to satisfy those of the opponent. Superficial harmony is created to maintain the relationship. In reality, this means 'giving in' to the other party at the expense of finding a proper and lasting solution.

5. Collaboration

Ideas are taken from both parties and ways are found to develop them all, without detracting from the overall goal. Abilities, values and knowledge are recognised. Decision making based on consensus is used to resolve the conflict.

5.4.5 Guidelines for conflict resolution

If conflict occurs on an interpersonal or intergroup level, certain guidelines can help to stay focused on solving the problem and dealing with conflict constructively.

- **Define the problem:** Both parties have to focus on the real cause/s of the problem, not the symptoms or results. When the problem is well defined, the real cause/s will often surface.
- **Collect information:** Gather the necessary facts and information needed to clarify the situation. Several questions must be answered: What happened? Where? When? How? Who is involved and affected? What policies and procedures are applicable?
- **Focus on the problem:** Concentrate on the problem and the issues involved. It is essential not to get personal by focusing on the other party involved. Personal attacks and blame should be avoided at all cost.
- **Communicate with the other party:** Communicate your thoughts and feelings to the other party. It is important to keep the lines of communication open throughout the conflict situation.
- **Understand the other party:** Make a concerted effort to understand the other party's point of view by listening very carefully. If you can put yourself in that person's shoes, the chances are very good that the conflict will be resolved.
- **Consider all options:** One should not jump to a final solution for the problem, but consider various ideas, approaches and options surrounding the conflict situation.

▓ **Agree on the solution/s:** Once options have been considered, a solution to the conflict can be jointly selected to achieve the desired goals.

▓ **Implement the solution:** Both parties should discuss how the solution will be implemented. This may include timetables for implementing the solution, and the responsibilities of both parties to honour and maintain the agreement.

▓ **Follow-up action:** Even the best solutions can fail unless all the parties involved in the conflict make the attempt to follow the solution through. Decide how the agreement will be maintained and how regularly the parties will meet to review the ongoing implementation of the solution and follow-up actions.

5.5   TEAMWORK

Teamwork has a crucial role to play in ensuring the success of engineering work. The nature, size and complexity of such work make it impossible for engineers to function effectively only as individual professionals. Efficient teams are therefore needed to enhance the success of the work. Without teamwork it is extremely difficult to implement modern engineering management, as you will realise when you study the chapter on project management (chapter 13).

5.5.1 Advantages of teamwork

From the above discussion it is clear that teamwork has multiple advantages for organisations and managers.

▓ Complex problems are solved more efficiently because a greater variety of knowledge, skills and experience are utilised.

▓ People work well in teams, and problems in human relations and communication can be prevented by effective teamwork.

▓ Teamwork improves job satisfaction because team members participate in decision making.

▓ If team members understand each other's work, they can fill in for one another when needed.

▓ The quality of group decisions in good teams is better than that of individual decisions, due to the increased level of input.

▓ There is a higher level of commitment among team members to the implementation of goals and tasks that contribute to goal attainment and productivity.

From this list it can be deduced that good teamwork will build trust, co-operation, synergy and interdependence in a group. The following slogan summarises the benefits of teamwork: 'Together Everyone Achieves More.' It is very appropriate that the first letters of the words in this slogan form the word 'team'.

5.5.2 Self-managing work teams (SMWTs)

The central principle behind self-managing teams is that the teams themselves, rather than managers, take responsibility for their work, monitor their performance, and alter their performance strategies as needed to solve problems and adapt to changing conditions (Wageman, 1997:49–61). In addition to the advantages of teamwork listed in section 5.5.1, the running of an organisation's day-to-day activities by means of SMWTs is said to:

- enhance the company's performance, because those closest to the customer and best able to respond to customer demands have the authority to meet those demands;
- enhance organisational learning and adaptability, because members of self-managing teams have the latitude to experiment with their work and develop strategies that are uniquely suited to tasks; and
- enhance employees' commitment to the organisation, because self-managing teams offer wider participation in and ownership of important organisational decisions.

See also the section on SMWTs in chapter 2.

5.5.3 Stages in team development

Teams don't immediately start to function if you simply put people together and expect them to perform well as a team. According to Tuckman (in Oakland, 1995:277), there are four stages of team development: forming, storming, norming and performing, as illustrated in figure 5.1.

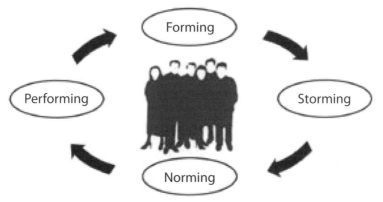

Figure 5.1 Stages in team development

- **Forming:** In the first phase, the team members come together and the team is formed. There is an awareness of the need for teamwork, but the members haven't worked together as a team before.
- **Storming:** Now they start to dialogue and this often results in conflict. The members don't have any guidelines or rules to follow in their behaviour and actions.
- **Norming:** Members start to co-operate. They establish ground-rules, e.g. that all members can agree and disagree and that there will be open communication.
- **Performing:** The members have mutual goals and function productively as a team. The team objectives are achieved successfully.

5.5.4 Team roles

Engineers and technicians should realise that each member of a team will fulfil different roles and responsibilities. People are not the same: their personalities, beliefs, values and experiences differ. Some members will be good at communicating ideas, others will be good at analysing problems, some will support fellow members, while others will be good at implementing the final plan or acquiring the necessary resources. The challenge for the engineer is to create a positive environment for teamwork so that each individual can contribute optimally in his/her unique way.

For SMWTs to function effectively, each team member has to fulfil seven roles, to a greater or lesser extent, depending on his/her level of maturity (Rayner, 1992; Fisher, 1993). The aim is for each member to develop each of these roles in order to become a fully contributing team player. The SMWT member roles are:

- problem solver;
- trainer;
- decision maker;
- customer advocate;
- team player;
- resource provider; and
- skilled worker.

➤ The *problem solver* understands and utilises problem-solving techniques to regularly identify and solve problems.
➤ The *trainer* trains others in job/skill areas and continually shares knowledge with others.
➤ The *decision maker* provides input and makes decisions on issues that directly influence the work area.
➤ The *customer advocate* strives to meet the needs of the customer better.
➤ The *team player* demonstrates good interpersonal skills and supports other team members.
➤ The *resource provider* has a diverse and ever-expanding set of skills that continually broadens the knowledge base.
➤ The *skilled worker* demonstrates all the necessary skills and knowledge to perform the job well, and continually strives to improve skill sets and assure total quality.

5.5.5 Attributes of effective teams

The engineer, as a member or a leader of a team, should actively encourage and support effective teamwork. To do this, he/she must ensure that the following attributes of effective teams are developed.

- The team must have clear objectives and goals, agreed to by the team members.
- Members must be open with and confront one another directly.
- Members should trust and support each other.
- There is co-operation and open information sharing.
- There is constructive conflict to generate new ideas and creativity.
- Information is collected quickly and quality decisions are made by means of consensus.
- The leader is part of the group and exhibits flexibility in his/her leadership approach.
- The team regularly reviews its performance to improve teamwork.
- Sound interpersonal relationships are maintained.
- There are ample opportunities for individuals to enrich and develop themselves.
- Members provide mutual help so that each can successfully accomplish goals.
- The leader must create a balance among the needs of the task, the team and the individual.

5.6 DIVERSITY MANAGEMENT

Most managers rarely receive the training they need to work effectively across cultures and diverse peoples. Today's engineers must be proficient in transferring knowledge from their own culture and simultaneously discovering what it means to work across cultures and other diverse groups.

Diversity management requires attention to the total environment and a cross-cultural understanding of key internal and external demands and resources. For example, factors such as how to resolve a conflict situation, how to use power, how to solve a business problem and how to negotiate a business deal vary by culture and gender. Cultural values and beliefs lie beneath the surface. Information, knowledge and skills are needed to view the world from another culture's perspective.

Diversity management can be defined as a comprehensive and holistic process for creating and maintaining an environment in which all employees feel comfortable, recognised, valued and appreciated despite differences such as race, gender, culture, religion, disability or sexual orientation.

In the previous chapter, the Employment Equity Act was discussed. It is evident that this act will fundamentally change the way in which people are managed in South African organisations. More blacks, women and people with disabilities are and will be employed. In addition, the Broad-based Black Economic Empowerment Act stipulates that companies should implement measures to improve black ownership of companies. But what does the concept 'diversity' entail? It relates to the diverse composition of the South African workforce. Here is an alphabetical list of some of the differences among people encompassed by the term diversity. You may be able to add to this list from your own experience:

- class;
- disability;
- ethnicity;
- gender;
- language;
- marital status;
- political affiliation;
- race;
- religion; and
- sexual orientation.

In the past, some forms of diversity were considered problematic in many South African organisations. The result was that discrimination occurred when certain groups were considered 'better than' or 'inferior to' other groups, based on the differences between these groups. Table 5.2 on the next page summarises the manifestation of discrimination in a company and solutions from a diversity management viewpoint.

Table 5.2 Manifestation of discrimination and diversity management solutions

Type of discrimination	Examples of workplace manifestation	Diversity management solutions
'Ableism'	The building does not cater for people with disabilities; a person in a wheelchair cannot move around everywhere.	The building and facilities are redesigned so that people can access the entire building and its facilities.
Ageism	An 18-year-old employee who joined the company recently is excluded from decision-making structures because the more experienced managers believe that she does not have the necessary insight and experience to provide input.	Employees are encouraged to say what they think and to contribute to decision-making structures. In fact, a lack of experience is seen as valuable for challenging and improving existing systems and methods in a creative way.
Classism	The tea room may only be used by managers and engineers. Administrative staff, technologists and labourers are not allowed in the tea room.	All facilities are open and available to all staff irrespective of organisational level or class.
Ethnocentrism	A Zulu employee avoids contact with a Venda employee because he believes that the Venda are inferior.	The company has an environment conducive to diversity and no group is seen as superior or inferior.
Heterosexism	Heterosexual employees make jokes about a lesbian colleague.	Company policy states that disciplinary action will be taken against heterosexism.
Racism	A black employee does not get the same training opportunities as a white employee.	Equal access to training opportunities is provided for all employees.
'Religionism'	An important meeting is scheduled for a Friday afternoon, when Muslim staff members will be unable to attend due to religious commitments.	Decision making regarding meetings, fuctions and catering requirements accommodates religious differences.
Sexism	An all-male selection panel rejects a female employee's application for a senior position because its believed that she cannot handle stress because she is female.	The selection panel should preferably be representative of both genders and selection decisions should be based on the criteria for the job, not the applicant's gender.

From table 5.2 it is clear that diversity management seeks to enable people of different groups not only to work together successfully, but also to value and take advantage of their differences. Such companies can even be more productive than they would be if the differences did not exist. According to Coghill *et al.* (1998), less diverse companies in the USA underperformed by 8 per cent. Diversity management is seen as an opportunity to improve competitive advantage, not only regionally and nationally, but also from a global perspective.

5.6.1 Benefits of diversity management

Successful diversity management can yield the following benefits for organisations.

- Alternative viewpoints and ideas of people from different groups can enhance creativity, problem solving and decision making.
- Diversity management helps a company to become more flexible. Given the fact that more sensitivity and openness are required in diversity management, employees learn more easily to adapt to changes in the company and the external environment. Employees and managers become more open to challenging old assumptions and beliefs.
- Managers become more effective and objective when making decisions about people. The elimination of stereotypes helps managers to be more professional, consistent and fair when selecting and developing employees, assessing their performance, and dealing with disciplinary cases and grievances.
- Diversity management can help an organisation to improve competitiveness by meeting the needs of diverse customers.
- Diversity management reduces destructive conflict between individuals and groups. This means that less energy is diverted from their work and they are therefore more productive.
- Better relationships develop among managers and employees and with customers. Teamwork and collaboration are thus improved.
- A higher level of trust and motivation develops among members of the organisation. They share information and communicate openly about their feelings, differences and similarities. There is a greater identification with the company's goals because employees develop a sense of inclusion in and ownership of the company's success.
- Diversity management can also contribute to the implementation of affirmative action and employment equity plans in accordance with the Employment Equity Act and Broad-based Black Economic Empowerment Act. A more open and inclusive organisation culture is created that can play a role in reducing the probability of discrimination-based lawsuits and non-compliance with employment equity and black economic empowerment legislation.

5.6.2 Unleashing the spirit of *ubuntu*

One of the most important challenges for managers is to unleash the spirit of the African concept *ubuntu* in the work environment. *Ubuntu* is based on the principle of humanity – the more respect you show for others, and the more you realise that your success depends on your ability to work with others, the more successful you will be. The application of the following principles related to *ubuntu* can help engineers, scientists and technicians to manage people more effectively in organisations by encouraging them to:

- use a more participative management style based on the principles of involvement and empowerment;
- consult staff on important matters and reach agreement based on consensus in order to ensure buy-in and support by all employees;
- get employees focused on the values of the company, e.g. respect, customer orientation, equality, excellence, quality and empowerment;
- analyse the talents of their employees and develop programmes to ensure that the talents of all employees are optimised in the company;
- treat each individual with respect and dignity;
- use effective teamwork to enhance employee motivation and the achievement of goals and objectives in the organisation;
- stay focused on organisation goals and reduce the impact of negative power plays and politics in the organisation;
- create a culture of openness and regular communication, not only top-down but also bottom-up so that all lines of communication are opened;
- encourage new ideas and innovation; and
- create, build and maintain effective relationships in the workplace.

5.6.3 Guidelines for successful diversity management

Engineers can use the following guidelines to implement effective diversity management.

- Communicate openly with all employees about cultural, employment equity and other diversity issues. Develop a relationship of trust and understanding by disclosing the elements of your own culture and values, and invite others to do the same. In addition, you should have a strategy for learning about the cultural values of people in your workgroup.
- Create a work environment where multiculturalism is respected and valued. Ensure that the working environment is open to all cultures and that aspects such as posters, symbols, language signs and food at functions do not represent one dominant culture.
- Racist and sexist behaviour and language should not be tolerated. Make it clear to all staff members that racist, sexist or religious jokes and inappropriate behaviour are not allowed, and that disciplinary action will be taken should these occur.
- When involved in meetings and decision-making structures, try to ensure that the principle of representativeness and inclusiveness in terms of diverse groups are adhered to.
- Initiate diversity workshops and seminars so that workgroups can acquire knowledge and skills to effectively manage diversity.
- Ensure that you understand the company's employment equity policy and black economic empowerment strategy. Liaise regularly with the human resource manager or other senior managers on the implementation and monitoring of this policy and strategy. Inform staff of the benefits of compliance with the legislation, not only from a legal point of view, but also from a business perspective, e.g. acquiring more government contracts.
- When recruiting, selecting and orientating new employees, emphasise the importance of diversity management and employment equity to gain their commitment to a diverse organisation.

▓ Actively seek to meet the diverse needs and expectations of all employee groups, e.g. by providing a disabled employee with the necessary resources and child care facilities for employees.

▓ Provide additional training, development and mentoring services to employees from designated groups so that they can acquire the skills needed to function effectively in the work environment.

▓ Offer assistance to the human resource manager when diversity audits or surveys are conducted.

▓ Ensure that internal and external publications and advertising media represent the diversity within the company, and that of the market and society.

▓ Prior to establishing contact with people from other countries and cultures, obtain and study as much information as possible about the cultures to ensure effective relationships and business partnerships.

5.7 CONCLUSION

Effective people management is an essential part of the responsibility of South African engineers. Not only must engineers be able to maintain sound human relations in the workplace, but they also have to acquire people skills to deal with conflict on a daily basis. Furthermore, engineers have to ensure that effective teamwork forms the basis of all engineering projects, in order to achieve optimal performance and outputs. Moreover, the diverse local and international marketplace and work environment requires specific skills in order to understand and appreciate differences among people. By developing these people management skills, engineers will become better managers of people and other resources. The end result is a satisfied workforce and, ultimately, a productive organisation.

Self-assessment

5.1 Why is it essential for engineers to be able to manage people?

5.2 Indicate the importance of human relations skills for engineers.

5.3 Discuss the guidelines engineers should follow to promote sound human relations in the workplace.

5.4 Identify the causes of conflict in the work situation.

5.5 Differentiate among major types of conflict.

5.6 Which conflict management strategy would you use when two of your project management team members are in conflict with one another?

5.7 Describe the different conflict management strategies. Which strategy do you use most of the time when confronted with a conflict situation?

5.8 Suppose you are involved in a conflict situation with another engineer. Indicate the guidelines you would follow to resolve the conflict.

5.9 List the advantages of teamwork.

5.10 Identify the attributes of effective teams.

5.11 Indicate the benefits of diversity management for engineers.

5.12 Explain how you can use the concept of ubuntu to improve people management in your organisation.

5.13 You are appointed as an engineering manager at an electronics company. Your team consists of seven members, of which two are Zulu, two are Venda, one is Indian and one Afrikaans-speaking. State the guidelines you would follow to manage this diverse group effectively.

References

Coghill, C.C., Beery, C.E. & Quick, J.G. 1998. 'Managing cross-cultural conflict incidents.' Unpublished paper delivered at the ASTD International Conference, San Francisco, 1 June.

De Beer, H. 1998. *Development Paradigms: From Paternalism to Managing Diversity*. Randburg: Knowledge Resources.

Fisher, K. 1993. *Leading Self-Directed Work Teams: A Guide to Developing New Team Leadership Skills*. New York: McGraw-Hill.

Human, L. 1993. *Affirmative Action and the Development of People*. Cape Town: Juta.

International Institute for Management Development (IIMD). 1998. *World Competitiveness Report*. Lausanne: IIMD.

Lee, G.L. & Smith, C. (eds). 1992. *Engineers and Management: International Comparisons*. London: Routledge.

Mbigi, L. & Maree, J. 1995. *Ubuntu: The Spirit of African Transformation Management*. Randburg: Knowledge Resources.

Meyer, M. 1998. 'Quality management: The essential component is teamwork.' *People Dynamics*, 16 (4), April.

Meyer, M. & Kirsten, M. 2005. *Introduction to Human Resource Management*. Cape Town: New Africa.

O'Connor, P.D. 1994. *The Practice of Engineering Management: A New Approach*. Chichester: John Wiley.

Oakland, J.S. 1995. *Implementing Total Quality Management: Text and Cases*. Oxford: Butterworth-Heinemann.

Rayner, S.R. 1992. *Recreating the Workplace*. Essex Junction: Oliver Wight.

Reece, B. & Brandt, R. 1993. *Effective Human Relations in Organizations*. 5th edition. Boston: Houghton-Mifflin.

Wageman, R. 1997. 'Critical success factors for creating superb self-managing teams.' *Organisational Dynamics* 26 (1), pp. 49–61.

Williams, R. 1997. *So You Think You Want Teams? Confessions of an Implementer*. Productivity Fact Sheet No 27. Pretoria: National Productivity Institute.

Websites

American Society for Training & Development: www.astd.org.
ASTD Global Network South Africa: www.astd.co.za.
Department of Labour: www.labour.gov.za.
HR Future: www.hrfuture.net.
HR Highway: www.hrhighway.co.za.
Institute of People Management: www.ipm.co.za.
Management Today: www.management-today.co.za.

National Productivity Institute: www.npi.co.za.
Society for Human Resource Management: www.shrm.org.
South African Board for Personnel Practice: www.sabpp.co.za
Workinfo: www.workinfo.com.

6 Engineering Contracts and Law

Alistair Glendinning

Study objectives

After studying this chapter, you should be able to:
- understand the basis of the law of contract
- list and explain the requirements for a valid contract
- apply the principles regarding offer and acceptance
- apply the principles that determine the presence or absence of consensus
- evaluate the capacity of a party to enter into a contract
- explain the principles relating to the interpretation of contracts
- explain the various terms that may be part of a contract
- assess the consequences of breach of contract
- explain the transfer and termination of rights
- apply the principles of the law of contract to engineering and construction contracts, and know the different parties traditionally associated with such contracts

The aim of this chapter is to introduce the reader to the general principles of the law of contract, and to apply these principles to engineering and construction contracts.

6.1 INTRODUCTION

South African law of contract has developed from Roman Law, which initially was concerned with a number of different types of contracts that developed into a general concept of contract, through to Roman-Dutch Law, which treated every agreement made seriously and deliberately as a contract.

The law of contract, which forms part of the law of obligations, is based on the simple principle that one must honour one's promises. It is essential for the smooth running of society that its members must keep their promises, and thus the law of contract has been described as the endeavour of public authority to establish a positive sanction for the expectation of good faith that has grown up in the mutual dealings of men and women of average right-mindedness.

A building or construction contract may be defined as a contract in terms of which one party, namely the builder or contractor, agrees to perform construction or engineering work for another, the client, who in turn normally undertakes to render counterperformance in the form of payment of a sum of money. With building contracts, many parties are involved, resulting in complex contractual relationships that require careful consideration. Thus in any project, one might find the following parties present:

- the client (or employer);
- the principal agent (or engineer);
- the contractor;
- the clerk of works (or resident engineer);
- sub-contractors; and
- other prime contractors or consultants.

6.2   OBLIGATIONS, THE CONCEPT OF CONTRACT AND REQUIREMENTS FOR VALIDITY

Not all agreements between persons are contracts. Only those agreements that create obligations are considered contracts. Thus, a contract is an agreement that creates obligations/ legal ties (rights and duties). The obligation is a juristic tie or relationship that exists between the parties to the contract, and entails a right to performance from one party and a corresponding duty on the other to perform. Thus we are dealing with a legal relationship between two or more legal subjects.

Certain agreements are not contracts because the parties do not *intend* them to be. Such agreements are mere social agreements and have no legal consequences. At most they create moral duties, and if one of the parties does not keep to the agreement, he/she cannot be forced to do so. Examples of such agreements are when two people agree to go to the theatre together, or where work colleagues agree to form a lift club to drive together to the office.

In order to determine whether an agreement constitutes a valid contract, it is seen that an agreement will be a contract only if the parties *intend* to create a legal tie or bond (obligation) between them and if, in addition, all *other requirements* that are prescribed in terms of law are met. These requirements are as follows.

- The parties must reach agreement (consensus).
- The parties must have the necessary contractual capacity (regulated by the law of persons).
- Performance must be certain or ascertainable, i.e. there must be certainty as to the obligations created by the contract.
- Performance must be possible at the time the contract is entered into.
- The conclusion of the contract and the performance and object of the contracting parties must be lawful.
- The formalities (if any) must be complied with.

The above are absolute requirements. Thus, should any of them be found to have not been met, the contract will be rendered invalid. Conversely, compliance with all the requirements must necessarily give rise to a valid contract.

For example, Roger offers to build a swimming pool for Claire at a cost of R45 000 and Claire accepts this offer. Presuming that all the above requirements have been met, this would constitute a valid contract between the parties, in terms of which both parties will be bound by the contract and will be obliged to perform in accordance with it. Roger is obliged to build Claire a swimming pool, and Claire is in turn obliged to make payment in the amount of R45 000 to Roger.

6.3 PERFORMANCE, PERSONAL RIGHTS, CREDITORS AND DEBTORS

An obligation creates a *right* to performance in favour of one party, called the *creditor,* and it places a *duty* to perform on the other party, called the *debtor.* It gives the creditor the right to *claim* that the debtor shall perform by giving or paying something, doing something or refraining from doing something. At the same time, it places a *duty* on the debtor to perform in terms of the agreement by doing something, by refraining from doing something or by giving/paying something. There are as many obligations as there are indivisible performances owing under the contract. In the example above, two obligations arose, namely the obligation for Roger to construct the swimming pool and the obligation for Claire to pay for the swimming pool.

Thus it is seen that the meaning of *performance* is human conduct, which may consist in someone's doing something, or not doing something, e.g. if Roger constructs the swimming pool for Claire, then Roger has performed (he does what he agreed to do) in terms of the contract.

The *creditor* is the person who may *claim* performance from the other party. (He/she may claim performance because the contract created certain rights in his/her favour.) The *debtor,* on the other hand, is the person who has a *duty* to perform. It often happens that the parties to a contract are simultaneously creditors and debtors, e.g. if Alex purchases a house from Mbogeni for R500 000 both are creditor and debtor.

■ Alex is a *creditor* because he has the *right to claim* delivery and registration of the house from Mbogeni. He is also a *debtor* because he has the *duty* to pay the purchase price of R500 000 to Mbogeni.

■ Mbogeni is a *creditor* because he has the *right to claim* the purchase price of R500 000 from Alex. He is also a *debtor* because he has the *duty to* deliver and effect registration of the house in the name of Alex.

Often contracts are concluded between more than two persons. Further, in a complex multipartite contract, it may not be possible to divide the parties into 'sides' due to the fact that two parties may be on the same side for the purposes of one term of the contract, but on opposite sides for the purposes of another, e.g. consider purchase and construct contracts entered into between a developer, a building contractor and a future home-owner, in terms of which the building contractor and developer are co-creditors as regards payments to be made as the construction progresses, whereas the building contractor will be a debtor to the developer in so far as the construction of the infrastructure of the development is concerned.

A right created in terms of a contract is often referred to as a *personal right,* because it can only be enforced against a particular person, namely the other person to the obligation. The creditor may claim performance only from the debtor and not from anyone else, while only that particular debtor (and not someone else) is compelled to perform. This does not imply that the personal right need only be observed by the debtor – third parties must observe this right in a negative manner. Thus they must not infringe the right by, for example, persuading the debtor to deliver to a third person, as opposed to the creditor. Conversely, the debtor must observe the right in a positive way, i.e. performing only as against the creditor.

6.4 AGREEMENT

In determining whether or not a contract has come into existence, it is necessary to determine whether there is consensual agreement by two or more parties. A person cannot contract with him-/herself alone, though our courts have recognised the validity of contracts entered into by one person acting in different capacities (*Ex parte Oberholzer* 1951 1 SA 554 (A)). That said, in general, two or more parties are required to conclude a contract, with this important proviso when considering contracts between partnerships – partnerships having identical partners cannot contract with one another, whilst partnerships only having a common partner may validly contract with one another.

The most common method for ascertaining whether there has been agreement is to identify an offer and acceptance of that offer, it being acknowledged that a binding contract is as a rule constituted by offer and acceptance.

6.4.1 True agreement contrasted with quasi-mutual assent

In *Conradie v. Rossouw* 1919 AD 279, where the court had to consider what constitutes a valid contract, it was held that:

> If two or more persons, of sound mind and capable of contracting, enter into a lawful agreement, a valid contract arises between them enforceable by action. The agreement may be for the benefit of one of them or both. The promise must have been made with the intention that it should be accepted; according to Voet the agreement must have been entered into *serio ac deliberato animo*. This is what is meant by saying that the only element that our law requires for a valid contract is *consensus*.

When considering the concept of agreement by consent, or true agreements, or a meeting of the minds, it needs to be stressed that this concept is grounded in philosophy as opposed to law. Thus, in *SAR & H v. National Bank of SA Ltd* 1924 AD 704, where the question that was raised was: Could the correctness of a letter, purporting to be a correct record of a prior oral contract, be challenged after a significant passage of time? It was held that in cases where there was an admittedly existing contract:

> The law does not concern itself with the workings of the mind of parties to a contract, but with the external manifestations of their minds. Even therefore if from a philosophical standpoint the minds of the parties do not meet, yet, if by their acts their minds seem to have met, the law will, where fraud is not alleged, look to their acts and assume that their minds did meet and that they contracted in accordance with what the parties purport to accept as a record of their agreements.

It was thus ruled that it was too late to challenge the correctness of the letter. Further, the decision in the English law case of *Smith v. Hughes* (1871) LR 6 QB 597 607, which has been accepted in South Africa, provides that:

> If, whatever a man's real intention may be, he so conducts himself that a reasonable man would believe that he was assenting to the terms proposed by the other party, and that other party upon that belief enters into the contract with him, the man thus conducting himself would be equally bound as if he had intended to agree to the other party's terms.

This is known as quasi-mutual assent.

Thus, in order to determine whether a contract exists, one would first look for the true agreement of two or more parties, with such agreement objectively revealed by external manifestations. Thereafter, in the event of disagreement as regards the existence of agreement, one would apply the doctrine of quasi-mutual assent, and whether objectively one party had given the other reasonably to understand that they were in agreement.

6.4.2 The offer

An offer is a statement of intention in which one party (the offeror) discloses to what performance and on what terms he/she will consent to bind him-/herself to the person to whom the offer is addressed (offeree). In order for an offer to be regarded as valid, the following requirements need to be complied with.

- The offer must be definite and complete, in that it must contain sufficient information to enable the offeree to determine exactly the content of the offer. There is no agreement when the parties have not yet, during the course of their negotiations, reached agreement on any outstanding terms, no matter how minor, due to the fact that the offer is, at that point in time, not the whole offer.
- The offer must be a firm offer, such that the mere acceptance thereof will result in a contract. In this regard it is noted that an advertisement or tender constitutes an invitation to do business, and not an offer – the offer is made by the person who answers the advertisement, or who completes and submits the tender. An advertisement that contains a promise of an award does, however, constitute an offer.
- The offer must come to the attention of the offeree. This requirement flows naturally from the fact that agreement is a conscious or declared mutuality of consent.
- An offer must, as a rule, be directed at a definite person or persons. Thus an offer made by a sub-contractor to the main contractor cannot be accepted by the client. Exceptions to this rule arise in the case of a promise of a reward, or in the case of a simple auction without conditions, where the bidder makes the offer, which the auctioneer considers and either accepts or rejects.
- Any formalities prescribed by legislation must be complied with. Thus the legislature prescribes that an offer (and subsequent acceptance) for the purchase of land has to be in writing.

Traditionally, when tenders are called for, this does not constitute an offer. This is because that person, not knowing in advance who will respond to the tender, would not normally intend to commit in advance to a contract with an unknown tenderer. Furthermore, the call for tenders is generally not directed to specific persons, and thus does not comply with the definite persons requirement. A tender completed and submitted by a tenderer constitutes an offer that may or may not be accepted.

Tender documents traditionally contain detailed information on aspects such as project specifications, standard specifications, tender drawings and schedules of quantities. In terms of case law, it has been held that when such information is prepared for those who are requested to submit tenders, this information does not constitute an implied warranty that the work can be effectively executed according to the said plans and specifications (*Robertson v. Maurice Nichols (Pty) Ltd* 1938 NPD 34).

As regards the accuracy of the plans, specifications and bills of quantities, our courts look at the true intention of the parties – it considers what the contractor has undertaken to do. Thus, in the case of a lump-sum contract to perform specific work, it has been held that quantities are not part of the contract (*Sharpe v. San Paulo Railway Company* (1873) LR 8 Ch App 597). In the case of a re-measurable contract, the schedule of quantities forms part of the contract, with the contractor entitled to receive payment for work actually done.

6.4.3 Expiry of the offer

After an offer has expired, the offeree can no longer accept it.

▧ An offer expires if it is revoked or withdrawn by the offeror. It can be revoked at any time prior to acceptance. The revocation of the offer, like the offer itself, is an expression of intention, and thus needs to be brought to the notice of the offeree before it can have any effect. An offeror may bind him-/herself by contract not to revoke his/her contract, and in such case we are concerned with an *option contract* in terms of which both parties agree that the option grantor (the offeror) will keep open the offer for acceptance by the option holder (the offeree). A question that has not yet been conclusively decided by our courts is whether or not an offer in terms of which the offeror unilaterally gives notice that the offer is irrevocable for a certain period is legally binding on the offeror. In *Building Material Manufacturers Ltd v. Marais* 1990 1 SA 243 (O), it was ruled that the offer was irrevocable from the outset because it was never contemplated by the parties that acceptance of its irrevocability was a requirement.

▧ An offer does not remain valid indefinitely, but rather lapses after expiry of the time that the offeror has prescribed for acceptance, or, in the absence of such prescribed time, the offer remains open for a reasonable period. What amounts to a reasonable period depends on the particular circumstances of each case (*Dietrichsen v. Dietrichsen* 1911 TPD 486).

▧ The offer lapses on the death of the offeror or the offeree. In exceptional cases it is submitted that an offer can be accepted by the executor of the offeree, or be regarded as remaining valid by the executor of the offeror.

▧ A well-established principle is that if the offeree rejects an offer, it lapses. A mere enquiry about the terms of the offer is not regarded as a rejection of the offer. When the offeree makes a counter offer, the original offer lapses, as the counter offer is regarded as a rejection of the original offer. A conditional acceptance, where the offeree indicates a qualified acceptance, also constitutes a counter offer, with the same effect on the original offer. One can think of cases where a counter offer will not incorporate a rejection, e.g. if an offeree states: 'I would like some time to consider your offer, but in the meanwhile you may wish to consider my counter offer on the following terms and conditions.'

6.4.4 Acceptance

Before acceptance can give rise to the formation of a valid contract between offeror and offeree, certain requirements for a valid acceptance must be satisfied.

▧ The acceptance must be unconditional and unequivocal. Consent is possible only where the entire offer, and nothing more or less, is accepted. Thus, if the offeree attaches a condition to the acceptance, the 'acceptance' constitutes a counter offer that may or may not be accepted by the original offeror. A purported acceptance is equivocal when it is neither positive nor unambiguous, e.g. where the offeree states that she would like to purchase your house, but must first ascertain whether or not she can afford this, with this 'acceptance' not being valid acceptance.

- The acceptance must be made with the intention of being legally bound.
- Normally, an acceptance may be made in writing or orally or even tacitly, e.g. by a hand movement. However, if the *law* prescribes that a contract has to be in writing, i.e. a formality prescribed in accordance with the law, the acceptance must be in writing, e.g. an acceptance of an offer to purchase land needs to be in writing in accordance with prescribed legislation. The offeror may also stipulate in the offer that it may be accepted in a specified way only, e.g. in writing. Any acceptance that does not comply with this stipulation will not constitute a valid acceptance of the offer and therefore no valid contract will come into existence.
- The offer must be accepted by the person to whom the offer was addressed. However, should the offer be made in general, such as in the case of an offer for reward addressed to the general public, or an offer to employees of a construction company to purchase townhouse units at a reduced price, it may be accepted by the general public or anyone within the particular group to which the offer was directed, respectively.
- The acceptance must be in reaction to the offer, as a person cannot accept an offer of which he/she was not aware. Thus in *Bloom v. American Swiss Watch Co.* 1915 AD 100, where the company offered a reward for the receipt of information that would result in the arrest of individuals who had stolen jewellery from the company, it was held that Bloom was not entitled to the reward as he had furnished information whilst unaware of the reward that was offered. The reason for this is that if the second party does not know what the first party is proposing, there cannot be a consensus of the two minds.
- The acceptance of the offer must come to the attention of the offeror (the mere intention to accept the offer is insufficient).

6.5 CIRCUMSTANCES AFFECTING CONSENSUS

6.5.1 Consensus (agreement)

Above it was explained that consensus is the basis for every contract, with this consensus constituted by a meeting of the minds of the contracting parties. This approach is known as the intention theory, or will theory, or consensual theory, with consensus thus seen to be the primary basis for contractual liability.

Thus, if there are circumstances that affect consensus, the existence of the contract may also be affected. Depending on the circumstances affecting consensus, the contract will be void (no contract arises) or voidable (a valid contract may be cancelled by the prejudiced party).

If the consensus of one of the parties is legally invalid, e.g. one of the parties has been declared a prodigal and thus lacks the capacity to act, there can be no contract. If consensus is absent for any reason, e.g. if there is a material mistake, the contract is, in principle, void. In contrast, where there is consensus, but it has been achieved in an improper manner, e.g. as a result of duress, misrepresentation or undue influence, the contract is voidable. Such a contract is valid but, owing to the fact that a party's consent has been obtained in an improper manner, the injured/prejudiced party has the choice of upholding the contract or canceling it.

6.5.2 Mistake/error

Mistake in the context of the law of contract carries a far more restricted meaning than in the everyday sense. Mistake in contract means that one or both of the parties have the incorrect impression, which incorrect impression affects the validity of the contract between them.

There are two possibilities as regards mistake. Either the mistake is material, due to the fact that consensus was lacking (e.g. a contractor purchases a loader in terms of an installment sale transaction, whereas he was under the belief that the loader was purchased in terms of a lease agreement). Alternatively the mistake is non-material, due to the fact that it only affects the decision to contract, e.g. Steve purchases a vehicle from Joe, who he mistakenly believes is a prominent cricketer. As Steve visited the motor dealership to purchase the vehicle, his mistake did not influence his decision to purchase, with the mistake therefore being non-material.

The fact that a mistake was material, with consensus thus lacking, may lead to unjust results under certain circumstances, due to the fact that a party may seek to rely on his/her own material misunderstanding in order to escape liability. Thus our courts accept that parties will be held to their declarations of intention unless circumstances are such that the mistake is deemed to be reasonable (see the doctrine of quasi-mutual assent, above).

Conventional approach to material and non-material mistake

South African law, in distinguishing between material and non-material mistake, divides mistakes in contracts into various types:

1. *error in negotio*, which is mistake regarding the nature of the contract being entered into (Veronica intends to purchase whereas Dana intends to let), which mistake is material;
2. *error in corpore*, which is mistake regarding the identity of the subject matter of the contract (Roger intends to construct a swimming pool for Claire, whilst Claire believes that Roger will construct a swimming pool and an external entertainment area for her), which mistake is material;
3. error in motive or mistake regarding the reason for entering into the contract. A person's motive (reason) for concluding a contract does not have any bearing on the obligations created under the contract, and is therefore non-material, with consensus therefore not absent. An example would be Patrick purchasing material for a nursery due to his mistaken belief that his wife was pregnant;
4. *error in persona*, which is mistake regarding the identity of the other party. In most cases such a mistake is material, with Pothier (*Oblig, s* 19) stating that: 'the consideration of the person with whom I contract is an ingredient of the contract I intend to make.' Thus where Ellen books a struggling musician whom she mistakenly believes belongs to a successful band due to a strong similarity in appearance, Ellen's mistake is material. This must be contrasted with the purchase of the motor vehicle by Steve above, which was not material; and
5. *error in substantia*, which is mistake as regards an aspect, quality or characteristic of the subject matter of the contract, which in general is non-material, e.g. Bradley decides to purchase a house thinking that a famous actress previously lived there, which was not the case.

The above list is not necessarily exhaustive, but the point to be noted is that any error that excludes consensus is regarded as material. The purpose of classifying mistake into different categories is to assist one in the determination of whether the parties were in agreement, with the real question always whether there was consensus or dissensus.

The mistake has to be reasonable

Once the mistake is found to be material, the next question is whether it is reasonable. A mistake that is not justifiable in the eyes of the law, i.e. an unreasonable mistake, will be enforced in accordance with the declarations of the parties to the contract, notwithstanding the fact that this differs from a party's real impression.

A mistake is regarded as reasonable if the reasonable person (the normal, careful person) under similar circumstances would have made the same mistake. Thus a mistake owing to the negligence of the person who wants to rely on it is normally not be regarded as being reasonable, as the reasonable person is not negligent.

Mutual mistake

Where both parties are at cross-purposes, due to the fact that each party has an incorrect impression that affects the validity of the contract, there exists a situation of mutual mistake, which needs to be tested against the doctrine of quasi-mutual assent. Should A's interpretation of what has been agreed be reasonable, whereas B's interpretation is not, then A can rely on the doctrine whilst B cannot. If both parties' interpretation is reasonable, then the doctrine does not apply, with the contract becoming void due to absence of consensus.

6.5.3 Rectification of a contract

Should either party consider that a written document does not truly reflect the terms of agreement between the two parties to the contract, he/she may apply to have the document rectified. It should be noted that what is rectified is not the contract itself, but the document in question, because it does not reflect the true intention of the parties. One needs to note that a void contract cannot be rectified by a court.

6.5.4 Misrepresentation

Culpable misrepresentation (either fraudulent or negligent) is a wrongful statement of fact made by one party (or his/her agent) to the other party (or his/her agent) prior to finalising the contract, which misrepresentation persuades the latter party to conclude the contract. The misrepresentation may be express or tacit (take place by means of conduct), and in some cases deliberate silence may amount to misrepresentation. Innocent misrepresentation is a pre-contractual false statement of fact made innocently by one party, and which induces the other party to enter into the contract, or to agree to terms that, had he/she known the truth, he/she would not have agreed to. One needs to remember that not all pre-contractual statements that turn out to be false are misrepresentations. Puffing consists in the praising or commending of the representor's performance, and does not constitute a misrepresentation, e.g. a seller who describes his car as 'being in mint condition'. Should he state that the engine's compression ratio is as per specification, and this later is found to be false, then this statement goes further than mere praise and as a result is wrongful.

Should a misrepresentation have been made, then the contract is voidable. The prejudiced/deceived party has the choice to either uphold the contract (to continue with it) or to rescind it (to cancel it). This decision needs to be made within a reasonable time after the deceived party first becomes aware of the misrepresentation. Upon making the election, such election cannot be changed. Should the contract be cancelled, then each party is required to return the performance he/she has received.

In the case of culpable misrepresentation, the prejudiced party may further claim damages, irrespective of whether the contract is upheld or canceled. Damages is the monetary equivalent of the damage the aggrieved party has suffered, thus putting him/her in the financial position he/she would have been in had the misrepresentation not been made.

Culpable misrepresentation is a particular form of delict, which arises if the following requirements are present.

■ A false statement of fact is made to the one party (or his/her agent) by the other party (or his/her agent), i.e. an act.

■ Wrongfulness is apparent, i.e. if, according to the legal convictions of the community, the defendant has wrongfully infringed the interests of the plaintiff (this element is missing in the case of innocent misrepresentation).

■ Fault is apparent, i.e. the misrepresentation was either committed intentionally or negligently by the misrepresentor (the party making the misrepresentation);

■ The misrepresentation must have caused or induced the other party to conclude the contract or to agree to terms that he/she would otherwise not have agreed to.

■ The result of the misrepresentation is the conclusion of the contract or the suffering of patrimonial damage.

6.5.5 Duress

Duress arises when a prospective contractant is forced or compelled by means of an unlawful threat or intimidation by the other party (or someone acting on his/her behalf), which causes fear in the prospective contractant, to conclude the contract. Duress is a delict, and thus has the same requirements as misrepresentation.

Due to the fact that the consent of one party is obtained in an improper manner, the contract is voidable. The aggrieved party has the choice of upholding or canceling the contract. Damages, if proved, can also be claimed.

6.5.6 Undue influence

Undue influence may be defined as improper or unfair conduct by one of the contracting parties by means of which the other contracting party agrees to conclude a contract with the former contrary to the latter's independent will. A party who alleges that he/she has been persuaded through undue influence to enter into a contract needs to prove that the following elements were present in order to succeed with the claim:

■ that a relationship of trust existed between the parties concerned, such as the relationship between an attorney and client, doctor and patient, religious advisor and disciple, or husband and wife;

■ that the influence weakened that person's powers of resistance and made his/her will pliable; and that the other party used this influence in an unconscionable manner to persuade the victim to consent to a transaction that was to his/her detriment and that he/she would not normally have entered into when acting with his/her independent free will (e.g. where an adult child tells his elderly parent that he will prevent her from having access to her friends unless she donates her property to him).

Undue influence may be treated as a form of delict, with the same requirements as misrepresentation. A contract concluded as a result of undue influence is voidable and the victim has the choice of abiding by the contract or canceling it. A claim for damages based on undue influence has not yet been considered by our courts, with it being submitted that actual patrimonial loss will need to be proven for such claim to succeed.

6.6 CONTRACTUAL CAPACITY

A party to a contract must have the necessary contractual capacity. Every party entering into a contract is presumed to have legal capacity to do so unless the contrary is proven, with the onus of proof on the person seeking to prove lack of contractual capacity. Contractual capacity can be divided into three categories.

- **Full contractual capacity:** Persons with full contractual capacity can conclude contracts without assistance from a third party.
- **Limited contractual capacity:** Persons with limited contractual capacity need assistance/permission from a third party designated by law for this purpose.
- **No contractual capacity:** Persons with no contractual capacity cannot conclude contracts at all, not even with the assistance/permission of a third party.

6.6.1 The state

To be valid, a state contract must be entered into in the normal course of government administration, and must be authorised (either directly or indirectly) by the responsible minister, e.g. the minister of public works in the case of contracts with the National Department of Public Works.

State liability under contract is determined by the State Liability Act 20 of 1957, which provides that:

Any claim against the State which would, if the claim had risen against a person, be the ground of an action in any competent court, shall be cognizable by such court, whether such claim arises out of any contract lawfully entered into on behalf of the State or out of any wrong committed by any servant of the State acting in his capacity and within the scope of his authority as such servant.

6.6.2 Corporations

The Companies Act provides that where a transaction is concluded by a company agent on behalf of the company, and the company (or its agent) lacks the capacity or power to act solely due to the fact that the transaction is beyond the scope of the company's business, the transaction is *valid*, e.g. where the company is involved in civil engineering construction and the directors purchase a holiday home on behalf of the company.

Where there are other reasons why the directors do not have authority to conclude a particular transaction, e.g. where the articles of association of the company specifically limit their authorisation, the company will not be liable where the directors have exceeded the limits of their authority. Where the articles provide for a specific director's (or directors') potential authority based on the authorisation by the general meeting, board of directors, etc., an outsider is entitled to assume that authorisation has been obtained (this is the so-called Turquand rule) unless the outsider knows, or should have known, that authorisation has not been given.

6.6.3 The age of a person

A person's contractual capacity is influenced by age. The law distinguishes between minors and majors. An unmarried person under the age of 21 years who has not been declared a major by the high court is called a minor. A minor under the age of seven has no contractual capacity, with the minor's guardian required to act on his/her behalf.

Minors between seven and 21 years of age have limited contractual capacity and thus need the consent of a guardian to conclude a contract in terms of which duties are incurred. The guardian's consent is not needed if the minor concludes a contract in terms of which only the other party incurs duties, with the minor only acquiring rights.

Where a minor concludes a contract without having obtained the prior consent of a guardian, the minor's guardian can still give consent, either expressly or tacitly, after the conclusion of the contract. A ratified contract becomes enforceable against the minor as if the minor had the necessary assistance of a guardian when entering into the contract. The minor can also ratify the contract on attaining his/her majority.

Any contract concluded without the necessary consent of a guardian is not enforceable against the minor. The minor can enforce the contract against the other party (assuming the latter is a major) provided that the minor is also willing to perform. If the other party has already performed but the minor has not, that party cannot enforce the terms of the contract against the minor (unless the minor's guardian ratifies the contract), though he/she would have a claim based on the fact that the minor was enriched at his/her (the major's) expense. The major would only be able to claim actual expenses incurred and not profit. The minor's enrichment needs to be evaluated at the time the claim is instituted and not at the time of contracting. Thus, should the minor have, in the meantime, lost what was initially received, he/she would no longer be liable. If the value of what was received has decreased, the other party is only entitled to the diminished value of the enrichment.

6.6.4 Marital status of a person

In South Africa, individuals have a choice upon entering into a marriage to regulate the patrimonial consequences of the marriage in accordance with one of the following systems:
- in community of property (in the absence of a choice, marriages are automatically regulated in accordance with this system);
- out of community of property subject to accrual; and
- out of community of property in terms of an ante-nuptial contract.

Marriages in community of property

When two individuals marry in community of property, their separate estates (consisting of their own assets and liabilities) are combined into one joint estate, with the two individuals being equal owners of an undivided share of the estate, irrespective of their individual contributions, whether at the commencement of the marriage or during the subsistence of the marriage.

The combined estate is managed jointly by the husband and the wife, with either of the two parties able to conclude contracts without the consent of the other spouse, with this contract binding on the joint estate. However, in certain cases, the law prescribes that the consent of the other spouse has to be obtained, e.g. where a spouse binds him- or herself in a contract of surety, such consent of the other spouse is required.

Marriages out of community of property

With marriages out of community of property, each spouse retains his/her individual estate upon entering into the marriage, and thus retains full contractual capacity with regard to his/her personal estate. Consent of the other spouse is thus not required when concluding contracts.

6.7   FORMALITIES

In South African law the concluding of a contract does not, in general, require compliance with prescribed formalities (*Globe Advertising Co. v. Johannesburg City Council* 1903 TS 333). Thus, in principle, contracts may be concluded orally, or even by conduct alone. However, the obvious advantage that arises with the conclusion of written contracts is that the possibility of disputes arising as regard to whether or not there was a contract, or to the particular terms of the contract, is minimised.

6.7.1 Formalities prescribed by law

Formalities that are prescribed by statute in South Africa cover the following types of contract.

- Contracts for the alienation of land (the sale, exchange or donation of land) must be in writing and signed by or on behalf of the parties.
- Contracts of suretyship must be in writing and signed by or on behalf of the surety.
- Executory contracts of donation must be reduced to writing and signed by the donor or by a person acting on his/her written authority granted by him/her in the presence of two witnesses.
- Any credit agreement must be recorded in writing and signed by or on behalf of every party thereto.
- For ante-nuptial contracts, long leases (with duration of ten years or longer), mortgages and assignment of patents to be effective against third parties, they must be formally registered.

Any variations with regard to the above types of contracts must also be in writing and signed by the parties/surety/their duly authorised representatives.

6.7.2 Formalities prescribed by the parties

Parties to a contract may validly prescribe formalities that become binding on themselves and from which neither of them may unilaterally depart. Thus the parties may intend that an oral agreement will only become valid and enforceable after it is reduced to writing and signed by the parties or their duly authorised representatives. Alternatively, the parties may intend that the written record of their oral agreement is required solely as a means to provide proof as regards the terms and conditions of the oral agreement, in which case the oral contract becomes binding immediately.

Our courts always presume the latter intention, and thus should a party to a contract wish to prove that no valid agreement arose prior to the reduction of the agreement to writing, that party would bear the onus of proving this claim.

When the parties include as a term in the contract that any variation of the contract and of that clause must be in writing, the parties cannot thereafter vary that clause, or any other, orally. This decision was arrived at in the case of *SA Sentrale Ko-operatiewe Graanmaatskappy bpk v. Shifren* 1964 (4) SA 760 (A), and was based on a policy decision based on the preference of commercial certainty and avoidance of litigation, as opposed to the principle of sanctity of contract (due to the fact that it should be theoretically possible for parties to orally alter the non-variation clause by a later mutual declaration).

6.7.3 The parol evidence rule

This rule provides that: 'When a contract has once been reduced to writing, no evidence may be given of its terms except the document itself, nor may the contents of such document be contradicted, altered, added to or varied by oral evidence' (*Lowrey v. Steedman* 1914 AD 532). The rule applies to *all* written contracts, even if the document is used for purposes of proof only.

This rule exists because it is assumed that, once a contract has been reduced to writing, the written document represents a final and conscious record of the agreement reached between the parties, and it would further lead to inconvenience and uncertainty were parties allowed to go beyond the document to prove a contrary intention.

There are several exceptions and limitations to this rule, with the following circumstances of relevance in the case of engineering and construction contracts.

- Extrinsic evidence is used to prove the existence, or otherwise, of the contract. Thus, in *Beaton v. Baldachin Bros* 1920 AD 312, it was held that evidence could be led to show that the document in question was not a contract, but instead was given for a different purpose. A party who succeeds in this regard can thereafter establish another verbal agreement as the true agreement.
- A party can lead evidence to prove that, apart from a written agreement, an independent oral contract also exists.
- Extrinsic evidence is admissible as regards the identity of the parties to the contract, to identify the subject matter and to show the meaning of technical expressions. Evidence may also be led to establish the meanings that words may have acquired on the basis of local customs/trade usages.
- Extrinsic evidence is used to clarify and explain ambiguities in the contract document.

6.7.4 Rectification

Should a party consider that a written document does not truly reflect the terms of a written agreement, he/she may bring an action for rectification before a court (or an arbiter if the arbitration clause covers this type of dispute) to have the document rectified, i.e. to have the document amended in order that it conforms precisely to the intention of the parties to the contract (*Weinerlein v. Goch Buildings Ltd* 1925 AD 282). A void contract cannot, however, be rectified by a court/arbiter.

Neither the parol evidence rule nor the rule that no evidence may be given to alter the clear and unambiguous meaning of a written contract operates to exclude evidence in support of a claim for rectification. The party who requests the court to rectify the written document bears the onus to prove what the parties' true intention was, and further that the written document does not reflect this intention. It is noted that this onus is not easily discharged. A party claiming rectification must set out the exact words of the amendment he/she wishes to be made to the contract, because the purpose of an action for rectification is to reform a written document in a specific fashion.

6.8   PERFORMANCE MUST BE OBJECTIVELY POSSIBLE

That performance must be objectively possible during the conclusion of the contract is a requirement for a valid contract, with a contract being void if performance is objectively impossible. Since the agreement is void, the parties will be required to return to each other what has been obtained through performance of a void contract. If what has been given as performance has been destroyed without fault of the party who received the performance, its value may not be reclaimed.

Objectively (or absolute) impossible performance is impossible for everyone. Thus if a horse dies before the conclusion of a contract, it is objectively impossible to sell the horse. In contrast, subjective impossibility is impossible for the particular debtor concerned, although not necessarily for other people. Thus whilst it is subjectively impossible for Kevin to sell Amanda's motor vehicle to Veronica, as Kevin cannot validly deliver the vehicle, Amanda may sell and deliver the vehicle to Veronica. It is thus seen that with subjective impossibility, it is merely inconvenient or difficult to perform, but not objectively impossible, e.g. in the example above Kevin could purchase the vehicle from Amanda and thereafter sell it to Veronica.

In contrast to objective impossibility, subjective impossibility does not render the contract void. Thus if Kevin does not perform, he would be in breach of contract, and will be liable for damages suffered by Veronica.

We distinguish initial objective impossibility from supervening impossibility of performance, which refers to the situation where the performance becomes objectively impossible after the conclusion of the contract, with this subsequent impossibility not occasioned by the fault of either party, e.g. after a racehorse is sold, it is struck by lightening and dies. In such a case, the contract is terminated with all existing rights and duties extinguished, unless the creditor, by agreement or by the operation of law, bears the risk of performance becoming impossible, e.g. a contractual clause allowing for the passing of risk to the purchaser of a motor vehicle on the signing of the contract document, whilst delivery takes place only two months later.

6.9 THE CONCLUSION OF THE CONTRACT, ITS PERFORMANCE AND ITS OBJECT MUST BE LAWFUL

A contract is unlawful when:
- its conclusion,
- the performance to be rendered, and
- the object for the conclusion of the contract,

is forbidden by law, whether statutory or common, or is contrary to public interest, policy or good morals. Good morals can be regarded as what society regards as good behaviour, with public policy being the public opinion of the community at any particular time and place.

Statutes can either expressly or implicitly prohibit the conclusion of certain types of contract, or they can declare the content to be void. Thus, statutory law forbids, or only allows subject to stringent conditions, the sale of poisons, firearms, rough diamonds and intoxicating liquor. Any contracts in contravention of a statute, which statute further imposes a criminal sanction in regard to such contract, are regarded as null and void.

However, not all agreements contrary to legislation are invalid, with the wording of the particular statute necessary to determine whether the contract is void or merely unenforceable. With unenforceable contracts, the contract is not void, though neither party may bring an action on the contract, e.g. most wagering and gaming contracts.

Considering actual performance, any contract to commit a crime, e.g. robbery or murder, would be contrary to a rule of law, with the contract thus void. Considering good morals, an agreement to give one's children to a third party in return for a financial reward would be to the detriment of the children and thus contrary to good morals, with the agreement as a consequence void. When considering public policy, restraint of trade agreements are of particular importance.

In terms of a restraint of trade agreement, the individual concluding the agreement is prevented from exercising his/her trade/profession or engaging in the same kind of business venture as his/her employer for a specified period and within a determined area. In principle, contracts of restraint of trade are valid, lawful and enforceable. However, should the period of restraint of trade or the specified area within which the restraint operates be unreasonable, then the restraint agreement will be against public policy and therefore unenforceable (though not void). The party who alleges that he/she should not be bound to the restraint agreement would bear the onus of proving this claim.

If the object of the agreement is unlawful, then the contract is void. An example would be where Conrad purchases a firearm from Linda in order to murder his wife. In such a case the contract of sale of the firearm would be void.

In the case of a void contract, neither party can claim enforcement of the contract. Thus in the example above, Conrad would not be able to claim delivery of the firearm, not would Linda be able to claim the purchase price. Should one of the parties have already performed, then the question that arises is whether that party should claim from the other based on the other's enrichment (the action would be brought on the unjustified enrichment of the party who had received performance, as opposed to an enforcement of the contract). In the case where both parties were aware of the unlawfulness and were therefore equally guilty, the courts will not grant redress, based on the equal guilt of the parties. In contrast where the parties are not equally guilty, e.g. Linda was not aware of Conrad's intention, then Linda will be able to recover her performance from Conrad, i.e. the firearm would be returned to her.

The above is based on the *par delictum* rule, which provides that where two parties are equally guilty, the one who is in position is in the stronger position. This rule is a rule of public policy, and as public policy also demands that justice must be done, it will not be applied in cases where simple justice between person and person demands that it does not apply (*Jijbhay v. Cassim* 1937 AD 537).

6.10 THE PARTIES TO A CONTRACT, AND PARTIES COMMONLY ASSOCIATED WITH ENGINEERING AND CONSTRUCTION CONTRACTS

6.10.1 Co-creditors and co-debtors

Contracts are usually concluded between one debtor and one creditor, though it is also permissible that several debtors bind themselves as against several creditors, e.g. Richard and Joseph agree to construct a house for Peter and Jenny, in which case Richard and Joseph are co-debtors who need to construct the house, with Peter and Jenny being co-creditors.

Simple joint liability

In the case where the parties to the contract have made no stipulations, then each debtor to the contract is simply jointly liable for his/her own share only. However, for simple joint liability to arise, the required performance must be divisible. Thus, in the example above, the construction of a house is not divisible, so simple joint liability cannot arise. If the performance required by Peter and Jenny is to pay the contract price of R800 000, this performance is divisible, with Peter and Jenny each liable for R400 000. (In the case of partners and fellow signatories to a negotiable instrument, e.g. cheques, the general rule does not apply, with the parties presumed to be jointly and severally liable.) With co-creditors, the parties are presumed to be jointly entitled to the performance and may thus claim a proportionate share.

Joint and several liability

Joint and several liability only applies if the parties have expressly contracted to be bound jointly and severally as debtors. In such case, the creditor can hold any of the debtors fully liable for the entire debt, or he/she can recover the entire debt in any portion from the various co-debtors. In the event of one of the co-debtors performing fully to the creditor, all the other co-debtors are absolved.

A joint and several debtor who has been obliged to pay in excess of his/her equal share can claim a contribution from his/her co-debtors. Thus if Bertie has paid R6 000 to Patricia, he may claim R2 000 respectively from his co-debtors, Carol and Stuart. Where co-creditors are entitled jointly and severally, each creditor may claim the entire performance or a part thereof. Payment of the whole debt to one co-creditor constitutes discharge of the debt as against all creditors.

Joint (common) liability

In such a case, the debtors are jointly liable only, and the co-creditors may only claim performance jointly against all debtors.

6.10.2 Parties commonly associated with engineering and construction contracts

The engineer

Engineers received statutory recognition in 1968 with the promulgation of the Professional Engineers' Act 81 of 1968. The registration, discipline and control of and reservation of work for engineers, engineering technicians and engineering technologists is now regulated by the Engineering Council of South Africa.

Engineers normally act as agent for the employer with regard to engineering contracts, with the scope of work of the engineer often regulated by the standard form of contract used for the works. In South Africa, the most common forms of contract used are the General Conditions of Contract for Works of Civil Engineering, 6th edition, 1990 (superseded by the 7th edition, 2004), the COLTO conditions of contract, the FIDIC suite of contracts, and the New Engineering Contract.

In terms of the General Conditions of Contract, 1990, the engineer is given authority to act as the employer's agent for purposes of the following:

- to administer the contract as agent of the employer, to supervise the works, and to be satisfied with the quality of materials and workmanship;
- to explain and adjust ambiguities that may arise in the contract documentation, and to certify any monies that may be due to the contractor as a result of such ambiguities;
- to establish basic reference pegs and benchmarks on the site in sufficient time to enable the contractor to comply with his/her obligations;
- to retain original drawings and to supply the contractor with copies thereof, along with revised drawings if so required;
- to approve the contractor's designs and drawings, if so required;
- to consider contractual claims based on unforeseen adverse physical conditions;
- to approve and, if required, to alter the order in which the construction works are to be carried out, and to receive the contractor's programme and particulars as regards resource planning, and to suspend the works if so required;
- to advise the contractor as to what information is required as regards the contractor's employees;
- to approve the contractor's site agent;
- to examine and measure the works, and to inspect excavations before permanent works are placed thereon;
- to order the removal of inadequate or negligent workmanship;
- to require the rectification of errors in the setting out of the works;
- to deliver to the contractor test results as regards materials and workmanship;
- to give instructions regarding facilities to be afforded to other contractors employed by the employer;
- to give instructions to the contractor as regards the dealing with fossils, coins, or articles of value and antiquity;
- to approve the contractor's security and lighting arrangements, and to direct the contractor as to any additional requirements for the protection of the works and for the safety and convenience of the public;
- to instruct the contractor to carry out repair works, if required, at the contractor's own expense;

- to order the contractor to effect repairs for work that was damaged or for materials lost due to 'excepted risks', and to certify payment due in respect of such repairs/replacement;
- to withhold payment certificates pending receipt of the contractor's guarantees;
- to notify the contractor as regards urgent remedial work that was undertaken by third parties due to the contractor's unavailability to effect such works;
- to give consent in writing for the contractor to remove his/her plant from the site;
- with regard to occurrences on site that cause damage to property, or injury or death to persons, to make enquiries as to the cause and result of such occurrences;
- to order variations to the contract and to determine the value of such variations in terms of the contract;
- to grant an extension of time due to delays caused by the employer or due to delays beyond the control of the contractor, and to certify an amount to cover costs that have arisen due to such delays;
- to allow an extension of time due to the requiring of the contractor to execute additional work;
- to grant the contractor permission to work outside of normal working hours;
- to require the contractor to accelerate the execution of the works, and to approve the proposed acceleration steps, should the engineer consider the rate of progress to be unduly slow;
- to receive the contractor's claim certificate; to measure the work completed by the contractor, including variations; to prepare a monthly certificate in respect of such work; and to give directions as regards provisional and prime cost sums;
- to issue a certificate of completion in respect of the works or part thereof, and to issue a final certificate as regards payment due to the contractor;
- to give instructions regarding the rectification of defects that arise during the defects liability period, to consider whether this period needs to be lengthened, and to determine whether the contractor is entitled to receive payment for any work executed during this period;
- to make valuations if the contract has been determined; and
- to consider claims made by the contractor.

The resident engineer

As the engineer who is responsible for the design and execution of the works is often not able to give his/her undivided attention to the works as they progress, a resident engineer is normally appointed to handle the day-to-day problems such as detailed scheduling and supervision. In terms of the General Conditions of Contract, 1990, the resident engineer must watch and supervise the works, and examine materials to be used and the workmanship employed in connection with the works.

The architect/quantity surveyor

Architects are regulated by the South African Council for Architects, with an architect being a professional person whose function is to design and supervise the erection of buildings. This includes the preparation of concept drawings, detailed plans accompanied by the necessary specifications and contract documents (most often the Joint Buildings Condition of Contract and specifications are used, with the architect/employer's agent defined as the

Prime Agent). An architect is thus somewhat analogous to an engineer in the case of building contracts (except that structural engineers are required to execute the work associated with the structural elements of the building).

Quantity surveyors are regulated by the South African Council for Quantity Surveyors, with a quantity surveyor being a duly qualified professional person who is required to take out in detail the measurements and quantities from architectural plans in order to produce a schedule of quantities, which details all the work to be carried out. The contractor would use this schedule in order to calculate the cost for the execution of the works.

The professional construction project manager/project manager

The recently constituted South African Council for Project and Construction Project Management Professionals (SACPCPMP) will, in the near future, require that any construction project be managed from the employer's side by a construction project manager, and from the contractor's side by a project manager. It can thus be expected that many of the management duties of the architect/engineer be taken over by the construction project manager in the future, though it is noted that many architects and engineers will be registered with both their councils and the SACPCPMP.

6.11   TERMS OF A CONTRACT

A term in a contract is a provision that imposes on one of the contracting parties one or more contractual obligations to perform some or other act, or to refrain from performing some or other act, or which qualifies the contractual obligations of the parties, i.e. the term specifies the time when or the particular circumstances in which the obligations become enforceable or are terminated.

6.11.1   Terms of contract in general

The terms of a contract fall into one of three classes:
- the *essentialia* of a contract, which are terms that the law requires as essential to place a contract in a certain category, e.g. a purchase price is essential in a contract of purchase and sale;
- the *naturalia* of a contract, which are terms that naturally form part of a contract, though which the parties can expressly change, e.g. it is a *naturalia* of a contract of purchase and sale that the item sold will be free of latent defects, though the parties may alter this by providing that the item sold is *voetstoets*, i.e. sold 'as is'; and
- the *incidentalia* of a contract, which are additional terms that the parties themselves make part of a contract, e.g. in a construction contract, that disputes will be resolved by means of the referral of the matters to an arbitrator.

6.11.2   Express, implied and tacit terms

Express terms

These are terms that the parties to a contract incorporate into the contract by means of communicated declarations of intent, whether oral or in writing.

Implied and tacit terms

Implied terms are terms not expressed by words, but that are incorporated into contracts by operation of law. Tacit terms are also not expressed in words, but are based on the parties' true intention, or the intention as assigned by law. In building and engineering contracts, these terms tend to be used interchangeably.

In construction contracts, it is implied that the owner will co-operate with the contractor, e.g. give possession of the site to the contractor, etc. It is further implied that the contractor will do the work in a proper and workmanlike manner, the materials used will be fit for their purpose (unless the employer or his/her agent prescribes the materials to be used), and that the contractor will conform with all applicable legislation and regulations.

It is noted that if a particular matter is dealt with by an express provision in a contract, neither party can claim that the matter is further regulated by an implied term (*Alfred McAlpine and Son (Pty) Ltd v. Transvaal Provincial Administration* 1974 (3) SA 506 (A)).

6.11.3 Conditions

A condition is a contractual term that renders the operation (both the arising and the termination of contractual obligations) and consequences of the contract dependent on the occurrence or non-occurrence of a specified uncertain future event. The uncertainty relates to whether or not the event will actually occur.

Conditions are traditionally categorised as suspensive or resolutive conditions. A suspensive condition is a contractual term that has the effect of suspending the operation of obligation/s in terms of the contract until the condition has been fulfilled. With a positive suspensive condition, the operation of the obligation would be suspended until the uncertain event takes place, whereas with a negative suspensive condition, the operation of the obligation would only come into existence if the uncertain event does not take place, e.g. Kirsty agrees to sell Bongani her house if she is transferred to Cape Town. The contract is binding on the parties on the conclusion thereof, though Kirsty only has to perform if she is transferred. In the event that she does not get transferred, then the condition would not have been fulfilled, with the suspended obligation coming to an end. Should she be transferred and then refuse to accept the transfer, then the suspensive condition will be deemed to have been fulfilled, based on the debtor's deliberate attempt to prevent fulfilment.

Resolutive conditions are contractual terms that render the continued existence of a contract dependent on the occurrence (or non-occurrence) of a specified uncertain future event. The normal consequences of the contract become operative and enforceable upon the conclusion of the contract, though if the condition is fulfilled, the contract is dissolved, with the contractual rights and obligations therefore falling away, e.g. Gareth agrees to give Kerry his grandmother's engagement ring, conditional on their getting married. Thus if Kerry terminates the engagement, she would be required to return the ring.

6.11.4 Time clauses

Time clauses are differentiated from conditions in that a time clause determines a specific time or period within which the contract will either become operative or be dissolved. The moment must be specified to provide certainty as to the exact moment that will result in the fulfillment or otherwise of the time clause. Time clauses may either be resolutive or suspensive, with a resolutive time clause providing for the termination of an obligation on a certain date or on the occurrence within the prescribed time of a certain future event. In contrast, a suspensive time clause postpones the operation of the obligations until the future defined moment or event.

6.11.5 Supposition

This is a contractual term that renders the operation of a contract dependent on an event that has already taken place, or on an existing state of affairs at the time of contracting, with the parties uncertain as to the true state of affairs. Thus, if Paul sells Tammy his car provided the vehicle has traveled less than 150 000 kilometers, with neither party sure of the true state of affairs due to the fact that Paul loaned Stiduso the vehicle three months earlier, obligations are created immediately if the vehicle has traveled less than 150 000 kilometers. Should the vehicle have traveled 155 000 kilometers, then the contract creates no obligations.

6.11.6 Warranties

A warranty is a contractual term whereby a contracting party assumes absolute liability for proper performance. Any aspect of a party's performance may be guaranteed, including facts of the past, present or future, or that the performance will be fit-for-purpose, e.g. with a turnkey project for the design and construction of a refinery, the debtor may warrant that the plant will have an output of 50 000 tons of steel per day. With insurance contracts, the insured is often required to warrant that his/her statements are both accurate and truthful.

6.11.7 Exemption clauses

Exemption clauses limit the liability of one of the parties to the contract, e.g. liability for latent defects in the item sold, for misrepresentation or for breach of contract. Exemption clauses may appear to operate unfairly against one party to the contract, though in general, due to the principle of sanctity of contract, in the absence of fraud or duress, the courts will enforce the terms. These terms will be interpreted narrowly by the courts, and will not be applied where the clause is contrary to public policy, or where it purports to condone fraudulent conduct or intentional breach of contract (*Wells v. South African Alumenite Company* 1927 SA 459 (A)).

6.11.8 Non-variation clause

Most contracts that are reduced to writing contain a clause that provides that any purported subsequent variation or alteration will be of no force and effect unless the alteration/variation has been reduced to writing and signed by the parties. In the absence of such a clause, parties to a written agreement are free to vary it orally (verbally).

6.11.9 Cancellation clauses

Generally, contracts may only be canceled due to a serious or material breach of the contract. However, a contract may include a cancellation clause entitling a contracting party to summarily cancel the contract due to the other's breach, no matter how trivial the breach may be. The former party would then be entitled to cancel the contract without notice. Should the cancellation clause require notice, then the former party, before he/she can cancel, must give notice to allow the other party to remedy the breach within the specified period. This provision must be strictly complied with, and failure to do so will result in the purported cancellation being invalid.

6.11.10 Penalty clauses

With breach of contract, in order to claim damages, the injured party bears the onus of proving the exact extent of his/her loss. As this onus is often difficult to discharge, the parties to a contract often include a penalty clause in their contract. With a penalty, the parties agree that the party who commits breach of contract must render a specified performance (such as the payment of a sum of money) to the aggrieved party, who can claim the penalty without having to prove that any damages were suffered.

In South Africa, the imposition of penalties is regulated by the Conventional Penalties Act 15 of 1962, which provides that the aggrieved party cannot recover both penalties and damages, and further that where penalties are expressly stipulated, the aggrieved party loses his/her right to claim for damages.

The party who is required to pay penalties (the debtor) may apply to a competent court to reduce the penalty payable due to the fact that the penalty is out of proportion to the actual damage suffered by the aggrieved party. The onus to prove that the penalty is out of proportion to the actual damage suffered lies with the debtor (*Smit v. Bester* 1997 (4) SA 937 (A)).

6.12 INTERPRETATION OF CONTRACTS

In the interpretation of contracts, primary rules of interpretation are always applied firstly. In the event of ambiguity remaining, secondary rules are applied, and in the event that these rules do not resolve the ambiguity, the tertiary rules of interpretation are applied as a last resort.

The primary rules of interpretation were summarised in *Coopers & Lybrand v. Bryant* 1995 3 SA 761 (A) as follows.

1. The intention of the parties must be determined from the language used in the contract, as well as from background circumstances, which is evidence of an identificatory nature. Thus, if Tasneem signs a contract to build a house in Johannesburg, evidence of what *erven* are registered in Tasneem's name can be used to determine which stand the contract refers to.
2. The words used in the contract must be given their ordinary, grammatical meaning, unless such meaning results in an absurd interpretation, or in an interpretation that the parties obviously never intended. With engineering and construction contracts, words will usually be given their technical meanings, based on the technical nature of such contracts. It is noted that the unambiguous meaning of a word or phrase must not be departed from simply because it operates unfairly upon one of the parties to the contract.
3. The contract will be interpreted as a whole, which involves considering the context in which a word or phrase is used in terms of its interrelation with the contract as a whole.

After applying the above rules, should the meaning of every word and phrase be clear, no further rules of interpretation will be applied. In the event of ambiguity remaining, the secondary rules of interpretation provide as follows.

1. The equitable interpretation that leads to fairness between the parties is favoured in the event of ambiguity, which is in accordance with the principles of contracting in good faith.
2. One must consider the meaning that best fits the nature and purpose of the agreement.
3. An interpretation that renders the contract valid is favoured over any other interpretation that renders the contract void.
4. It is presumed that every word used in the contract was intended to have some effect, or be of some use.
5. One must consider the manner in which the parties executed the contract in order to determine their original intention.
6. One must consider surrounding circumstances in order to determine the parties' original intention.
7. In cases where a general word is followed by more specific words, the general word is limited by the specific word or words.

Tertiary rules of interpretation, which are used as a last resort, provide that:
1. the interpretation that leads to the least inconvenience to the debtor is favoured over any other interpretation; and
2. the clause is interpreted against the party in whose favour it was inserted, or against the party (or his/her agent) who drafted the clause.

6.13 BREACH OF CONTRACT

Breach of contract occurs when the one contracting party culpably interferes with the rights of the other contracting party. The law of contract recognises five distinct ways in which breach of contract may occur, namely default by the debtor (*mora debitoris*), default by the creditor (*mora creditoris*), malperformance, repudiation and the prevention of performance.

6.13.1 Default by the debtor *(mora debitoris)*

Mora debitoris arises where the debtor neglects or fails to perform timeously whilst performance remains possible, with the debt in question due and enforceable, i.e. the time for performance must have arrived, the debt must not have prescribed, and no prior performance must be required by the creditor. If the debtor cannot perform in time because of circumstances beyond his/her control, then in the absence of a guarantee, the debtor will not be in *mora*.
   Two forms of *mora debitoris* are distinguishable.

■ *More ex re* occurs where the time for performance is determined by the contract and the debtor fails to perform on or before the prescribed date, with the delay attributable to the fault of the debtor.

■ *Mora ex persona* occurs where the contract does not stipulate a specific date for performance, with the debtor having to perform within a reasonable time after the conclusion of the agreement. The creditor will have to fix a time for performance by making a demand for performance, which demand must be reasonable, taking into account all relevant circumstances, with the debtor in *mora ex persona* if he/she fails to perform when the time arises. Should the time in the demand by the creditor not be reasonable, then the demand is ineffective, and the creditor will be required to make a demand anew before *mora* can arise. It has been held that the demand must be in writing (*West Rand Estate v. New Zealand Insurance Co.* 1926 AD 173).

6.13.2 Default by the creditor *(mora creditoris)*

Where the co-operation of the creditor is needed for the debtor to perform, the creditor will be guilty of breach of contract, in the form of *mora creditoris,* if he/she fails to co-operate timeously whilst the performance remains possible. This failure to perform properly can occur when the creditor fails to accept the proper performance that has been tendered by the debtor, or if he/she fails to perform an act that is necessary to enable the debtor to be able to perform his/her obligations, e.g. in order to enable Roger to construct a swimming pool for Claire, she is required to perform by opening the boundary gate to allow Roger access to the site.

6.13.3 Positive malperformance

Positive malperformance occurs when a contracting party does not comply with the terms and conditions of the contract, either by performing something in a manner that does not comply with the terms of the contract or by doing something that he/she undertook not to do, e.g. in the first instance Roger agreed to build a swimming pool with a circumference of 25 metres, but the actual circumference of the swimming pool is only 22 metres. An example of the second type of positive malperformance is when Matthew starts an engineering consulting practice in Johannesburg that is contrary to a restraint of trade agreement concluded with his former employer.

6.13.4 Repudiation

Repudiation occurs when one contracting party conducts him-/herself in such a manner that the other party reasonably concludes that the former will not render performance, or will not render further performance in accordance with the terms and conditions of the contract (*Culverwell and Another v. Brown* 1990 (1) SA 7 (A)). It should be noted that a mere delay by a contracting party to perform does not necessarily indicate repudiation, but rather the conduct of the guilty party must be such that the other party is reasonable to presume that the former has no intention to perform (either at all or properly). Examples of repudiation include a denial of liability, an offer of compromise, a refusal to perform, etc.

In order to determine whether or not a party has repudiated a contract, one must objectively determine whether that party has acted in such a way as to direct a reasonable person to the conclusion that he/she no longer intends to fulfil his/her part of the contract (*Tucker's Land and Development Corporation (Pty) Ltd v. Hovis* 1980 (1) SA 645 (A)).

Our courts have accepted that repudiation constitutes an infringement of an existing obligation. The effect of repudiation is not to automatically end a contract, but rather to allow the creditor the choice to accept or reject (ignore) the repudiation. If the repudiation is accepted, the creditor is said to have rescinded the contract, i.e. terminated the contract. If the repudiation is rejected, the contract remains in full force and effect, with the innocent party entitled to claim that the other party must fulfil his/her obligations.

6.13.5 Prevention of performance

Prevention of performance arises where performance is made impossible by either contracting party after the conclusion of the contract. Examples can be where the seller of a motor vehicle negligently destroys it, or where a creditor refuses to accept milk that subsequently sours.

Thus it is seen that both debtor and creditor can commit breach of contract in the form of prevention of performance, though in both cases the impossibility must be due to the culpability/fault/blameworthiness of the guilty party. If neither party is to blame for the impossibility, it will be a case of supervening impossibility of performance (provided the impossibility is objective), in which case the contract will be terminated.

6.14 REMEDIES FOR BREACH OF CONTRACT

With a breach of contract, there are three remedies that may be available to the innocent party, with the availability of each remedy dependent upon the form, and the severity (seriousness), of the breach.

6.14.1 Specific performance

Specific performance is the performance of that which the parties agreed to when entering into the contract. An innocent party is always entitled to claim specific performance, with the court having a discretion as to whether to grant such an order or not (*Farmers' Co-op Society v. Berry* 1912 AD 343). Each case must be judged according to its own circumstances, with specific performance not granted in cases where the performance has become impossible, where it would cause undue   hardship to the debtor, where the courts cannot enforce the decision (e.g. where performance must take place in Kenya), where the debtor is no longer in a position to fulfil his/her obligations, or where it concerns the freedom of the individual, e.g. breach of a promise to marry.

In a reciprocal contract, for the innocent party to claim specific performance, he/she must have performed in terms of the contract. It should be noted that a claim for specific performance does not exclude a claim for damages where the innocent parties has suffered damages as a result of the breach.

6.14.2 Cancellation/rescission of the contract

Cancellation or rescission of a contract is a juristic act in terms of which the consequences of a prior valid contract are terminated. Rescission is only available in exceptional circumstances where:
- the breach is material, i.e. of a sufficiently serious or important nature; or
- the parties have incorporated a cancellation clause in their contract, with the cancellation of the contract effected exactly in terms of the provisions of the cancellation clause; or
- where restitution, i.e. a return of the performance, is possible.

Where the above circumstances exist, the innocent party has the choice to uphold or rescind the contract. For the innocent party to be entitled to cancel the contract, the right of rescission must be exercised:
- clearly and unambiguously;
- within a reasonable time after the breach of contract came to the attention of the innocent party; and
- with immediate effect.

Once this choice to rescind/uphold the contract has been made, the innocent party cannot change his/her mind. The choice to cancel must be communicated to the other party (orally or in writing), unless the contract prescribes requirements for cancellation, which must be strictly complied with.

With rescission, the contract dissolves, with each party obliged to restore what has been received in terms of the contract. Rights that have already crystallised are enforceable, e.g. the right to receive a salary for those completed periods of work. The guilty party is further liable for any damages suffered.

6.14.3 Damages

'Damages' is defined as the monetary equivalent of damage awarded to a plaintiff with the object of eliminating, as fully as possible, his/her past as well as future damage, with damage being the diminution, as a result of the damage-causing event (in this case the breach of contract) in the utility or quality of a patrimonial or personality interest that the law recognises as being worthy of protection. In this case, we are interested in the innocent party being compensated for financial losses occasioned by the other party's breach of contract.

Damages are claimable irrespective of whether the innocent party cancels or upholds the contract (claims specific performance). The object of an award of damages is to place the innocent party in the same financial position he/she would have been in if the contract had been performed properly. To succeed with a claim for damages, the innocent party is required to prove that he/she has suffered actual patrimonial (financial) losses. In terms of the law of contract, no damages can be claimed for pain and suffering, inconvenience, injured feelings, etc. (Such amounts are claimable as a claim for satisfaction based on injury to personality, though this aspect falls outside the scope of this chapter.)

The financial losses must have been caused by the breach of contract and must reasonably have been foreseen or contemplated by the parties when the contract was entered into. The innocent party must try to mitigate his/her losses by taking positive steps, e.g. where a sub-contractor has defaulted, the main contractor arranges for another sub-contractor to complete the works in order to reduce the penalty being charged by the client for late completion of the project.

6.14.4 Typical damages due to delay caused by an employer with regard to construction contracts

Additional on-site expenses

A delay by the employer will result in the contractor having to take additional time to complete the works, and as such will cause him/her to incur additional on-site expenses such as extra supervision, security expenses, small plant and tool hire, service charges, etc. Often such amounts are included in the time-related preliminary and general item of the schedule of quantities, with the amount claimable based on a pro-rata share as opposed to actual costs expended, depending on the terms of the contract.

Escalation on material, labour, fuel, plant and sub-contractors

In the absence of a suitable escalation clause, the contractor will claim actual additional expenses incurred due to the effects of inflation on material, labour, fuel, plant and sub-contractors.

Additional expenses due to loss of productivity

Delays by the employer may result in the contractor not being able to construct the project in accordance with the original contract programme, e.g. the contractor may have planned to have balanced cut-to-fill operations, so any delay in handing over of sections of road will necessarily affect productivity due to changed haul routes necessitated by the use of material from quarries. It should be noted that this type of loss is often very difficult to quantify in practice. As the onus to prove damages lies with the contractor, when a delay arises, it is recommended that detailed calculations and representations be submitted to the engineer and updated on a regular basis thereafter.

Additional costs of financing and insurance

Delays in the construction process may adversely affect the cash flow of the contractor, resulting in his/her needing additional financing. In addition, insurance premiums are often based on contract durations, and thus delays result in the levying of further premiums. These costs are claimable in the event of the delay arising due to the default of the employer.

Loss of profit and off-site overhead

In principle, any construction delay will affect the ability of the contractor to contract to execute work for other employers, and thus any delay results in a loss of profit to the contractor. This type of damage is seldom awarded in practice, due to the difficulty in proving that the contractor would have made profit elsewhere. Thus this type of damage is not presumed to flow naturally from the breach of contract.

6.15 THE TRANSFER AND TERMINATION OF OBLIGATIONS ARISING FROM A CONTRACT

6.15.1 Cession

Rights arising from a contract can be transferred by the holder thereof (the cedent) to another (the cessionary) by a cession agreement. Generally no formalities are required for a valid cession, nor is the debtor's consent. However, the debtor's consent is needed if the right is ceded in part or split up and ceded to a number of cessionaries, or if the contract requires the permission of the debtor. The effect of a cession is that the cedent loses the ceded right/s to the cessionary, who can enforce those rights directly as against the debtor.

6.15.2 Delegation

Delegation is the transferring of contractual duties from the original debtor to a third party. The permission of all the parties to the contract is required, as through delegation the original duty is extinguished and replaced with a new duty created between the creditor and new debtor. Where one party to a contract has rights and duties, these rights and duties can be *assigned* to a third party with the consent of all parties to the contract.

6.15.3 Discharge

A contract is terminated naturally by its discharge, i.e. performance in accordance with what the parties envisaged when they entered into the contract. Discharge is normally a bilateral juristic act where the co-operation of the creditor is required, though it may also be a unilateral juristic act where no co-operation is required from the creditor, e.g. where Courtney agrees to chop down the tree on her property so as to stop leaves blocking her neighbour's pool filter.

Discharge can be effected by means of tender, delivery or payment. Performance must furthermore be carried out at the place indicated on the contract. If no place has been indicated, one must determine the place of performance from the intention of the parties, e.g. a meal must be delivered in the restaurant it was ordered. When an amount of money must be paid, the creditor must go to the debtor to collect the money. In the event that a place of performance cannot be determined from the intention of the parties, the general rule is that performance must take place at the place where the contract was concluded.

As regards the time of performance, in the case where the contract has fixed the date of performance, the debtor must perform on or before the fixed date. In the case where the date was fixed in favour of the creditor, in the event of the debtor performing earlier, he/she must pay interest up to the fixed date, e.g. where money is loaned for a one-year period, and the debtor wishes to make earlier repayment, additional interest will be payable (subject to the provisions of the Usury Act). Where no time for performance has been set, the debtor must perform immediately, or within a reasonable time, depending on the circumstances of the contract, e.g. the delivery of perishables must be performed immediately.

6.15.4 Release

This is a further agreement between the creditor and the debtor in terms of which the creditor releases the debtor from his/her obligations in terms of the original contract. No consideration is required from the debtor for such release.

6.15.5 Novation

This is an agreement between a creditor and a debtor to an existing obligation in terms of which the original debt between them is extinguished and a new obligation is created in its place, e.g. Paul was required to deliver a BMW to Sharon, and the parties agree that he will deliver a Volvo in its place. It is noted that novation cannot take place in respect of an agreement that was void for some reason.

6.15.5 Set-off

Set-off is the extinction of debts owed reciprocally by the parties, e.g. Thomas owes Swazi R1 000, whilst Swazi owes Thomas R780. Thus Thomas owes Swazi R220 after set-off. The requirements for set-off are that the debts must be similar in nature, the debts must be liquid (have an exact monetary value), the debts must be due, and the debts must be between the same parties.

6.15.6 Compromise

If the parties are in dispute as to the performance that is due, they may agree to settle the dispute by way of a compromise agreement, e.g. Lara claims Chloe owes her R25 000 whilst Chloe maintains that the debt is R12 000, with the parties agreeing to a compromise in terms of which Chloe will admit to a debt of R20 000, which Lara accepts as the amount owing to her. The compromise then forms the basis for the rights and duties between the parties.

6.15.7 Prescription

The passage of time, by way of prescription, influences obligations. Prescription starts to run as soon as the debt is due and payable, and in terms of the Prescription Act 68 of

1969, contractual obligations prescribe (become unenforceable through the lapse of time) after three years, after which the debtor will be released from his/her duties in terms of the contract. The issuing of summons against the debtor interrupts prescription.

6.16 DIFFERENTIATING BETWEEN THE CONTRACT FOR THE LETTING AND HIRING OF WORK, AND THE CONTRACT FOR THE LETTING AND HIRING OF SERVICES

In Roman Law, there were three species of *locatio conductio* (letting and hiring) contracts, namely:
- contracts for the letting and hiring of a specific thing, e.g. a horse;
- contracts for the letting and hiring of work; and
- contracts for the letting and hiring of personal services in return for remuneration.

The differentiation between the second and third type of contract, i.e. between contracts of work and contracts of service, is important in the sphere of engineering and construction contracts, with the former applicable where an independent contractor, e.g. an engineer, agrees to complete a piece of work in accordance with the contract between the parties. With the latter, one is concerned with an employer–employee relationship, in terms of which the employee would be under the control and direction of the employer.

Thus in the case of a contract of work, the contractor would perform work for the owner, though the contractor is not an employee of the owner, nor is under the direction and control of the owner. The contractor can further select the individuals who will undertake the actual work, and may therefore not execute the works in his/her personal capacity. Engineering and construction contracts are traditionally put out to tender by an owner, with the contract formed by the owner accepting a particular tender. No formalities are prescribed by law for the formation of construction and engineering contracts.

With contracts of work, the main duty of the contractor is to complete the works, in a proper and workmanlike manner, and within the prescribed period or a reasonable period if no actual period is prescribed, in accordance with all other terms of the contract and specifications. The main duty of the owner is to remunerate the contractor for work done in accordance with the terms of the contract. The terms of the contract could specify payment of a lump sum for work done, payment based on the measurement of actual work completed, or payment based on the actual cost to the contractor, plus a mark-up based on a percentage of the actual cost or a pre-determined sum.

With a contract of service, an engineer or architect would undertake work whilst acting in the capacity of employee. In such case, the engineer or architect would be under the control and direction of the employer, and would be required to undertake the work in his/her personal capacity. One would thus be dealing with an employment contract, with an employee defined in the Labour Relations Act as:

> ... any person, excluding an independent contractor, who works for another person or the State and who receives, or is entitled to receive, any remuneration, and any other person who in any manner assists in the carrying on or conducting of the business of an employer.

In practice, it is important to distinguish between the two types of contract, as the former is regulated in accordance with the prescripts of the law of contract, whilst the latter is regulated by the prescripts of the law of contract as modified by the prescripts of labour law (which includes the Labour Relations Act, the Employment Equity Act and the Basic Conditions of Employment Act).

Self-assessment

6.1 Explain what a contract is.

6.2 List the requirements for a contract to come into existence.

6.3 Explain what performance is.

6.4 Briefly explain whether a valid contract was concluded in the following cases:
   a) Peter makes an offer to George to repair equipment for R4 500. George tells him that he is not prepared to pay more that R4 000.
   b) George informs Peter one day later that he is willing to pay R4 500.

6.5 Briefly explain the requirements for a valid offer.

6.6 Marty and Susan agree that their verbal contract must be reduced to writing. Explain whether a valid contract will be concluded before it has been reduced to writing.

6.7 Explain the following:
   a) conditions
   b) guarantees
   c) cancellation clauses
   d) penalty clauses.

6.8 List the ways in which breach of contract can occur.

6.9 Briefly distinguish between *mora debitoris* and *mora creditoris*.

6.10 A sells his car to B for R20 000 and delivers the car to B. B undertakes to pay the purchase price to A within a reasonable time. Three days after conclusion of the contract, B informs A that he cannot perform his obligation of paying the purchase price. Discuss the form of breach committed by B.

6.11 Distinguish between rendering performance impossible and initial impossibility of performance.

6.12 Discuss the remedy of cancellation for breach of contract.

6.13 What should the economic result be of a reward for damages in a case of breach of contract?

6.14 Briefly explain the principles relating to the contract for the letting and hiring of work.

References

Basson, A. et al. 2005. *Essential Labour Law.* 4th edition. Cape Town: Labour Law Publications.

Christie, R.H. 2001. *The Law of Contract in South Africa.* 4th edition. Durban: Butterworths.

Gibson, J.T.R. 1997. *South African Mercantile and Company Law.* 7th edition. Cape Town: Juta.

Havenga, P. et al. 2000. *General Principles of Commercial Law.* 4th edition. Cape Town: Juta.

Hutchison, D. et al. 1991. *Wille's Principles of South African Law.* 8th edition. Cape Town: Juta.

McKenzie, H.S. 1988. *The Law of Building and Engineering Contracts and Arbitration.* 4th edition. Cape Town: Juta.

Van Aswegen, A. et al. 1996. *General Principles of the Law of Contract.* Pretoria: Unisa Press.

Visser, P.J. & Potgieter, J.M. 1993. *Law of Damages.* Cape Town: Juta.

7 Operations Management

David Kruger

Study objectives

After studying this chapter, you should be able to:

- define the term 'transformation process'
- describe the nature of the transformation process
- define manufacturing and service operations
- discuss and differentiate between the manufacturing and service industries
- discuss and explain the operations design process
- discuss the different means of defining and measuring capacity
- explain the importance of capacity planning
- explain what a flexible manufacturing system (FMS) is
- explain and discuss the term 'just-in-time' (JIT)
- discuss the goals of the just-in-time system
- discuss all the just-in-time building blocks
- discuss the benefits of the just-in-time system
- define and discuss the term 'inventory' and the major reasons for having inventory
- list the main requirements for effective inventory management
- discuss the A-B-C method of inventory control
- define the term 'productivity'
- discuss the different ways of improving productivity

The aim of this chapter is to provide an introduction to operations management.

7.1 INTRODUCTION

The most important reason for any entity to be in business is to earn a profit. This is true whether an organisation produces goods or renders a service. If no profit is earned, the business is usually of a charitable nature. To be profitable, any manufacturing or service industry must manufacture goods, supply a service or mine the raw materials necessary for the production of various goods. This chapter applies specifically to the operational stage (as discussed in chapter 1) of a facility such as a mine, factory or plant.

Organisations in the same area of business often compete for the same customers. It is therefore essential that special care and attention should be given to the following.

- **Empathy:** Treat customers with consideration and give attention to the needs of each individual.

- **Reliability of service:** If the organisation/service provider makes a promise, that promise should be fulfilled completely and without any hesitation.
- **Assurance to customers:** The courtesy that staff show towards customers is non-negotiable. Employees should be able to inspire trust and confidence when dealing with customers.
- **Tangibles:** This refers to the physical appearance of the place(s) where service will be rendered. It also involves staff courtesy and knowledge when dealing with customers.
- **Adaptability:** Organisations should be willing to change in order to meet new customer requirements.

7.2   THE TRANSFORMATION PROCESS

No matter what the type or size of the business, it must follow the steps in the transformation process. The transformation process can be described as follows: equipment, i.e. machinery directly related to producing goods and/or services, is used by trained human beings to produce tangible or intangible goods. The machinery and workers are located in facilities such as factories, buildings, hospitals, plants and mines. It is very important to distinguish between the different types of inputs. Firstly there are the transformed inputs. These are the materials that are changed during any operations of the complete transformation process. Secondly there are the transforming inputs. These are all the resources that may be required by the operations in the complete process to transform inputs. The resources themselves will not be transformed.

Each of these processes has its own characteristics and name. Each type of transformation process will have characteristics unique to that specific process. The inputs are also known as the eight Ms of operations management. These are: money, materials, machines, manpower, management, markets, methods and messages. Therefore it is necessary that the inputs (the 8 Ms) should follow the same process when they pass through the transformation process. Omitting any of these inputs will result in an incomplete transformation process. That in turn will result in the production of poor quality goods or the rendering of an inferior service. Figure 7.1 illustrates the transformation process.

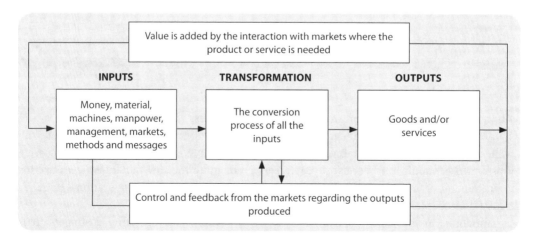

Figure 7.1 The transformation process – from inputs to outputs

The transformation process can be physical – something physical has to be done to produce the finished product or service. It can however also be in the form of an exchange, such as in the retail trade when money is exchanged for goods or services. The process can also be informational. Ordinary communication where information is traded for money serves as an example. Telkom has different services where, for the price of a local call, information about a person or business can be obtained. The transformation process can also be locational, such as the transporting of passengers. The process can also be physiological, such as the care of sick people. Lastly, the transformation process can be psychological, such as going to see a movie or a live show (see table 7.1 for examples of this process). For the transformation process to be successful, value must be added to the inputs. The value added is the difference between the costs of all the inputs and the price fetched by the finished product or service. It is vital that the transformation process adds value as the product or service progresses through the process, i.e from raw material to finished product. If no value is added to the product, then only cost will be added. This will result in high costs and low or no profit. It is not unusual that if this is a continuous process, the organisation will eventually fail. No organisation can sustain losses indefinitely.

The efficiency and effectiveness of the transformation process is of the utmost importance. For this reason continuous feedback from the whole process is required. Feedback is the function that will inform the organisation whether it is performing efficiently and effectively. It is important to measure and compare results achieved with standards previously set, in order to exercise control (see chapter 2 for more information on the controlling function).

Table 7.1 Examples of transformation processes

OPERATION	INPUTS	TRANSFORMATION PROCESS	OUTPUTS
Hospital	Doctors Nurses Patients	Beds Medicines Operations	Test results Healthy patients Research
Canned goods	Fresh foodstuffs Machines Other equipment Labour	Prepare the food Can and cook the food	Canned goods
Police	Officers Crime information Public Computers	Prevent crime Solve crime Arrest criminals	Lawful society Public feel safe
Food store	Foodstuffs for sale Staff Cash registers Customers	Display goods Give advice Sell goods	Goods and customers are brought together

7.3 MANUFACTURING VERSUS SERVICE INDUSTRIES

Both manufacturing and service industries have unique characteristics that differentiate them. These are listed in table 7.2 on the next page. Another difference between the two can be seen in *how* things are done rather than *what* is done. Manufacturing is more focused on the product it produces, while a service industry is more focused on the act that constitutes its end product, e.g. sales staff are taught how to treat customers so that they will return. In contrast, manufacturing will concentrate on supplying high-quality and reasonably priced items to retain a customer base. It is important to note that there is a place for both types of industries in any economy. During the last decade most of the jobs were created in the service industry. Table 7.2 illustrates the main differences between the manufacturing and service industries.

> **Example 7.1 Examples of companies participating in the manufacturing and service industries**
>
> **Manufacturing:** Motor vehicles (Toyota, Ford and BMW), bottling of beverages (Coca-Cola and SAB Miller), canning of foodstuffs (Koo and Heinz), manufacturing of household appliances and other electrical appliances (LG and Defy).
>
> **Service:** Financial services (Nedbank, FNB, Standard Bank and ABSA), cellular phone service providers (MTN, Vodacom and Cell C), doctors, attorneys and retail outlets (Edgars, Woolworths and Pick 'n Pay).

Table 7.2 Differences between manufacturing and service industries

MANUFACTURING	SERVICE
A physical (tangible) product that may be durable. The product may be used for many years.	An intangible product. This type of product is usually perishable and can be used once only.
Goods produced can be stored in a warehouse or at home.	The final product cannot be stored.
No direct contact between consumer and producer.	Given the nature of the service industry, there must be direct contact.
Output is usually produced in large plants or factories.	Facilities for the service industry are usually much smaller.
Large capital outlays are required.	Very labour-intensive operations.
Easy to determine and measure quality.	Much harder to determine and measure quality.
Not essential to be on the consumer's doorstep. Goods can be transported.	Essential to be close to the consumer of the service rendered.
Output produced can be resold numerous times.	Output cannot be resold to a third party.
Before consumption of the output can take place it must be produced.	Provision and consumption take place simultaneously.
The product produced can be patented.	It is difficult to patent the output.
Due to the low variability, production tends to be efficient and smooth.	The activities tend to be slow and awkward. Output is variable.
Productivity measurements tend to be straightforward.	Productivity measurements tend to be more difficult.

7.4 OPERATIONS DESIGN

Many factors must be taken into account when a product or service is designed. Such factors can be regulations published by the government, new technologies or pressure from the competition in the marketplace, and, most importantly, what the customer requires.

The following guidelines can be used when a new product or service is designed.

- *Determine the target market and costs associated with the design.* For this reason it is important to have a multifunctional design team. The team should have as wide a representation as possible. The marketing department will provide the important input of what the customer wants – it will give the team possible sales figures, profit targets and the competition's pricing structure. The logic behind this is to ensure that, once the new product has been manufactured, it will not cost more than what it can be sold for. (See chapter 11 for more information on target markets and chapter 5 for more information on teams.)

- *Keep the operations and parts necessary to manufacture the new product to a minimum.* The same applies to a service industry. When the number of parts has been decided on, it is important to analyse each part individually. The reasoning behind this is to ensure that only the best methods and materials are used in the manufacturing process.

- *Focus on what customers require.* The importance of this step cannot be overemphasised. Neglect this stage and the company may end up with a product or service nobody wants. This is a continuous process, as customers' needs may change over time. Product and service design cannot be undertaken and must never be attempted in isolation. It is of the utmost importance that everybody concerned with manufacturing is involved in the design stages. Each department should have a representative on the design team to ensure the manufacturability of the final designed product. Therefore not only the customers' needs must be analysed. The product required by customers may not be easily manufacturable. Suppliers both within and outside of the business must have input during the design stage. These people supply the materials and sub-assemblies that will be used to manufacture the product. During this stage, the suppliers and all the other stakeholders will be able to determine whether the product will reach the production stage.

- *Design the product/service for process capability.* Here the design team's responsibility is twofold. Firstly, the design team must ensure that the product can be delivered or manufactured using existing processes, machines, labour, materials and facilities. Secondly, the design team must be familiar with the capabilities required to manufacture the new product. There are many ways to measure the capabilities required – for example, educational qualifications needed, level of training and cross-training required, documentation available on the newly designed process and the experience of each team member in designing new processes and products.

- *Standard procedures, materials and processes must be put into place to ensure easy manufacturability.* Standard procedures should enable those involved in the manufacturing process to understand what is required of him/her. Care should be taken to ensure that products meet standards.

- *Design the product/service for ease of use.* This is especially applicable in the do-it-yourself market. If a complex level of interaction between the product and the user is required for it to operate, it is unlikely to sell well. The same applies to service industries. Assume, for example, that a software company develops a new computer program that requires a high level of knowledge and understanding to operate it. The average person is unlikely to buy such a product, as he/she does not have the time to study the new product intensively before operating it successfully. If a competitor markets the same program in a simpler form, it is more likely to be a success. (See chapter 18 for more information on product development.)

▨ *Design the new product so that it can be assembled and disassembled with ease.* In the high-tech environment of today, it is important that items are easy to assemble. The opposite also holds true. Once the product has reached the end of its useful life, it must be easy to take apart and recycle the components. (See chapter 19 for more information on product life cycle costing and assessment.)

▨ *Design for straight-line assembly.* The assembly of a new product at the same work centre at different stages of the manufacturing process should be avoided. The manufacturing process must be developed in such a manner that the product travels in a straight line. It should never be allowed to double back to a work centre it has already passed through.

▨ *Avoid special features.* Develop the process to be as simple as possible. Try to avoid the attachment of special features and fasteners, especially in the case of low-cost products. Special features and fasteners may complicate the assembly of the final product and more time will have to be spent on quality control.

▨ *Make designs robust.* Resist all temptations to take unsafe shortcuts. Extra costs will be incurred when the mistake that results from the shortcut has to be rectified. Be courteous to customers who experience problems with your product. Treat sensitive equipment with care.

▨ *Take legal issues and product liability into account.* Liability will be incurred if poor-quality goods and services are sold or rendered. Organisations can be held liable in a (civil or criminal) court of law if customers are harmed, injured or killed when consuming goods or services of inferior quality as a result of defects in the design of the product or service.

▨ *Take environmental issues into account.* Products and services must be environmentally friendly. The impact that the product or service may have on the environment and people when it is consumed must be considered during the product planning process. The main concern is how to ensure optimum satisfaction to the consumer without harming the environment. Many natural resources are non-renewable and must be used responsibly to ensure that some will be left for future generations. (See chapter 19 for more information on sustainable development.)

If the proper communications and consultations take place during these steps, a properly designed product will be produced. A minimum of comebacks or problems during manufacturing will be the result.

7.5 CAPACITY PLANNING

Capacity planning is one of the most important areas in operations management. What is capacity planning? Capacity can be defined as the maximum use of all the limited resources available to produce a set quantity of goods, in a given time frame that is fixed, if all the conditions for production are ideal. The capacity available can be seen as the upper limit of the amount of work a productive unit will be able to perform. The amount of work can be expressed in hours or units. A productive unit can be the whole facility (e.g. plant, mine or factory), a department within that facility, or a worker within the department or other functional unit (e.g. a machine shop). There are various ways in which the capacity available can be measured, e.g. the inputs or the outputs of the productive unit can be measured.

To enable forward planning by management, it is imperative to know what capacity will be available in the productive unit. If management do not know this, it becomes virtually impossible to determine whether the company will be able to fulfil its obligations to its customers. Before any attempt can be made to determine the availability of capacity, some important questions need to be answered.

Firstly, it must be determined what type of capacity will be needed; secondly, what amount of capacity will be needed; and, thirdly, when it will be needed. The correct amount of capacity can be assured through short-, medium- and long-term capacity planning. Short-term capacity is needed to achieve daily production targets. Medium-term capacity planning ensures that capacity exists to complete the production for the next fortnight or so. And long-term capacity planning should be done to ensure that enough capacity is available to implement the overall production plan. Long-term capacity planning may not exceed six months, depending on the type of facility that is involved. The further the planning horizon, the less reliable the planning becomes, as capacity can fluctuate on a daily basis. Facilities such as deep-level mines may take years to be developed and at very high cost. Such mines are usually operated at high overhead costs and should therefore produce high volumes to achieve low unit costs. Figure 7.2 illustrates where capacity planning fits into the overall plan of a company.

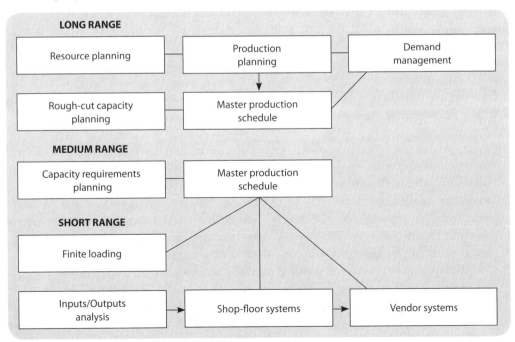

Figure 7.2 The relationship between capacity planning and other plans

7.5.1 The objectives of capacity planning

The capacity decisions taken by management will have an impact on the overall running of a business. It is therefore important that the objectives of capacity planning are clear and definite. Some issues and generic capacity-planning objectives are given below.

■ *The balance struck between capacity and demand will influence the cost structure.* Available capacity will be underutilised if more capacity is available than the demands placed on the system by customers. As a result, the cost per unit produced will increase.

■ *The balance struck between capacity and demand will influence revenue.* If capacity is fully utilised or even overutilised, it will ensure that no revenues are lost.

■ *If finished goods are manufactured for stock (inventory), then working capital will be tied up.* Working capital that could have been used elsewhere will be tied up in holding stock for as long as it takes to sell it. (See also chapter 14 for more information on cash flow.)

■ *Large fluctuations in demand may cause quality problems.* Quality problems may be experienced when less-experienced, temporary workers have to be employed during short periods of unusually high demand. As a result of the unfamiliarity of the workers with the manufacturing process, mistakes may be made.

■ *Enhancement of the speed a customer's order can be filled may cause stock/inventory problems.* In order too speed up the fulfilment of a customer's order, it may be drawn from stock. A drawback of this method is the extra cost that is involved with keeping stock/inventory. Another method of speeding up the fulfilment of a customer's order is to provide for extra capacity. This will prevent customer queues. Orders can then be filled as they are received.

■ *The closer the demands are to the available capacity, the more dependable the supply of goods will be.* The result will be fewer disruptions because of the non-availability of goods.

Effective capacity can be defined as the maximum rate at which a process can produce for extended periods when the production conditions are normal, taking maintenance requirements, rest periods, the production mix to be manufactured, how well employees are trained in their jobs, their work schedules and the methods used to schedule that work into account. This definition shows that all possible influences must be taken into account. If not, a false picture will emerge of how much capacity is really available. The following factors influence availability of production capacity.

1. **Process design:** The maximum output of any multi-stage production process will be determined by the operation that takes the longest time. It can also be called the slowest operation. Design the process in such a way that this is taken into account.

2. **Product design:** It is usually easier to manufacture a well-designed product. Available manufacturing capacity for well-designed products should be higher than for poorly designed products. The reason for this is that when manufacturing well-designed products, less time should be taken to set up machines and to rework scrap.

3. **Product variety:** As few as possible products or varieties of the same product should be manufactured if a cost leadership strategy is followed. The result will be fewer changeovers and better utilisation of labour and machinery. Another benefit is that employees and machines can be more specialised. A direct result of this is that more capacity will be available. The more flexible a process is, the less capacity will be

available. (See chapter 17 for more information on cost leadership and other business strategies.)

4. **Product quality:** This will have a huge impact on capacity availability. The method of production, testing and inspection of the product will determine the capacity requirements for a particular product. The higher the level of quality, the more capacity will be required. It stands to reason that if the requirements are less, more capacity will be available. It is therefore important that high-quality products are manufactured well to limit the amount of rework to be done. The result will be more available effective capacity.

5. **Production scheduling:** Minimising unproductive time, balancing workflow through the process and synchronising workflow can result in more effective capacity availability.

6. **Materials management:** Shortages of the required materials will result in production stoppages. Conversely, excess raw materials may encourage waste. A just-in-time (JIT) system can be installed to manage materials properly. This will enhance utilisation of available capacity. (JIT is discussed in section 7.7 of this chapter.)

7. **Equipment maintenance:** The constant breakdown of badly maintained machines can influence the availability of capacity negatively. For the duration of the breakdown of that machine, its capacity is lost. The result will be that less product can be manufactured. Equipment should therefore be properly maintained to prevent possible production losses due to their non-availability. (See chapter 10 for more information on maintenance.)

8. **People management:** The number of high-quality products produced is directly proportional to the calibre of the people working in the process. Factors such as their level of training, motivation, number of days absent from work and the design of the product they manufacture will directly influence the available capacity.

7.5.2 Capacity measurement

Available production capacity must be measured accurately. The accurate measurement of capacity can be influenced by various factors, the most important being the definition of the output measurement of the process. The second factor that influences the measurement of capacity is the assumption that conditions are perfect during the manufacturing process. Unfortunately, this is rarely the case. A method should be chosen that does not have to be updated at regular intervals. Capacity should therefore not be measured in terms of monetary value, as the selling price of the finished product will vary over time and will not provide a true reflection of the available capacity. Nor will it indicate whether capacity was maximised. It is advisable to have a unit of measure that will remain constant. Capacity measurement is simpler if only one product is manufactured. It becomes more difficult when several products are manufactured. The reason for this is that each product will have its own throughput rate and unit of measure. Examples of how production capacity can be measured are listed in table 7.3.

Table 7.3 Examples of how capacity can be measured

BUSINESS	INPUTS	OUTPUTS
Petrol station	Number of pumps	Quantity of fuel pumped
Airline	Number of seats	Number of passengers
University	Number of students	Number of students graduating
Hospital	Number of beds available	Number of patients treated
Cinema	Number of seats	Number of patrons
Repair facility	Number of jobs received	Number of jobs completed

There are three different ways of measuring capacity.

1. **Design capacity:** If everything is ideal during the manufacturing process, design capacity is the maximum output that can possibly be attained by that manufacturing process.
2. **Effective capacity:** If all the constraints, such as the number of products to be manufactured, breakdown of machines, scheduling and quality are taken into account, effective capacity will be the capacity necessary to attain the maximum possible output from that process.
3. **Actual output:** This is what was actually achieved by the process. In this case, the actual output can never exceed the effective capacity. In practice, it is often found that that the actual output is less than the effective capacity. Some of the reasons are a shortage of raw materials, breakdown of machinery and the manufacturing of poor-quality products.

Once these three definitions are understood, they can be used to calculate the effectiveness of the process. The effectiveness is measured as the efficiency and/or utilisation of the process. Efficiency can be expressed as the ratio of the actual output to the effective capacity. Utilisation can be defined as the ratio of the actual output to the design capacity. The following formula can be used to compute efficiency:

Efficiency = Actual output/Effective capacity **Equation 7.1**

Utilisation can be computed using the following formula:

Utilisation = Actual output/Design capacity **Equation 7.2**

It is misleading to concentrate only on the efficiency rate as a measure of how well capacity is being used. If this happens, a very distorted view will be obtained. Example 7.2 illustrates this point

Example 7.2

Use the following information from the repair workshop of a motor dealership and compute workshop efficiency and utilisation factors:

Design capacity : 100 vehicles per day

Effective capacity: 80 vehicles per day

Actual output: 72 vehicles per day

Then:

Efficiency = Actual output/Effective capacity = 72/80 = 90%

Utilisation = Actual output/Design capacity = 72/100 = 72%

If the effective capacity of 80 vehicles per day in example 7.2 is compared to the actual output of 72 vehicles per day, it looks quite good. However, if it is compared with the design capacity of 100 vehicles per day, then 72 vehicles per day as the actual output seems less impressive but a more meaningful comparison. To improve capacity utilisation, effective capacity must be improved.

To ensure continuous improvement, the focus should be on effective capacity. This may be influenced by the following factors.

- **External factors:** These are conditions that have been imposed from outside the organisation. An example is legislation regarding health and safety imposed by the government.
- **Facility factors:** These factors refer to how well facilities were designed for their intended purpose.
- **Human factors:** These factors refer to how well humans can function within the facility in its current form.
- **Operational factors:** These include issues such as bottlenecks, breakdowns and failures of equipment.
- **Process factors:** These may include factors such as technology, specifications and customer requirements over and above the usual specifications.
- **Product/service factors:** These will depend on the specifications required to produce the goods or render the service.

7.6   FLEXIBLE MANUFACTURING SYSTEMS (FMSS)

Any manufacturing company that wants to be at the forefront of its industry will have to seriously consider using flexible manufacturing systems (FMSs). There are several ways of operating an FMS. The most common method is a group of self-contained machines, operated by a centralised computer. The group of machines can consist of as little as three machines. The maximum number of machines in a group is usually 12. Between the machines in the group and the warehouse, materials are handled mostly automatically. Automated guided vehicles (AGVs) are used for this purpose. As the materials are needed, the central computer prompts an AGV. The vehicle goes to the material's storage place, withdraws the required material from the bin, and delivers it to the machine. Another characteristic of an FMS is that all tool changes on a machine are done automatically. The computer program operating the machine is written in such a way that the machine will select the correct tool for the job at hand. In an FMS environment, a variety of similar products can be manufactured. The reason for this is that the computer can be programmed to change over from one job to another automatically. An FMS can thus be described as an automated job shop that does intermittent processing of jobs.

Flexible manufacturing systems are successfully used in the manufacturing of gears, electrical components, clothing and many other types of manufacturing. Numerically controlled (NC) machines usually do the jobs. An FMS can be characterised by the following features.

- An automated loading mechanism loads and unloads materials and tools.
- Tool changes can take place at two machines simultaneously.
- A tool storage and changing system will be found at each of these machines.
- An automated materials-handling system moves materials when required.
- At each machine, the materials will be automatically loaded and stacked when completed or waiting to be loaded.
- A computer program controls the complete FMS.

NC machines are the simplest example of a flexible manufacturing system. Examples of more advanced systems are welding robots, spray-painting booths, cleaning equipment and monitoring equipment in complex plants. Welding robots and spray-painting booths are used extensively in the manufacturing of motor vehicles. Monitoring equipment is usually used in power stations and chemical plants. Automated cleaning equipment is usually found in bottling plants. In all the above examples, one operator can operate several machines or pieces of equipment. The operator's main function is to ensure that any stoppage that may occur is immediately cleared, to ensure the FMS operates smoothly. Certain advantages or benefits can be derived from a flexible manufacturing system. These are given below.

- Labour costs can be reduced significantly. Fewer people are required to operate more machines. Where previously an operator operated each machine, with an FMS, one operator can operate two or more machines.
- Better quality products can be produced. This is made possible through the programmability of the machines. Exact requirements can be programmed into the machine and every job will be done to the exact standard.
- The flexibility and efficiency of automation is combined in an FMS.
- The flexibility of the system eliminates the need for jobs to be moved between machines or work centres by hand. All this can be achieved by using automated guided vehicles.
- As a result of the repetitiveness of an FMS, improvements in the flexibility of the system and the productivity of the process can be obtained.
- Lower capital investment is required.
- Savings in machinery set-up times can be achieved (the machine is automatically set up before the job reaches the machine). The set-up of the job is written into the program that controls that machine.
- Almost no work-in-process is retained. Jobs and materials are delivered at the machine when required. No half-finished jobs clutter the workplace. Another advantage of delivering a job or material to a machine only when it is needed is that hardly any queues form of jobs or materials waiting to be loaded. Automated guided vehicles will deliver materials only when instructed to do so by the computer program.
- Being able to program the machine means that a one-off unique job can be done without affecting the efficiency of that machine.
- The machines in a group can be reprogrammed at random, without affecting capacity, quality or efficiency.

148

Certain disadvantages are associated with the flexible manufacturing system.

■ An FMS can handle only a narrow range of products.
■ Given the narrow range, only products that use the same machines can be manufactured by an FMS.
■ Much longer planning and development times are needed, given the complexity of the machines involved. The programs that must be written for each machine take a long time to produce.

Before management decides to implement an FMS, a thorough investigation must be done into whether such an undertaking will be viable. It is important to determine from the beginning whether an FMS will suit the core business of a company.

7.7   JUST-IN-TIME (JIT)

JIT can be seen as a system that is designed to fulfil any demand without delay. It will do so with the minimum of waste and only with high-quality products. In simpler language, it means producing the product only when it is required. This means that the product will not be delivered earlier or later than the date specified by the customer. JIT originated in Japan and was developed by the Toyota Motor Corporation of Japan. When the Western world saw all the benefits that could be derived from using the system, it too converted to it, but not all the Western businesses that implemented this system had the same success as the Japanese. The reasons for this are varied.

The most important characteristic of a JIT system is that there is no idle time. The meaning of 'no idle time' is that no inventory is kept in warehouses, no extra workers are employed and all the space in the facility is used optimally. The result is a new way of production, known as lean production, which can be seen as a refinement of the JIT concept. The term 'lean production' means that there are no extras built in for unforeseen happenings during the production process. In many South African companies, a certain amount of 'fat' is built into the production process to provide for higher-than-anticipated demand, scrap and rework. Businesses that have implemented JIT will have certain advantages over those that still use older methods of production. The following advantages can be derived from implementing a JIT system.

■ Costs will decrease.
■ Fewer defective products will be manufactured, which will result in higher customer satisfaction. (This is one of the reasons why costs will decrease.)
■ Flexibility of the production system will increase.
■ New and more improved products can be made available to consumers much faster.

Once a JIT system has been installed, it must be managed properly and, most importantly, continuous improvements must be made to the system, i.e. the system must be fine-tuned constantly. The main purpose of a JIT system is to eliminate waste in the production process, i.e. anything that does not add value to a product or service. Examples of waste include the following:

■ an operator standing or sitting watching a machine while the manufacturing process is in progress. That person's presence does not add value to the product. The operator is there just in case the machine breaks down;

- the reworking of sub-standard quality products. No value is added when products have to be reworked because quality is low or non-existent;
- the moving around of completed or half-completed products over long distances. No value is added, as valuable time is wasted;
- operators looking for lost or misplaced tools;
- machines or work centres waiting for raw materials or parts for the manufacturing process. No value is added when the machine and the operator are idle;
- the holding of inventory. The space required could be put to better use, and holding excessive inventory costs money;
- the breakdown of machinery. Better maintenance plans should be put into operation to eliminate this kind of waste;
- continuous counting and recounting of existing stock. This is one of the biggest sources of waste in any business. Nothing constructive is produced. For this very reason, no inventory exists in the JIT system; and
- the overproduction of products. Many companies do this to prevent the non-availability of their products: 'fat' is built in, just in case of stock-outs.

The goal of the JIT system is to eliminate waste and ensure a smooth production process. For a successful JIT system, the following criteria must be met.

- *Waste must be eliminated.* If not, resources that could have been used in production are kept busy with unproductive work.
- *The system must be flexible.* It must be developed in such a way that any mix of products can be handled without this being problematic. The balance and rate of throughput must be maintained at a steady rate.
- *Disruptions must be prevented.* Disruptions have a negative influence on the production process. They interrupt the smooth flow of products through the production process. When this happens, one of the cornerstones of the JIT system is absent. Factors such as poor quality, continuous changes in the production schedule, late receipt of materials and the breakdown of machinery will disrupt the smooth flow of products through the production process.
- *Delivery times of raw materials and the set-up times of machines must be reduced.* The longer these take, the less value is added to the final product. This will also impact negatively on the flexibility of the production process.
- *The minimum inventory must be kept.* Ideally, there should be no inventory in a JIT system. The more the inventory that is held, the more the space that is required. Inventory increases the cost of the final product.
- *A good supplier network must be established.* To eliminate the holding of stock, a reliable supplier network must be built up. The smaller the number of suppliers, the more reliable deliveries will be. Suppliers must constantly be kept informed of when and what quantities of goods are required. It will be a great help if the suppliers are situated close to the factory.
- *Only quality supplies must be sourced.* Quality is a prerequisite for a successful JIT system. Ensure that only high-quality parts and raw materials are received from suppliers. In this way, the manufacturing of zero-defect products becomes more viable.

Many benefits can be derived from a properly installed JIT system. These are discussed below.

- *All levels of inventory (work-in-process, bought-in goods and finished goods) will be reduced.* Properly installed, JIT will trim inventory down to only the materials needed for that day's production.
- *The quality of the goods produced will improve.* Rework and scrap will be reduced and eventually eliminated.
- *All the costs involved will be reduced.* This includes the costs of rework, scrap and inventory holding, and that of constantly setting up machines.
- The time taken to manufacture a finished product will be reduced.
- The productivity of both the operator and the machine will improve because of better utilisation.
- There will be more flexibility in the production process.
- Better relationships will be built with suppliers.
- The scheduling and controlling of jobs will be simplified.
- *Due to fewer interruptions in the manufacturing process, capacity will be increased.* This will result from better quality materials, better employee training and fewer machinery breakdowns.
- *Employees can be used more productively.* Employees can now become actively involved in the solving of production problems. The people working in the process are usually more aware of where a problem is likely to occur and how to solve it. Listen to their input.
- Indirect labour (progress chasers, storepeople and planners) can be reduced.
- More types of products can be produced.

To become more competitive and remain so, more and more companies will have to seriously consider converting to a JIT system. Those that choose not to convert risk becoming irrelevant in the manufacturing environment. Their costs will be high, capital will be tied up in unproductive stockholding and their quality will be almost non-existent. It is for these reasons that the Japanese and other Asian countries dominated the world markets from the late 1970s, through the 1980s and into the early 1990s. It is only lately that Western countries have been able to compete on an equal footing with Asian countries. These countries have learned their lessons well.

7.8   INVENTORY MANAGEMENT

To ensure that inventory holding costs are kept to a minimum and on-hand stock is maximised, it is important that inventory is properly managed. If a company keeps inventory, it is investing large amounts of money in that stockholding. Holding inventory does not add value to the production process. The tied-up capital could have been used to replace old machinery or invest in newer production technologies. In some instances, the investment in inventory can be as high as 25 per cent of the total assets of a company. If it is so expensive to hold inventory, why do companies bother to do it? In many instances it is the way in which the production process was set up. Even though the costs are high for holding inventory, it acts as a safety net during the production process. If a problem arises during production, demand can be satisfied from inventory. Doing this may prevent customers from buying a competitor's product due to a stock-out (See table 7.4 for examples of different types of inventory).

When dealing with the planning and control of inventories, it is important to distinguish between two types of demand: dependent and independent demand. *Dependent demand* is demand for the sub-assemblies or components used to manufacture the final product. This type of demand usually originates from within the company and can be seen as an internal demand. *Independent demand* is demand placed by a customer for the finished product manufactured by the company. This type of demand, as a rule, usually originates from outside the company. Therefore it stands to reason that these demand types will influence the demand for the product differently. In order to ensure that the production process runs smoothly, the following types of inventories are often held:

- *finished goods inventory*, i.e. the final product of the company. It is kept in a warehouse, awaiting a customer order. The products can also be in the warehouse waiting to be shipped to the customer after receipt of an order;
- components or finished good that are *in transit* between the manufacturer and the customer. The items are not at the manufacturer's premises, but have not yet been received by the customer. Because no payment and delivery has taken place, they will show as stock held;
- *office supplies, machine tools and the parts needed to repair broken-down machines*;
- *raw materials* required for the manufacturing of goods, and the purchased parts used in the manufacturing process; and
- partially completed goods in the manufacturing process. This type of inventory is known as *work-in-process* (WIP).

Table 7.4 Examples of different types of inventory

BUSINESS SERVICE	TYPE OF INVENTORY HELD
Doctor	Syringes, needles and dressings
Retailer	Goods for sale (e.g. groceries)
Motor spares shop	All types of motor spares
Manufacturing	Raw materials, WIP, finished goods
Petrol station	Fuel and lubricants
Fast-food outlet	Food ingredients

As one can see, in certain cases it is important to hold inventory, e.g. if a customer walks into a fast-food outlet and orders a hamburger, but the establishment does not have rolls, meat or the condiments available to fill the order, the customer will leave. This will result in a lost sale and loss of the customer's goodwill. No customer will support an establishment that cannot satisfy his/her demand. Inventories can be used for the following reasons:

- *to ensure smooth operation of the manufacturing process:* In this type of inventory, raw materials and work-in-process will be included. This type of inventory will ensure continuous production;
- *to provide buffer stock:* In this instance, stock of half-finished goods is kept at the machine to ensure that the machine will not be idle waiting for a job;
- *to prevent stock-outs:* Products are manufactured in excess of the existing demands. When a surge in the demand occurs, there will be enough stock in the warehouse to satisfy the unforeseen demand. This will lead to customer satisfaction;

- *to deal with seasonal demand:* Businesses with demand that is influenced by particular seasons use this type of inventory. During the off-season, enough products are manufactured and stored to ensure that all the demands can be met during the high season. An example of this is a knitwear company that will produce jerseys during summer, to ensure that enough garments are available during winter;
- *to create efficient order cycles:* This will ensure the most advantageous buying of raw and other materials. Buying in cycles ensures that goods are purchased at the best economic order quantities. Consequently, the cost of reordering will be minimised; and
- *to deal with increase in prices:* If there is a price increase in the pipeline, buy in larger quantities to offset or beat the imminent price increase. In this way, costs can be minimised.

As stated throughout this chapter, holding inventory will incur costs, but it is not only the cost of investment in inventory that must be taken into account. Various other costs will have an equal impact on the profitability of a company. Unfortunately, these costs are not always visible when determining the stock-holding costs. Some of these costs include the following.

- **Holding cost:** This type of cost is also known as carrying cost. This is incurred when inventory is physically stored in a warehouse. This cost is determined by factors such as quantities held, the value of the items stored, the time period stored, tax, depreciation in item value, obsolescence, the cost of insurance and many other costs. There are two basic ways of expressing holding cost. It can be expressed as a percentage of the selling price, e.g. if the holding cost per item is 10 per cent of R50, this amounts to R5. Alternatively, it can be expressed as a rand value per item, e.g. the holding cost per item is R1,50.
- **Ordering cost:** This type of cost is also known as the order preparation cost. Costs will be incurred each time inventory has to be replenished. The first part of the cost will be incurred when an order is placed and the second part will be incurred when the goods ordered are received into stock. The ordering costs are usually expressed as a fixed amount per order – for example, R100 per order.
- **Shortage cost:** This type of cost is also known as the customer service cost and is only incurred when demand exceeds supply. To determine this type of cost is very difficult, as it is done subjectively. Factors taken into account at one company may not be taken into account at another. The costs that are included on a regular basis are loss of customer goodwill, charges for the late delivery of goods, loss of sale and a host of other subjective costs. To illustrate the subjectiveness of these costs, take the loss of customer goodwill. How can this be determined? At most it will be an educated guess.

As a result of the high costs involved in holding stock, it becomes critically important to manage that stock properly. A general rule is that the higher the monetary value of an item, the higher the degree of attention given to that item. The lower the monetary value, the lower the degree of attention given to that item. This is only a general rule of thumb. Because of the high investment in inventory, a more scientific method had to be found to do the analysis to determine how to manage inventory properly. The technique that came into being is popularly known as the *A-B-C classification system.* Other names used for this technique are the Pareto principle or the 80/20 principle. Inventory is classified into three main classes or categories, taking into account the monetary value of each item. Category A items contain all the high-monetary-value items. A small number of items, around 20 per cent, are responsible for the largest slice of the costs. These items are usually responsible for 80 per cent of the costs. That is why it is known as the 80/20 principle: some 80 per cent of the cost is caused by 20 per cent of the items in stockholding. These items should be controlled strictly. Category B items are moderately important. Category C items are the least important. Items included in this category are usually of low monetary value, e.g. nuts, bolts and washers, and need the least attention.

Management will concentrate on controlling category A items. These items are usually of high value and can be sold outside the company at a profit. When discrepancies are found in stocktaking, it is usually in this category. Pilferage of these items usually takes place if employees see that control is lax.

7.9   PRODUCTIVITY IMPROVEMENT

As competitiveness becomes more important, so too does the increase of productivity. Low productivity in a company usually leads to that company becoming less competitive. This non-competitiveness will lead directly to an increase in the cost of manufacturing for that firm. It will cost the company more and take it longer to manufacture one completed product compared to the competition.

Productivity can be defined as the measurement of output in direct relation to the input needed to produce the desired number of end items. The following formula can be used to express productivity:

Productivity = Output/Input   **Equation 7.3**

Output is the product/service produced or rendered. Inputs may vary from company to company. The most important inputs are labour, materials and energy. The output that is measured during the production process will, to a large extent, depend on the type of job performed. Below are some examples of productivity measurements.

Average m² of bricks laid per day = m² of bricks laid per day/Hours worked per day
Average number of cars serviced in a day = Number of cars serviced per week/Number of work days per week
Average jobs done by a machine per hour = Number of jobs done per day/Hours worked per day

These formulas only measure a small part of the overall productivity. There are numerous other fractional ways to measure productivity. The main determinant of which formula to use will be 'what has to be measured'. First determine what has to be measured and the reason for measuring that specific entity. Then determine the objective that has to be reached by that measurement.

Example 7.3

Given the information below, determine the productivity in each case.

1. Six bricklayers laid 900 m² of bricks in one 8-hour shift.

2. The total production run for a machine in a 12-hour shift is 16 000 items. Out of the total production run, only 12 650 items can be used.

Answers:

1. Productivity = m² of bricks laid in one 8-hour shift/Total hours worked = 900/(6 x 8) = 19 m² of bricks laid per hour, per bricklayer.

2   Productivity = Number of quality items produced per shift/Hours per shift = 12 650/2 = 1 054 items per hour worked.

Why is it necessary to do productivity analysis? Most companies do it to work out their real productivity performance. Once that has been determined, their next objective is to improve on the current state of productivity. The productivity performance of a company will directly influence its competitiveness. The lower a company's productivity is compared to that of its competitors, the less it will be able to compete. If a company wants to remain competitive, it can take the following steps.

▥ *It is important that, throughout the company, for every operation, a productivity measurement is in place.* Without this, it will be difficult to determine how well the company measures up to productivity standards. Management and control of the operation will become very difficult.

▥ *Determine which of the operations are the most critical.* Critical operations are those that could influence the whole process negatively. A bottleneck is a sure sign of a critical operation.

▥ *Better methods must be employed to enable a company to achieve better productivity results.* Involve employees in the setting of productivity measures. Ask them to come forward with new ideas on how productivity can be improved. This will make them feel involved and work harder to attain the set standards. Better methods can be put in place by using benchmarking. Study the methods the competition employs to achieve high productivity and employ the best of these in the company.

▥ *Ensure that the objectives set to better productivity are obtainable and fair.* Nothing demotivates people more easily than not reaching goals. This is counterproductive.

▥ *Employees must see that management are part of the drive to attain a higher standard of productivity.* Management must encourage and support the productivity programme.

▥ *The improvements must be measured and the results published on a regular basis.* It is important for the morale of the employees to be informed about how they are doing.

To further enhance productivity in a company, certain factors must be taken into account. There are two main categories of factors that influence productivity. Firstly, there *are outside or external factors*, which management cannot change or influence. Management have to work within the framework of these factors – e.g. current labour legislation – and can only endeavour to minimise the impact external factors will have on the productivity of the company. Secondly, there are two sets of *internal factors* that influence productivity. Management may experience difficulty changing or influencing the first set of these factors. Examples include the way in which the company is organised, human resource policies and the work ethic of employees. To attempt to change these factors, the essence of the company will have to be changed. Management will have to realise that they need to manage productivity within this framework. The other set of factors that will influence productivity internally is within management's power to change and influence. By installing newer and better machines to enhance productivity, there will be fewer breakdowns and better quality. Better processes and work methods can be introduced and maintained and, lastly, management can help to achieve higher productivity by installing proper productivity measurement criteria.

Productivity has far greater impact on a company than most people realise. The prosperity of a nation can be directly linked to its productivity. The lower its productivity, the fewer new job opportunities become available. A country where low productivity is rife becomes less competitive than a country where productivity is high. The result is that the country with low productivity will price itself out of the marketplace because it will cost that country more to produce products than a country with a high productivity index. Employees should realise that, when salaries increase, productivity should increase roughly by the same percentage. Example 7.4 illustrates what will happen if productivity does not increase in direct proportion to salary increases.

Example 7.4

Productivity measurement before salary increase:

A company produces 2 000 items per day. Labour cost is R230, material cost is R100 and overheads are R170.

Productivity = Quantity produced/(Labour costs + material costs + overheads) = 2 000/(230 + 100 + 170) = 4

So, four items can be manufactured for every rand spent.

Productivity measurement after increase:

Assume that the labour cost increases from R230 to R400, and all the other figures remain the same.

Productivity = Quantity produced/(Labour costs + material costs + overheads) = 2 000/(400 + 100 + 170) = 2,9

After the salary increase in example 7.4 only three items can be manufactured for every rand spent. Because productivity did not increase, the result was a decrease in the number of items produced. In other words, it cost more to produce less. Wealth cannot and will not be created in this manner. A high productivity index is needed for the creation of wealth and new job opportunities. It is therefore everybody's responsibility to ensure that the highest standards of productivity are maintained.

Self-assessment

7.1   Discuss the transformation process and give relevant examples.

7.2   Distinguish between the manufacturing and service industries. List the differences in a table.

7.3   Discuss the guidelines for effective operations design.

7.4   Define 'capacity'.

7.5   List and discuss the objectives of capacity planning.

7.6   Discuss the different types of capacity measurements available, with examples.

7.7   Discuss the characteristics of a flexible manufacturing system.

7.8   List the advantages of a flexible manufacturing system.

7.9   Discuss the term 'just-in-time'.

7.10 Discuss the advantages of lean production as part of a just-in-time system.

7.11 List the different types of inventory that can be held by a company.

7.12 Discuss the reasons for holding stock.

7.13 List the costs associated with holding stock.

7.14 Discuss the steps that can be taken to improve productivity.

References

Chase, R., Aquilano, N. & Jacobs, F.R. 2001. *Operations Management for Competitive Advantage*. 9th edition. New York: McGraw-Hill.

Dilworth, J.B. 2000. *Operations Management: Providing Value in Goods and Services*. Fort Worth: Dryden Press.

Hanna, M.D. & Newman, W.R. 2001. *Integrated Operations Management: Adding Value for Customers*. Upper Saddle River: Prentice-Hall.

Heizer J. & Render, B. 2004. *Principles of Operations Management*. 5th edition. Upper Saddle River: Pearson Education.

Heizer, J & Render, B. 2004. *Operations Management*. 5th edition. Upper Saddle River: Pearson Education.

Hill, T. 2000. *Operations Management: Strategic Context and Managerial Analysis*. Houndsmill: Macmillan.

Nicholas, J. 1998. *Competitive Manufacturing Management: Continuous Improvement, Lean Production and Customer-Focused Quality*. New York: Irwin/McGraw-Hill.

Schroeder, R.G. 2000. *Operations Management: Contemporary Concepts and Cases*. New York: McGraw-Hill.

Slack N., Chambers, S. & Johnston R. 2001. *Operations Management*. 3rd edition. Harlow: Pearson Education.

Stevenson, W.J. 2004. *Operations Management*. 8th edition. New York: McGraw-Hill.

Waters, D. 2002. *Operations Management: Producing Goods and Services*. 2nd edition. Harlow: Pearson Education.

Gaither, N & Frazier, G. 2002. *Operations Management*. 9th edition. Cincinnati: South Western/ Thomson Learning.

8 Total Quality Management

Winton Myers

Study objectives

After studying this chapter, you should be able to:
- define total quality management
- understand the seven fundamental principles of total quality management
- gain an overview of total quality management practices and techniques
- apply the total quality management principles, practices and techniques to manufacturing and service organisations

The aim of this chapter is to introduce students to the basic principles of total quality management (TQM).

8.1   INTRODUCTION

Total quality management (TQM) is a management process that ensures that products and services are designed, developed, produced, delivered and supported to fully meet customer expectations – the first time, every time.

Formally defined, TQM is an interlocking arrangement of procedures and practices that ensures that all employees in every department are adequately trained and directed to continuously implement aligned improvements in quality, service and total cost, such that customer expectations are met or exceeded (Bellefeuille, 1993).

TQM is not only a way of thinking, but, more importantly, it is also a way of doing. It involves all people at all levels of an enterprise, leading them to focus their energies on the customer and on constantly refining the processes that deliver customer value. An important goal of TQM is to sustain the habit of continuous quality improvement, while the ultimate target is perfect quality – often called zero defect.

In his book *Quality is Free* (1979), Philip Crosby states that we must define quality as 'conformance to requirements' if we are going to manage it. There are seven fundamental principles of TQM.

1. Understand and answer the voice of the customer.
2. All people in an enterprise must be totally involved in quality improvement.
3. Continuously strive for zero defect.
4. Design and build quality into the product.
5. Focus on the process.

6. Suppliers are partners in quality.
7. Quality is free.

8.2   CASE STUDY: GRANITE ROCK COMPANY

Granite Rock Company is a 100-year-old, family-owned company based in Watsonville, California, USA. It produces rock, sand, gravel aggregates, ready-mix concrete and asphalt for sale to commercial and residential builders and highway construction companies. The company also retails building materials made by other manufacturers and runs a highway paving operation. It employs 400 people, distributed among branch offices, several quarries and 15 batch plants.

Granite Rock operates in a competitive marketplace in which customers demand high quality and most competitors are owned by multinational construction material companies with considerable financial backing.

Because the products Granite Rock sells are commodities, it is difficult to differentiate one supplier from another and customers typically buy from the lowest bidder. One factor that separates Granite Rock from its tough competition is quality. Since 1980, Granite Rock has increased its market share significantly. Productivity also has increased, with revenue earned per employee rising to about 30 per cent above the national industry average.

Most of the improvement has been realised since 1985, when Granite Rock started its Total Quality Programme. The programme stresses satisfying two types of customers: the contractor who normally makes purchasing decisions and the end-point customer who ultimately pays for the buildings or roads made with Granite Rock materials. By emphasising the hidden costs associated with slow service and sub-standard construction materials, such as rework and premature deterioration, the company is convincing a growing number of contractors of the value of using its high-quality materials and unmatched service.

As a result of its investments in computer-controlled processing equipment and widespread use of statistical process control, Granite Rock can assure customers that its materials exceed specifications. Also, customers can be confident about on-time delivery. Granite Rock's record for delivering concrete on time, a key determinant of customer satisfaction, rose from less than 70 per cent in 1988 to 93.5 per cent in 1991.

Charts for each product line help assess Granite Rock's performance relative to competitors on key product and service characteristics, ranked according to customer priorities. The company's objective is to obtain a 10 per cent lead over its nearest competitor for each performance indicator. Co-ordination across divisions is fostered by ten Corporate Quality Teams that oversee and help align improvement efforts across the entire organisation. Although senior executives chair committees, members include managers, salaried professional and technical workers, and hourly union employees. Teams carry out quality improvement projects as well as many day-to-day activities and operations. In 1991, nearly all workers took part in at least one of the company's 100-plus quality teams.

As part of Granite Rock's effort to reduce process variability and increase product reliability, many employees are trained in statistical process control, cause-and-effect analysis, and other TQM and problem-solving methods. This workforce capability helps the company exploit the advantages afforded by investments in computer-controlled processing equipment. Its newest batch plant features a computer-controlled process for mixing batches of concrete, enabling real-time monitoring of key process indicators. With the electronically controlled system, which Granite Rock helped a supplier design, the reliability of several key processes has reached the Six-Sigma level. Six Sigma is a statistical term indicating a defect rate of 3.4 per million, which is discussed later in the chapter.

Applying statistical process control to all product lines has helped the company reduce variable costs and produce materials that exceed customer specifications and industry- and government-set standards. For example, Granite Rock's concrete products consistently exceed the industry performance specifications by a considerable amount.

A primary reason why Granite Rock is so successful at meeting customer needs is that the company takes great pains to find out what customers want, what they like about Granite Rock, and what needs improving. Each time customers do business with Granite Rock, the company asks them to fill out 'quick response' cards that give them the opportunity to 'comment on the service and products you received today.' Every year since 1987, the company has conducted an annual survey that allows customers to compare the company with its competitors. Every three to five years, more detailed surveys are conducted.

Customer complaints are handled through product/service discrepancy reports that require analysis of the problem and identification of the root cause. Ultimate customer satisfaction is assured through a system where customers can choose not to pay for a product or service that does not meet expectations. Costs incurred in resolving complaints are equivalent to 0.2 per cent of sales, compared with the industry average of 2 per cent. In 1992, Granite Rock won the prestigious Malcolm Baldrige National Quality Award, and in 1998, the company won the National Asphalt Pavement Association's Quality in Construction Award.

8.3   THE VOICE OF THE CUSTOMER

TQM starts with gaining a thorough understanding of what customers need. There are many approaches to understanding customers and their requirements. The main ones to use in TQM are as follows.

- **Informal customer research:** Customer requirements can be analysed informally by talking to customers and gathering information through the company's or the industry's informal network. Valuable information can be gained by developing close relationships with customers and even incorporating them in product development teams.
- **Formal customer research:** Formal approaches to identify, measure and understand customer expectations can be classified into qualitative and quantitative techniques.
- **Qualitative research:** Qualitative research involves free-format responses in which words and observations are used. It is used to develop insights into the nature and content of customers' expectations and perceptions and the forces that drive them.
- **In-depth interviews:** These are face-to-face interviews conducted on a one-to-one basis or in small groups. They resemble conversations more than formal, structured interviews. A detailed discussion outline must be designed prior to conducting the interviews. Questions should be general and open ended, i.e. the questions must not direct responses towards fixed, pre-determined choices but instead allow the respondent to state whatever thoughts occur. Time must be taken to explore and investigate important ideas.

- **Focus groups:** The focus group technique is the most useful and versatile qualitative research technique, particularly for analysing external customers. Focus groups comprise seven to twelve customers who share common characteristics. The group meets for about two hours to offer opinions, viewpoints and perceptions about group members' requirements and expectations. Focus group interviews are freewheeling discussions guided by a skilled facilitator. Due to their spontaneity and openness, these sessions bring both emotional and rational needs to the surface.
- **Quantitative research:** Quantitative research is used to develop statistically reliable information about customer needs from sample data that can be generalised to a larger population. It is useful if the customer population is large, and is mainly used to understand external customers. This research usually involves surveys conducted by mail and telephone.

8.4   EMPLOYEES' INVOLVEMENT IN QUALITY IMPROVEMENT

Satisfying customer needs through TQM should not be abdicated to so-called quality professionals. TQM is a company-wide responsibility; it is a way of doing business. Establishing a quality control department, recruiting a quality manager, increasing quality standards, employing more inspectors and implementing a rework or 'hospital' department in the factory will not improve quality on their own. All they will do is increase the cost of quality.

Improved competitiveness starts with the recognition that quality is strategically important and top management should lead a TQM initiative. This initiative should be focused on preventing quality problems from happening rather than detecting them and fixing the product after it has been manufactured. The only way to reach world-class quality standards is for everyone in the enterprise to be totally committed to and involved with quality improvement.

To implement TQM on an enterprise-wide basis, the following steps need to be taken.

- Set up a quality council or similar group to formulate and implement an enterprise-wide quality policy. This council should be a cross-functional team comprising members of senior management and the organisational functions that have an impact on quality, such as marketing, product development, manufacturing, logistics and after-sales service. The quality council should always show total commitment to TQM principles.
- The quality council should develop a formal, documented quality policy that must be communicated to all employees.
- All people in the organisation should be involved in quality and continuous improvement activities.
- Everyone, including senior management, should be trained in TQM principles and techniques, particularly customer needs analysis and the TQM tools.
- Small groups, called process improvement teams (PITs), should be set up at shop-floor level to improve processes and solve quality problems. These groups should be given the responsibility for quality in their work area rather than the quality control department.
- Sufficient time should be allocated to quality training and small-group activity.
- Quality improvements should be monitored, and people and groups rewarded for quality achievements. If incentives are in place, people will find ways to improve quality.

■ Quality specialists in the quality control department should *facilitate* quality improvement initiatives and constantly *coach* others in the organisation to implement the TQM principles and techniques.

8.5   CONTINUOUSLY STRIVE FOR ZERO DEFECT

Richard Schonberger (1982) identified seven rules that sustain the habit of continuous quality improvement while aiming for zero defects.

1. **Focus on the process:** This rule is discussed in greater detail later in the chapter.
2. **Make quality visible:** Set up measurable standards of quality and clearly display them for everyone to see. This tells everyone – the work group, management, customers and visitors – which quality characteristics are important and are being measured, the targets, current performance and what is being done to reach the targets. Quality indicators should be easy to understand and displayed at every work area.
3. **Insist on compliance:** Too often production is more important than quality. To achieve zero defects, it is important for everyone to comply with quality standards as the top priority. Senior management needs to make it clear that quality comes first and output second.
4. **Stop the line:** This rule reinforces commitment to quality. The worker on the line should be given the authority to stop the line the moment a defect is detected. The quality problem can then be corrected and further defects prevented.
5. **Correct your own errors:** The first four rules enforce the assignment of responsibility for quality to the people working in the process, producing the product or service. This fifth rule closes the loop. The rework of poor quality products and services must be done by the person or work group that produced the poor quality in the first place.
6. **Do a 100 per cent check until the process is capable:** While the process is still unstable, check every item made – not just random samples. This should be applied rigidly to finished goods and, where possible, to component parts. This rule should be applied by the person or work group making the product, checking it at source rather than relying on quality inspectors to do the checking at the end of the process. Once the process has been proved capable, random samples can be used to ensure that it remains stable.
7. **Improve on a project-by-project basis:** Have a continual series of quality improvement projects running in every work area, year after year. When some of the quality problems have been solved, work groups should look for others to solve. Quality improvement is a journey without end. Continuous improvement is often called *kaizen* (which is a Japanese word meaning 'improvement'). This requires ongoing improvement, regardless of how good the product or service may be.

8.6   DESIGN AND BUILD QUALITY INTO THE PRODUCT

The TQM process starts with the new product development process, as it is here that quality is built into the product. When an enterprise is committed to TQM, all participants in the new product development process are expected to provide quality service to each other, to others who will be involved with the introduction of the new product and, ultimately, to the customer who will use it.

With TQM in place, a new product team will constantly consider the impact of design decisions on the quality of the ultimate product and associated services. The allocation of new product development resources will centre around the core concept of meeting customer requirements rather than on narrower, traditional financial justifications of expenditure.

The *prevention* of future quality problems is a key responsibility of the new product development team. It costs much less to avoid potential quality problems than it does to fix them once they are already embedded in the product. It is better to spend an extra few rand on prevention than to lose hundreds of thousands correcting the product when it fails in the field.

The key to the application of TQM in product development is to assure quality through effective design control. The role of design activities is to translate customer needs into technical specifications for materials, products and processes. Designs should:
- be unambiguous and clearly define characteristics important to quality, such as fitness for purpose, reliability and serviceability;
- specify methods of measurement, testing and acceptance criteria during development and manufacture; and
- result in products and services that can be produced and give customer satisfaction at acceptable prices.

8.6.1 Quality function deployment (QFD)

One method that has been particularly successful in analysing customer requirements and building these into product designs is quality function deployment. QFD is a process that can help determine:
- what a customer's requirements are;
- what a customer's highest priorities are;
- the characteristics of a product or service that are likely to satisfy customer requirements; and
- which characteristics of a product or service will have the greatest impact on customer requirements.

This information can greatly increase the product's or service's ability to focus on areas that will ensure customer satisfaction.

QFD is a system for translating customer requirements into appropriate products and services at each stage, from research and product development, to engineering and manufacturing, to marketing/sales and distribution. It is a formal method of transforming customer requirements into technical requirements.

In QFD the voice of the customer (VOC) expresses the real requirements of the customer. The QFD method uses the VOC chart to provide the discipline and communication required to focus on answering two questions: What does the customer need? and How can we satisfy the customer's need?

Figure 8.1 shows a VOC chart for a household steam iron. This chart has been prepared using the following steps.
- **Identify the customer's needs:** Investigate the customer's needs using the techniques given in the section on understanding and answering the voice of the customer. These needs must be stated from the customer's point of view. List these in the VOC column of the chart.

▦ **Rate the importance of the VOC:** For each VOC, identify the importance the customer places on the need. Rate them on a scale of 1 to 3, with 3 being the most important, 2 being moderately important and 1 being least important.

▦ **Identify the product characteristics that will meet the customer's needs:** These are the technical counterparts that must be deployed to meet the customer's needs. They are listed across the top of the chart.

▦ **Develop the VOC matrix:** Wherever there is a relationship between the VOC and the product characteristic, the importance score is entered into the cell.

▦ **Prioritise the product characteristics:** By totalling the columns, the product characteristics can be prioritised according to their importance to the customer.

Product: Steam iron										
Important: **3 = Most important: 2 = Moderately important: 1 = Least important**										
					Product characteristic					
VOC Number	**VOC**		**Importance**	Lightweight	Self-cleaning	Rotating cord connector	Teflon-coated base	20-second warmup period	Intelligent controls	Transparent water reservoir
1	Saves time		3	3	3		3	3	3	
2	Right temperature for all types of fabric		3					3	3	
3	Plate clean at all times		2		2		2		2	
4	User can see when water is depleted		2							2
5	Heats up in a short period		2					2	2	
6	Cord does not twist and snag		2			2				
7	Optional manual temperature and steam control		2						2	
8	Easy to store		1	1		1				
		Total:	4	5	3	5	8	12	2	

Figure 8.1 VOC chart for household steam iron

It can be seen from figure 8.1 that intelligent controls and a 20-second warm-up period are the most important characteristics, and therefore require the most attention and resources in the development of the steam iron.

8.6.2 Example: Building quality into the product at Motorola

Stephen Rosenthal (1994) provides insight into how Motorola became world leaders in the manufacture of mobile communication products. Confronted with an onslaught of competition from the Far East and Europe, Motorola was committed to fighting for global markets for its electronics products. One of its main strategies in this continuing battle is improved quality – in products, services and, in fact, every phase of all jobs. In the mid-1980s, Motorola embarked on a programme to improve product quality 100 times in five years.

Even as the company moved toward this goal, which seemed almost impossible when first discussed, it was realised that the result would not be good enough to remain competitive. Further, the '100X' goal spoke only of product quality and did not cover all of the other aspects of the daily jobs of the people at Motorola.

Therefore, a new goal, designated Six Sigma, was formulated. Motorola defined Six Sigma as 'a measure of goodness – the capability of a process to produce perfect work.'

Six Sigma is measured in terms of *defects per opportunity for error*, where a defect is defined as any mistake that results in customer dissatisfaction. In order for a process to be at Six Sigma it must have less than 3.4 defects per one million opportunities. The strength of this seemingly simple concept is that by applying the term 'customer' to anyone who has a need for a product or a service, the measurement of quality can be extended to every function in the company.

For example, a customer can be the next person on the assembly line, someone buying food in a company cafeteria, an order administrator handling field orders or even the purchaser of a paging receiver. Under the rallying cry of 'Six Sigma', everyone at Motorola is considered to be both a customer and a supplier.

Six Sigma has grown into a quality and process improvement methodology that defines, measures, analyses, improves and controls processes, and claims to tie quality improvements directly to business results. A good overview of Six Sigma is provided by Greg Brue (2002).

In terms of product quality at this high level, it was soon realised that no amount of time or money could take a product designed in the past and make it reach Six Sigma. This quality level had to be designed in from the beginning.

8.7 FOCUS ON THE PROCESS

The traditional approach to quality control was to prevent poor quality products being delivered to customers. This involved inspectors examining products and either scrapping those that did not pass or sending the non-conforming products back for reworking.

One objective of TQM is to replace inspectors looking at products with people who take ownership of the processes that deliver customer value. To achieve this, it is necessary to continuously improve and control the processes that design, develop, produce, deliver and support products.

As early as 1961, A.V. Feigenbaum (1961) stated that 'the burden of quality proof rests not with inspection but with the maker of the part: machinist, assembly foreman, vendor, as the case may be.' Before looking at how to improve processes, we must examine what a process is and what causes it to produce poor quality.

A process is a repetitive set of interacting activities that uses resources to transform a defined set of inputs into outputs that are of value to a customer.

A process may be a manufacturing process or a business process. Manufacturing processes produce physical products (outputs) by transforming raw materials and components (inputs), using people, machines and facilities. Services such as insurance and banking are produced and delivered by business processes.

165

The elements of a process are called the four Ms:
- materials (raw materials or components);
- manpower (people);
- machines (tools, equipment and facilities); and
- methods (operating procedures).

All of the Ms are not always apparent in a process. Services, for example, usually involve few or no materials.

8.7.1 Variation: Special and common causes

When processes produce output and the quality of the output varies, this is called *process variation*. This is a key concept in TQM.

Output can vary a great deal or only a small amount, depending on how effective the process is. All processes have some inherent variation due to chance, which in TQM is called *variation due to common causes*. Other variations are not due to chance but due to special causes, such as an incorrect tool setting or a person not following procedures.

A key objective of TQM is to identify when a special cause or variation exists so that these can be eliminated. This requires separating special causes from common causes. Special-cause variation is also called *assignable variation* because it can be traced to a specific source.

Apart from identifying special causes of variation, process improvement also aims at reducing the range of variation due to common causes. The concept of variation was popularised by W. Edwards Deming (1986). He recognised that the responsibility for quality improvement could be assigned according to the type of variation. Those working in the process, the front-line workers, have the job of finding and eliminating special causes. Those managing the process are responsible for eliminating common causes, which are more difficult to pin down and usually involve working on the process as a whole and the investment of capital to improve the process. Deming's rule of thumb is that 94 per cent of all variation is due to common causes and only 6 per cent is due to special causes.

8.7.2 Example

Let us look at an example of a manufacturing process in order to understand process variation. An operator makes brass hose fittings on a machine. The part number of the fitting is A800. Hexagonal brass bar is cut into 50 mm lengths, then an 8 mm thread is cut on one end of the fitting and a barb is machined on the other end to fit the internal diameter of an 8 mm rubber or plastic hose.

These fittings have several *quality characteristics* that can be measured. The quality characteristics are achieved by the process that produces the fittings. The ability of the process to produce acceptable quality characteristics is called *process performance*. An analysis of customer needs will indicate which quality characteristics are important enough to formally control. For a product like the A800 fitting, customer-critical characteristics might be length, barb diameter and thread depth. Following Schonberger and Knod's (1991) suggested method for process improvement, the operator measures the quality characteristics of 30 consecutive fittings and notes the results. One of the characteristics measured and recorded is called 'fitting length'. Table 8.1 shows the recorded fitting lengths.

Table 8.1 Sample of 30 consecutive A800 fittings before process improvement

A800 hose fitting														
Sample of 30 consecutive fitting lengths in millimetres														
Before improvement														
1	2	3	4	5	6	7	8	9	10	11	12	13	14	15
49,9	50,3	50,1	49,7	50,0	49,9	51,1	50,5	50,1	50,2	49,8	50,4	50,2	50,1	50,2
16	17	18	19	20	21	22	23	24	25	26	27	28	29	30
49,8	49,5	50,0	49,8	49,9	49,6	49,7	51,6	49,5	49,5	49,4	49,8	49,6	49,5	49,4

The operator then plots the measurements on a run chart, the results of which are shown in figure 8.2. What can you conclude?

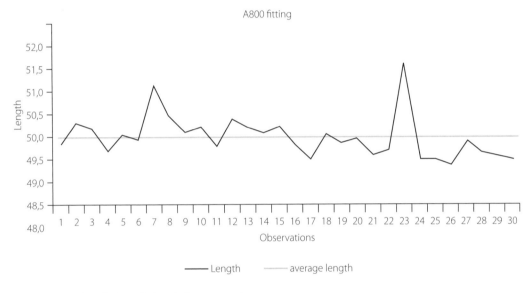

Figure 8.2 Run chart for A800 fittings before process improvement

The objective is to make fittings that are 50 mm long. Table 8.1 and figure 8.2 show, however, that not all the fittings are exactly 50 mm long. The varying length is an example of process variation. It would probably be difficult, if not impossible, to produce every fitting exactly 50 mm long using the available process. It is important, however, to reduce the variation in fitting lengths as far as possible.

In the example fittings, we find that the lengths vary from a minimum of 49.4 mm to a maximum of 51.6 mm. The process appears to be very unstable. It is now the job of the operator and the supervisor to investigate the causes of the variations.

Suppose the investigation reveals that a badly worn machine was being used to cut the brass bar lengths. This is a special cause of variation and must be corrected. Replacing the worn parts on the machine and maintaining it regularly may eliminate the problem.

Once the operator and supervisor feel that the special causes have been eliminated, the operator measures another sample of 30 consecutive fittings using the improved process. The results are shown in table 8.2 and figure 8.3.

Table 8.2 Sample of 30 consecutive A800 fittings after process improvement

A800 hose fitting														
Sample of 30 consecutive fitting lengths in millimetres														
After improvement														
1	2	3	4	5	6	7	8	9	10	11	12	13	14	15
49,8	50,2	50,1	49,8	50,0	49,9	50,1	49,8	50,2	50,0	49,8	49,9	50,2	49,8	50,1
16	17	18	19	20	21	22	23	24	25	26	27	28	29	30
49,8	50,2	50,0	50,2	49,9	50,0	49,8	50,2	49,9	49,8	50,2	50,1	49,8	50,1	50,2

The process appears to be stable, but we do not know whether it is in control. The process is said to be in control when all special causes of variation have been removed. When a process is in control there is still some variation resulting from common causes, but the process output is predictable, i.e. it follows an identifiable pattern. To determine whether the A800 fitting manufacturing process is in control, it is necessary to monitor the process once it appears to be stable. This is done through the use of control charts (discussed later in this chapter).

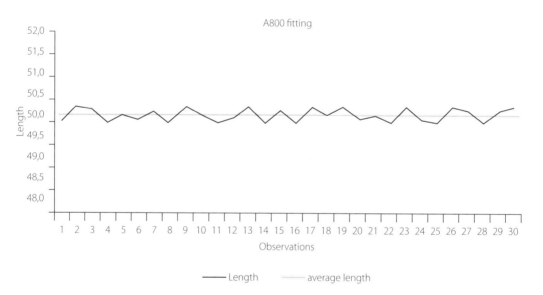

Figure 8.3 Run chart for A800 fittings after process improvement

8.7.3 Quality improvement tools

There are seven quality improvement tools that can be used individually or together to improve processes and thus the output of processes. These tools are a simple and long-established set of techniques for data-gathering and -analysis in a work situation. The emphasis of these tools is on prevention rather than cure; and they enable people who own and work in processes to do their own improvement and control. Peter Mears (1995) and John Bicheno (1994) give detailed information on these tools.

Tool 1: The process chart

The process chart is used to map the activities involved in the manufacture of a product or the delivery of a service. It helps identify value-adding and wasteful activities, thus revealing possible sources of quality and productivity problems. *As-is* process charts describe the process before it has been improved, and *to-be* charts describe what the process should look like after it has been improved.

Once the to-be process has been implemented, this chart serves as a record of how the process should be carried out. It provides complete documentation of the process flow.

A sophisticated form of process chart that is widely used in TQM is called an IDEF model. IDEF is an abbreviation for integrated definition for process modelling. An IDEF model represents the activities of a process, how they interrelate, the resources used to conduct each activity, and the result or output of each activity. The model consists of graphics and associated text supporting the graphics. It is an intuitive way to define, analyse and document the manufacturing and business processes of an enterprise. Process charts and models are developed using a team approach, preferably involving people who are directly involved in the process. In the case of the manufacture of the A800 fittings, the process should be mapped from the receipt of the raw materials through to the packaging and delivery of the finished product.

Tool 2: Pareto analysis

Pareto analysis is a simple yet effective tool for identifying and prioritising quality problems that need to be solved. It recognises that a few problem types account for a large percentage of the total number of problems that occur. It describes an effect that is often called the 80/20 rule, because 80 per cent of the problems that occur often comprise only 20 per cent of the problem types.

The Pareto method was first named by Joseph Juran (1988), who described it as a universal problem-solving method to identify the 'vital few' as opposed to the 'useful many'. It was named after Vilfredo Pareto, a 19th century economist who showed that the largest share of world income was held by a small number of people.

It is a graphical method that begins by ranking problems from highest to lowest. Then the cumulative number of problems is plotted on the vertical axis of the graph. Along the horizontal axis are arranged the problems in descending order. The resulting graph rises rapidly, then levels off to an almost flat plateau. Once the graph has been prepared, it is easy to identify the problems that need top priority.

Tool 3: The Ishikawa diagram

Once quality problems have been identified using Pareto analysis, they need to be solved. The Ishikawa diagram is named after Kaoru Ishikawa (1987), who first popularised the tool. It is also known as the cause-and-effect diagram or the fishbone diagram. The problem that needs to be solved becomes the 'spine' of the fishbone, and the contributing causes are the 'bones'. The diagram is constructed using brainstorming to identify possible contributing causes of a particular quality problem. To structure the diagram, use the four Ms (see page 166) as the four initial bones. The Ishikawa diagram is a concise, clear and graphical way of identifying the causes of problems that need to be solved to improve quality.

Tool 4: Histograms

The histogram has much in common with the Pareto diagram. It is used to show, graphically, the relative number of occurrences of a range of events. Using vertical bars, it plots frequency on the vertical axis against events arranged one after the other on the horizontal axis. Once the Ishikawa diagram has been prepared, data needs to be collected and classified according to each of the causes suggested. When data is shown on a histogram, the most likely causes become apparent.

Tool 5: Run charts and correlation diagrams

The next stage requires some experimentation to find out how causes can be eliminated. Run charts and correlation diagrams are helpful here as they are used to explore relationships between events and time, and between problems and causes. They are used to find out when and how problems arise, and how they can be solved. A *run chart* is simply a graph of the number of events plotted against time. An example is figure 8.2, which was prepared by the operator of the A800 fitting manufacturing process. It is used in an initial investigation of a process. Although this is a simple tool, it is very effective in the hands of a person trained to identify patterns in the data displayed.

A *correlation diagram* is used to identify specific cause and effect relationships. Let us assume that we have a process such as baking bread where a correct oven temperature range is critical for acceptable loaves of bread. The number of rejected loaves is a measure of process performance and is plotted on the vertical axis of a correlation diagram. The temperature is plotted on the horizontal axis.

This is called the *experimental variable*. The diagram would probably show that, as temperature rises, rejects fall. Then around a specific temperature range, rejects are at a minimum and when the temperature is increased further a greater number of rejects occur. The two are often used in conjunction with one another. First, a run chart is used to detect change over time and, once a pattern has been established, a correlation diagram is prepared to identify relationships.

Tool 6: Control charts

Control charts are used to control processes once they are considered stable. There are two main types of control chart: *variables* and *attributes charts*. A variables chart measures some characteristic that is variable along a scale, such as length or diameter. It measures something that can be measured. An attributes chart is used where there are only two possibilities: pass or fail, yes or no. With an attribute, a judgement is made rather than a measurement taken. The application of control charts in TQM is part of statistical process control (SPC).

1. Variables charts: Average and range charts

The two most important variables charts are *average* and *range charts*. Once processes are stable, samples are taken of the products produced by the process and their quality characteristics measured. Average charts plot the averages of samples taken, while range charts plot the ranges of the samples – the maximum minus the minimum values. Both charts are necessary.

Whereas the averages may be acceptable, the ranges might be unacceptably wide. Alternatively, the range of the sample may be acceptable, but located in the wrong place. In other words, it has an unacceptable average. Typically, from time to time, a sample of the most recently produced products is taken. The product characteristics are measured and the average and range values of the sample calculated. These two results are plotted on the charts. The charts indicate if the process is in control. When the process is in control, work continues; if not, work stops and an investigation is carried out.

Control charts must be set up for each process. The manufacture of the A800 fittings will be used to illustrate how the charts are constructed and used. Once the operator and supervisor are satisfied that the special causes of variation have been removed, the process is run and fittings are produced. At hourly intervals, samples of five consecutive fittings are taken and the quality characteristics measured. Average and range control charts are prepared for the fitting lengths. Table 8.3 shows the averages and ranges of 20 consecutive samples. Figures 8.4 and 8.5 show the same data in graphical form.

Table 8.3 Averages and ranges of lengths of A800 fitting samples

A800 hose fitting										
20 samples of 5 consecutive fitting lengths in millimetres										
	1	2	3	4	5	6	7	8	9	10
Average	49,90	50,14	50,18	50,14	50,30	50,14	49,98	50,10	50,02	49,86
Range	0,80	0,40	0,40	1,10	0,50	0,60	0,60	0,80	0,80	0,70
	11	12	13	14	15	16	17	18	19	20
Average	50,28	50,34	50,18	50,10	49,86	50,14	50,20	49,84	50,26	50,00
Range	0,50	0,40	0,50	0,30	0,70	0,60	0,50	0,80	0,30	0,80

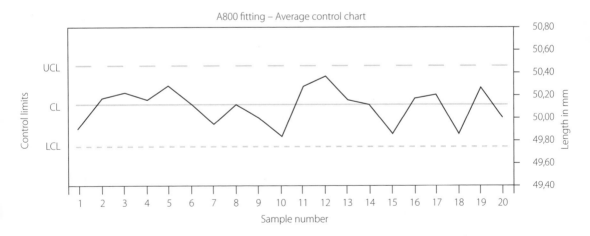

Figure 8.4 Average control chart for A800 fitting lengths

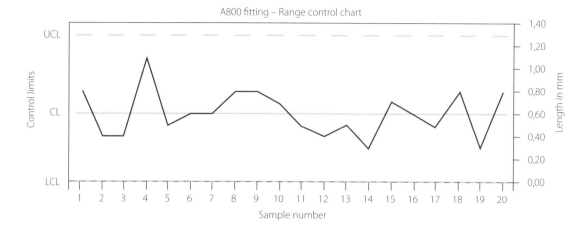

Figure 8.5 Range control chart for A800 fitting lengths

The average row of the table shows the simple average (mean), $\overline{X}$, of the lengths of the five fittings in each sample taken. For example, the average for sample 1 is:

$$\overline{X_1} = \frac{\sum x}{n} = \frac{49,9 + 50.1 + 50.3 + 49,5 + 49,7}{5} = 49,90$$

The range row shows the range, R, which is the longest minus the shortest of the five lengths in the sample. Thus for sample 1:

$R = 50.3 - 49.5 = 0.80$

We also need to know $\overline{R}$ the average (mean) range of the 20 sample ranges. In addition, we must compute the grand average, $\overline{\overline{X}}$, which is the average of the sample averages. The values $\overline{R}$ and $\overline{\overline{X}}$ become the centre lines (CL) on a control chart for the range control chart and the average control chart respectively. The calculations are as follows:

$$\overline{R} = \frac{\sum R}{k} = \frac{12,10}{20} = 0,60$$

$$\overline{\overline{X}} = \frac{\sum \overline{X}}{k} = \frac{1001,96}{20} = 50,10,$$

where k is the number of samples taken.

The next step is calculating control limits. Since we plot sample averages rather than individual piece measurements, we are actually dealing with a distribution of sample averages. The standard deviation (sigma) of this distribution of sample averages, also called the standard error, depends on the size of the sample taken.

Process control limits are usually set at three standard errors above and below the centre line; this is referred as the three-sigma limit. This means that approximately 99,72 per cent of all sample averages should fall between the upper and lower control limits if the process is in control. This is because the distribution of sample averages is regarded as approximately normal.

Table 8.4, from Schonberger and Knod (1991), provides a convenient set of factors to calculate control limits.

Table 8.4 Process control chart factors

Sample size	Control limit factor for average control chart	UCL factor for range chart	LCL factor for range chart	Factor for estimating process sigma ($\hat{\alpha} = \bar{R}/d_2$)
(n)	(A_2)	(D_4)	(D_3)	(d_2)
2	1,880	3,267	0	1,128
3	1,023	2,575	0	1,693
4	0,729	2,282	0	2,059
5	0,577	2,115	0	2,326
6	0,483	2,004	0	2,534
7	0,419	1,924	0,076	2,704
8	0,373	1,864	0,136	2,847
9	0,337	1,816	0,184	2,970
10	0,308	1,777	1,223	3,078

Earlier we determined that $\bar{R}$ = 0,60, $\bar{X}$ = 50,10 and the sample size was 5.

Control limits for the average control chart are:
Upper control limit $(UCL_{\bar{x}})$ = $\bar{\bar{X}}$ + $(A_2)(\bar{R})$ = 50,10 + (0,577)(0,60) = 50,45
Lower control limit $(UCL_{\bar{x}})$ = $\bar{X}$ − $(A_2)(\bar{R})$ = 50,10 − (0,577)(0,60) = 49,75

Control limits for the range control chart are:
Upper control limit (UCL_R) = $(D_4)(\bar{R})$ = (2,115)(0,60) = 1,27
Lower control limit (LCL_R) = (D_3) $(\bar{R})$ = (0)(0,60) = 0

Table 8.3 shows the averages and the ranges for the 20 samples and the control charts are shown in figures 8.2 and 8.3.

1. Chart interpretation

All the points are within the control limits and the process appears to be in control; the control charts are 'the voice of the process'.

If control charts continue to show that the process is under control, there may be no need to keep using them. However, measurement and plotting on control charts should still be done on an audit basis occasionally in order to ensure that the process does not stray out of control.

With only natural variation occurring, one would expect measurements to be spread more or less evenly on either side of the centre line. If there are runs of successive measurements above or below the centre line, or if there is a positive or negative trend in the values, these are indicators of possible problems. The interpretation of process control charts is a skill that can be developed. Particular chart patterns are indicative of specific problems that may be developing.

2. Attributes charts: The p and c charts

It is not always possible to measure variables. Some defects, such as scratches or cracks, are either there or they are not. The products either pass or fail. This is where *p* charts and c charts come in. A *p* chart (*p* being an abbreviation for percentage defective) is used where there are batches of a product and the percentage that is defective can be determined. A *c* chart is used where there are a number of possible types of defect associated with a particular product – for instance, the number of scratches or cracks. With attributes, only one chart is plotted, not two, as with the average and range charts. The basic principles of taking samples, setting up the charts and interpreting them remain the same.

3. Process capability

Customer requirements are stated formally as specifications. A specification has two parts: a nominal value and tolerances. In the case of the A800 fittings, the specification of the fitting's length could be stated as: 50 mm ± 0,5 mm. Specifications should reflect customer requirements and they exist independently of the process's ability to produce products that meet specifications.

Once the special causes of variation have been removed and the process is shown to be in control, quality is still not certain. It is now necessary to establish that the process can produce output within the tolerance of the specification.

This is called *process capability*. Process capability can be measured using a process capability index. If the capability index equals 1,0 the process just meets the specification. If it is less than 1,0 (but positive), only part of the process output meets the specification. If it is greater than 1,0 the inherent capability of the process is within specification. A rule of thumb is to aim for a capability index of at least 1,33.

Process capability refers to the match between the location of the upper and lower process control limits (UCL and LCL), and the specification limits. The location of the process control limits is due to the natural variation of the process that makes the product, whereas the designer of the product gives the specifications. These are two distinctly separate things, but they must relate in order to produce quality products.

Tool 7: Checklist

The last of the seven tools is used to maintain the level of quality attained by the application of the previous tools. A checklist is a list of items that need to be done and the order in which they should be done. It details the correct procedure to be followed. Progress can be determined by ticking off the items when they are complete. The procedure given in the checklist must be verified or audited at specific intervals.

8.8   SUPPLIERS ARE PARTNERS IN QUALITY

Increasingly, companies purchase raw materials, components and sub-assemblies rather than manufacturing them themselves. This means that the application of the TQM principle extends beyond the boundaries of an enterprise to its suppliers.

The starting point for integrating suppliers into a TQM programme is to treat them as partners. The traditional adversarial attitude adopted by buyers whose only interest is to bargain for the lowest price does not work if partnership is the aim. Partnership means taking co-responsibility for meeting customer expectations the first time, every time.

Working with suppliers with TQM as an objective means that you are not just buying the product, but also the capability of the supplier to meet your technical specification, and your quality and delivery requirements.

When you have a strong TQM programme in place and you have a supplier with the capability to meet your requirement, a transfer of TQM knowledge to the supplier will help you to meet your TQM objectives. By training the supplier in TQM principles and techniques, a long-term partnership is often forged.

The logical consequence of developing a sound TQM-based relationship with suppliers is that you adopt a policy of single sourcing for each product class you purchase. Although this policy has its risks, companies that have made a commitment to TQM are increasingly adopting it.

The advantages of establishing single sources of supply are as follows.

- You get better service. A supplier that has all your business for a product class will be better able to invest in the right people, equipment and methods to meet your needs. You will also be in a position to expect the best service.
- You reduce the amount of monitoring and expediting of purchased products.
- You can develop a more collaborative, information-sharing relationship with the supplier.
- You can get a better price. The more business you give a supplier, the lower the price should be. You should also be able to negotiate favourable long-term supply contracts with single-source suppliers.

8.9 QUALITY IS FREE

There is a cost attached to doing things wrong, called the cost of quality, or COQ. Cost of quality can be thought of as the cost of achieving conformance to quality standards plus the cost of non-conformance. It is the cost of rework, scrap, inspection, warranty claims, testing and similar activities to ensure conformance to quality standards. Costs related to quality are usually separated into at least three areas, as follows.

1. **Prevention costs:** These costs are associated with all the activities that focus on preventing defects or non-conformance with quality standards. In many organisations this includes all the people in the quality department who inspect the product and the cost of operators who do their own inspection. Also included in this group are activities to assure supplier conformance.

2. **Appraisal costs:** These costs are associated with measuring, evaluating or auditing products to assure conformance with quality standards and performance requirements.

3. **Failure costs:** These costs are associated with evaluating and either correcting or replacing defective products, components or materials that do not meet quality standards. Failure costs can be either internal failure costs that occur prior to the completion or delivery of a product or service, or external failure costs that occur after a product is delivered or a service is provided.

The basic relationship among the three types of cost is that money invested in prevention and appraisal can substantially reduce failure costs. Apart from reducing expenses, the reduction in external failures results in fewer dissatisfied customers, resulting in fewer product returns, fewer customer complaints and increased customer loyalty. Generally, in a well-managed enterprise, the cost of prevention and appraisal should be one-third of the cost of failure. In addition, the benefits resulting from the application of the principles of TQM can often outweigh the cost of prevention and appraisal, leading to the conclusion that 'Quality is free. It's not a gift, but it is free' (Crosby, 1979).

8.10 THE ISO QUALITY SYSTEM STANDARDS

The Geneva-based International Organisation for Standardisation (IOS) first published a series of standards for quality management and quality assurance in 1987, revised in 1994 and 2000. These standards became known as the ISO 9000 standards (ISO being derived from a variant of the Greek word *isos*, meaning 'equal').

The ISO 9000 standards are a written set of standards that define the basic elements of a management system an enterprise should use to ensure that its products and services meet or exceed customer needs and expectations. The standards are generic and can be applied to any kind of product or service. The standards do not specify technology to be used for implementing quality system elements.

The current set of ISO 9000 and related standards comprises the following.

- **ISO 9000:2000, Quality management systems: Fundamentals and vocabulary:** This standard establishes a starting point for understanding the standards and defines the fundamental terms and definitions used in the ISO 9000 family of standards.
- **ISO 9001:2000, Quality management systems: Requirements:** This is the requirement standard used to assess an organisation's ability to meet customer and applicable regulatory requirements and thereby address customer satisfaction. It is now the only standard in the ISO 9000 family against which third-party certification can be carried.
- **ISO 9004:2000, Quality management systems: Guidelines for performance improvements:** This guideline standard provides guidance for continual improvement of a quality management system.
- **ISO 19011:2002, Guidelines on quality and/or environmental management systems auditing:** This standard provides guidelines for verifying that a quality or environmental management system achieves defined quality objectives. This can be used to audit suppliers.
- **ISO 10005:1995, Quality management systems: Guidelines for quality plans:** This standard provides guidelines to assist in the preparation, review, acceptance and revision of quality plans.
- **ISO 10006:1997, Quality management systems: Guidelines to quality in project management:** This guideline standard helps to ensure the quality of both project processes and the project products.
- **ISO 10007:1995, Quality management systems: Guidelines for configuration management:** This standard gives guidelines to ensure that a complex product continues to function when components are changed individually.

- **ISO 10012:2003, Measurement management systems: Requirements for measurement processes and measuring equipment:** This standard gives guidelines for calibration systems to ensure that measurements are made with the intended accuracy.
- **ISO/TR 10013:2001, Guidelines for quality management system documentation:** This standard provides guidelines for the development and maintenance of quality manuals.
- **ISO 10015:1999, Quality management: Guidelines for training:** This standard provides guidance on the development, implementation, maintenance and improvement of strategies and systems for training that affects the quality of products.
- **ISO/TS 16949:2002, Quality management systems: Particular requirements for the application of ISO 9001:2000 for automotive production and relevant service part organisations' quality systems:** This is a sector–specific guidance standard for the application of ISO 9001 in the automotive industry.

Additional information on ISO quality standards is obtainable from the International Organisation for Standardisation, http://www.iso.org.

A major difference between ISO 9001:2000 and its predecessors is that it makes a process approach to quality management mandatory. It is very clear that organisations must identify and manage the processes that make up their quality management systems. They also need to manage the interaction between these processes, and the inputs and outputs that link these processes together.

A quality management system is a network of processes, and a process is made up of people, work, activities, tasks, records, documents, forms, resources, rules, regulations, reports, materials, supplies, tools and equipment – all the things that are needed to transform inputs into outputs. In general, a quality system includes all the things that are used to regulate, control and improve the quality of products and services.

The objective of an ISO 9000 quality system is to create and, on a continuous basis, to improve the processes by which the enterprise meets customers' needs, leading to customer satisfaction and resulting in success for the enterprise.

The standard recommends that 'appropriate levels of management define, where necessary, specialised quality objectives consistent with corporate quality and other corporate objectives.' Suggested quality objectives include:

- standards for 'key elements of quality: fitness for use, performance, safety and reliability'; and
- standards for quality of the process itself: its control, capability, performance, safety and reliability.

In pursuit of these quality objectives, the standard advises that certain quality activities be carried out, such as:

- defining customer needs in a measurable way;
- implementing preventive actions and controls to prevent customer dissatisfaction or adverse effects by the organisation on society and the environment;
- optimising quality relating costs;
- building collective commitment; and
- continuously reviewing requirements and achievements to seek ways to improve.

Fulfilment of all the above should result in organisational success that the standard defines as:
- meeting well-defined needs, uses or purposes;
- satisfying customers' standards, specifications and expectations;
- complying with statutory (and other) requirements of society; and
- providing products at competitive prices for a cost that yields profit.

The ISO 9000 quality system comprises methods designed to:
- plan the means by which the quality will be achieved;
- document these plans into procedures;
- communicate the procedures to everyone in the organisation;
- monitor the success of the efforts;
- modify the procedures, on a controlled basis, to improve them based on the feedback from monitoring; and
- prove to others the existence and efficacy of the quality system.

Table 8.5 outlines the documents that need to be produced to set up and run an ISO 9000 quality management system.

Table 8.5 ISO documents and responsibilities

Document	Responsibility
Quality policy	Quality council
Quality manual	Middle management
Procedures	Process improvement team (PIT)
Job instructions	PITs, supervisors and employees
Records and documentation	Employees

The quality policy, quality manual, procedures, job instructions and recorded data must be documented and ready for inspection at any time. They must be the basis of the quality practices of an enterprise.

Self-assessment

8.1 Identify the TQM principles used by Granite Rock Company. Describe how the company has applied them. Which TQM practice do you think has made the greatest contribution to the company's success?

8.2 Why do you think all people in an enterprise should be totally involved in quality improvement?

8.3 Give examples of how quality can be made visible to customers, work groups and management.

8.4 What is the meaning of the Japanese word *kaizen*? Why is *kaizen* important in TQM?

8.5 How can the new products development process contribute to product quality?

8.6 Why is the concept of process variation key to TQM? Explain this concept and give some examples of process variation that can be measured.

8.7 What is the difference between a common cause and special cause? Why is it important to differentiate between these two types of causes?

8.8   Name the quality improvement tools. Which of these tools could be used to analyse the quality problems experienced in the manufacture of the A800 fittings? Describe how you would apply the tools to analysing and solving the problems.

8.9   Do you agree that quality is free? Explain your answer.

8.10 Interview a person in an organisation that is pursuing or has achieved ISO 9001:2000 accreditation.

8.11 Identify which standard is being or has been implemented and the objectives of the programme.

8.12 Also determine the benefits resulting from the programme and the difficulties that have been experienced.

8.13 There are many organisations that are renowned for the quality of their products and services. Some examples are Toyota, Boeing, Panasonic, AVIS, Nuclear Technology Products, Tongaat-Hulett Sugar, Impala Platinum Refinery and Dekro Paints. Research one of these organisations and find out which of the TQM practices it has successfully applied.

8.14 Visit an organisation you normally do business with, such as a retail store, a bank or a motor vehicle repair workshop. Through informal research and direct observation of its customer-facing processes, such as customer assistance and speed of service, suggest ways in which it could apply TQM practices to improve its product or service.

References

Banks, J. 1989. *Principles of Quality Control*. New York: John Wiley.

Bellefeuille, J. 1993. *IEEE Spectrum*, 30 (9), p. 47.

Bicheno, J. 1994. *The Quality 50: A Guide to Gurus, Tools, Wastes, Techniques and Systems*. Buckingham: PICSIE Books.

Brue, G. 2002. *Six Sigma for Managers*. New York: McGraw-Hill.

Crosby, P.B. 1979. *Quality is Free: The Art of Making Quality Certain*. New York: McGraw-Hill.

Deming, W.E. 1986. *Out of the Crisis*. Cambridge, Mass.: Massachusetts Institute of Technology, Centre for Advanced Engineering Study.

Feigenbaum, A.V. 1961. *Total Quality Control: Engineering and Management*. New York: McGraw-Hill.

Ishikawa, K. 1987. *What is Total Quality Control? The Japanese Way*. Translated by David J. Lu. Englewood Cliffs: Prentice-Hall.

Juran, J.M. 1988. *Quality Control Handbook*. New York: McGraw-Hill.

Mears, P. 1995. *Quality Improvement Tools and Techniques*. New York: McGraw-Hill.

Rosenthal, S.R. 1994. *Effective Product Design and Development: How to Cut Lead Time and Increase Customer Satisfaction*. Homewood: Business One Irwin.

Schonberger, R.J. 1982. *Japanese Manufacturing Techniques: Nine Hidden Lessons in Simplicity*. New York: Free Press.

Schonberger, R.J. & Knod, E.M. 1991. *Operations Management: Improving Customer Service*. Homewood: Irwin.

Websites

Granite Rock Company: www.graniterock.com.

International Standards Organisation: www.iso.org.

9 An Introduction to Safety Management

Carl Marx

Study objectives

After studying this chapter, you should:

■ have a basic understanding of the theories and concepts of industrial safety
■ have an understanding of the complexities of safety and the law
■ know the responsibilities of the engineer regarding ethics
■ be able to evaluate a safety management system and a safety audit
■ have a basic understanding of qualitative risk assessments
■ be able to assess a risk assessment done by a safety professional
■ have a basic understanding of the processes necessary to do a proper accident investigation and deal with accident statistics

The aim of this chapter is to introduce you to safety management as it is applied in an industrialised environment. The chapter's scope is limited in that it only provides a basic insight into a very complex field of study. Should you be interested in obtaining a working knowledge of any of the specific topics discussed in the chapter, further study is strongly recommended.

9.1   INTRODUCTION

The occurrence of industrial accidents and injuries has been with us since the first industrial accident of the Industrial Revolution. These types of accident and injury in the modern industrialised environment have become unacceptable to the public, state agencies, public organisations and companies, as well as to the victims and their families.

In most countries, employers are legally responsible, apart from the humanitarian concerns, to reduce the risk of accidents and, in the event of an accident, the associated suffering as a result of the accident.

This chapter is about the tasks that any person in a supervisory role should carry out pertaining to the safety of subordinates. The focus of the chapter will be on introducing engineers to their important role with regard to safety. The safety management approach is grounded in an array of competitive moves and business approaches that skilful engineers depend upon to ensure the successful performance of companies. The managing of safety involves every function and department of a company such as purchasing, production, finance, marketing, human resources and R&D.

Specialisation of knowledge, division of labour and detailed job specifications are common managerial responses to the complexities of the modern business. Attention is focused on the minutiae of managerial problems and techniques for dealing with them. Yet, when the successes and failures of companies are analysed one usually finds a correlation between safety and financial successes. Successful companies normally realise that an investment in the safety of their workers pays handsome dividends.

The crafting and implementing of safety management practices form part of the core management functions. Amongst all things managers do, few affect company performance more fundamentally than how well its management/supervisory team incorporates safe work practices and procedures into the tasks that are carried out to produce the company results. It is said that a good safety performance in a company is the most trustworthy indication of good management practice.

It is accepted, however, that a good safety record does not guarantee that a company will be able to avoid periods of poor business performance. One must remember that even well-managed organisations can face adverse and unforeseen conditions. But the maintenance of a good safety record may be the one factor that can prevent total disaster.

9.2 THEORY AND CONCEPTS

In order to ensure that real value is added when applying the concepts and principles of safety in the workplace, it is important to first establish a conceptual foundation of industrial safety. One might argue that theories and concepts are useless and that the engineer would want to get to the part where he/she can learn some practically implementable practices and processes. However, it is important to realise that there is a very strong need for concepts and theory. They provide a framework for the paradigm required for someone to become a safety-conscious engineer. The theoretical framework assists in simplifying and structuring any problem for the generation of safe solutions. Theory and concepts are a very valuable foundation in safety education and training. They aid in generalising past experience and condensing it for broader implementation.

In order to prevent accidents it is necessary to understand the elements involved in an accident. Different authors have developed a number of different models. It is felt that the one described below will be sufficient in detail for the purposes of this discussion.

A graphical representation of the elements said to be present in an accident is given in Figure 9.1 on the next page. This model identifies the ten most critical performance areas, i.e.

- energy sources out of control;
- management system failure;
- training deficiency;
- latent design defects;
- inappropriate maintenance;
- imperfect procedures;
- unsuitable task directives;
- sub-standard physical conditions;
- unsafe acts; and
- barrier failures.

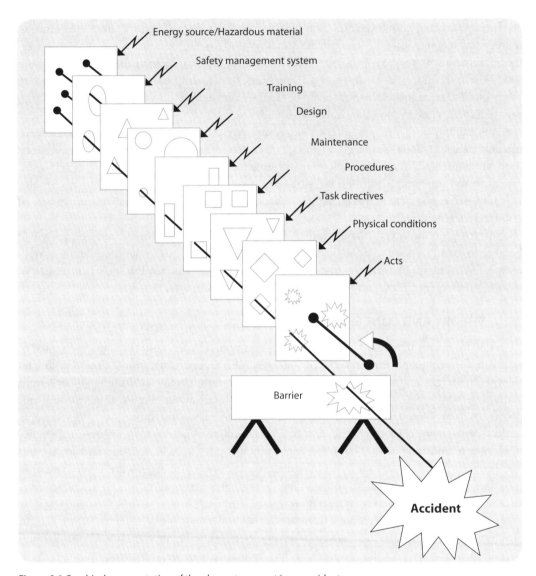

Figure 9.1 Graphical representation of the elements present in an accident

With reference to figure 9.1, the critical performance areas are represented as solid plates rotating at their own individual speed on a common axis. Each plate represents a critical element that contributes to the accident occurring. The solid plates represent a perfect condition in each of the elements. The randomly positioned holes in them represent unsafe acts, unsafe conditions, unsafe sections in procedures, latent design defects, ineffective safety training and a lack in safety leadership respectively. If at any point in time all the holes are aligned or arranged so that a line of sight passes through all the plates, then an accident will occur. In practice this means that an unsafe act alone cannot cause an accident. The unsafe act only forms part of a system where all the elements must have a defect in place in order for the accident to occur. The final condition required is that all the defects must coincide at the same time. This may explain why defects in the critical performance areas may exist for a long time before they result in an accident.

9.3   SAFETY AND THE LAW

There are a number of excellent publications dealing with safety and the law. For the purpose of this chapter it would suffice to say that any person in a supervisory position should ensure that he/she is aware of any statutory requirements placed upon the position he/she holds. In addition to statutory requirements pertaining to the position held, certain specific engineering duties also have specific statutory requirements placed on the engineer doing the design of a system or installation, for instance. Specifying these statutory duties falls outside the scope of this chapter. The importance of the duties can, however, not be overemphasised.

A legal liability does not always arise as a result of an intentional harmful act or of some proven oversight like negligence. The allocating of a liability as a result of injury may historically merely have been a peace-conserving substitute for the custom of an injured party taking revenge on the party believed to have caused the injury. It is a historic fact that the focus of the law has changed to ensure that the one who was negligent should compensate the one who has lost something as a result of the action of the negligent party, even if that action is otherwise blameless under the law.

In essence, any person in a supervisory position can cause his/her company to be prosecuted and in some cases can be held personally liable for both criminal and civil proceedings in the event of his/her failure to obey statutes and regulations. The following main points should be made with regard to safety and the law from a management point of view.

9.3.1 Common law

The name *common law* is derived from the medieval theory that the law administered by the king's courts represented the common custom of the realm, as opposed to the custom of local jurisdiction that was applied in local or manorial courts. In its early development common law was largely a product of English courts. In today's legal system common law is based on previous decisions by the courts and is also referred to as law of precedent.

The characteristic element of common law is that it represents the law of the courts as expressed in court decisions. The grounds for deciding cases in common law are acquired in precedents provided by past decisions of courts. With the exception of a few technical exceptions, a court is generally bound to follow earlier decisions of courts of equal or higher status.

9.3.2 Vicarious liability

Vicarious liability can be defined as the duty of a principal, like an employer, to take responsibility for the consequences of the acts or omissions of an agent, such as an employee. In other words, vicarious liability is when somebody can be held accountable for the actions and conduct of persons acting on his/her behalf. In this instance it is clear that a worker or contractor acting on behalf of an employer when carrying out a task may cause the employer to be held legally accountable for that action. Vicarious liability will be the basis for prosecution whether or not an act or omission was specifically authorised by the employer or not, as long as the act or omission by the employee was in the course of employment. To avoid vicarious liability, an employer must demonstrate either that the employee was not negligent in that the employee was reasonably careful or that the employee was acting in his/her own right rather than on behalf of the employer or his/her business interest.

9.3.3 Criminal liability

It is an established principle in law that behaviour must be criminalised before a person can be prosecuted and convicted. Duties placed on any person by acts of Parliament are enforceable and punishment may be passed by a court of law where failure to comply with these duties was found. Subordinate legislation in the form of regulations may also be promulgated under most occupational-safety-related acts. A contravention of a regulation is also a criminal act and may be prosecuted in a criminal court. The state always plays a part in a criminal prosecution case and it is very rare to find a private individual prosecuting another for a criminal act. The state will identify that there was an alleged non-compliance with legislation that may constitute a crime. Such an alleged crime may be committed by an individual or business and the state will prosecute the offender. The attorney-general prosecutes on behalf of the state through state prosecutors in the criminal courts situated throughout the country.

9.3.4 Civil liability

In the event of a supervisor not acting in the best interest of workers and as a result of this an accident occurs, the workers have the recourse to bring a civil case against the individual and/or the company for compensation, reparation or specific compliance. There is no collective list of circumstances that will result in an entity becoming liable in a civil case. Non-compliance with legislation will in most instances indicate negligence and it is therefore of vital importance that the law is complied with in full.

9.3.5 Administrative liability

In more recent years a move towards decriminalising safety-related offences has resulted in the introduction of an administrative fine system. In general this system will result in a company receiving an administrative fine as a result of a breach of safety-related statutes by a member of the management staff of the company. The issuing of such an administrative fine normally does not prevent criminal prosecution of the individual responsible for the contravention.

9.4   ETHICS

9.4.1 Understanding ethics

To understand ethics, the difference between morals and ethics must be understood. The term 'morals' normally refers to generally accepted standards of right and wrong in a society, while the term 'ethics' refers to more abstract codes that are specific to a particular profession. Both these terms involve standards of good conduct and the appraisal of specific actions by those standards. Moral and ethical statements should also be distinguished from laws. The fact that an action is legally allowable does not mean that it is morally and ethically permitted.

Engineering ethics are the moral issues, decisions and related questions about good conduct, character ideals and relationships of people facing engineers.

9.4.2 Engineering ethics

Some of the main ethical issues that an engineer is normally expected to adhere to can be summarised as follows.

- An engineer should always ensure the health, safety and welfare of workers and the public.
- If an engineer's assessment is contradicted in an environment that may endanger life or property, he/she must notify his/her employer or client and such other authority as may be appropriate.
- An engineer may only perform duties in the area of his/her competence.
- An engineer may only approve engineering designs that are in accordance with applicable industry standards. If his/her employer insists on unprofessional conduct, the engineer must notify the proper authorities.
- An engineer must be objective and truthful in professional reports or statements. All relevant and pertinent information must be included in such reports or statements.

9.5   SYSTEM SAFETY

The principle of system safety was first developed more than three decades ago. The process revolves around an approach where people, processes, plant and tools form an integral part of the duties required to carry out a task. System safety does not only take cognisance of the risk that each of these elements brings into the process, but also evaluates the multiplying effect caused by their interaction.

Although system safety was originally developed to design systems to prevent accidents, the system safety approach is increasingly utilised for the identification of system faults that may have been contributing to accidents. It is becoming increasingly important to analyse accidents from all angles in order to consider all means of improving overall safety. A system safety approach analyses an incident by considering the individual, equipment and procedures involved.

According to the US MIL STD –882B, the five basic steps of systems safety are as follows.

- Design for minimum risk.
- Incorporate safety devices.
- Provide warning devices.
- Develop procedures and training.
- Understand and expect residual risk.

System safety will only be successful in companies where a strong management commitment towards this approach exists, regardless of the industry.

9.6   SAFETY AUDITS

Safety auditing is a systematic method utilised to establish the fundamental safety performance of a company. An increased focus on auditing safety management systems – as opposed to physical condition inspections only – normally has a gradual and sustainable effect on the number of accidents occurring in an industrialised environment.

Safety performance management covers a wide range of issues, including:
- devolution of decision making;
- accountability for safety performance;
- measurement and benchmarking of safety performance; and
- responsiveness by employers to workers' safety initiatives.

The starting point of safety performance management is the setting of safety standards. These standards must be reflected in a safety policy and other detailed written standards. The written standards must include standards for at least the following critical elements to establish a safe working environment:

- physical condition standards (housekeeping standards);
- statutory compliance standards;
- standards for dealing with hazards;
- maintenance standards;
- training standards for task and safety training;
- standards for safe work procedures;
- standards for design; and
- competency standards for managers/supervisors.

Once standards have been set, safety performance must be reviewed through:

- safety programme evaluation and performance auditing;
- the use of tripartite mechanisms;
- the use of strategic management planning processes; and
- the creation of workplace forums for safety.

Performance audits should have at least the following objectives:

- to utilise a structured process of collecting information on the efficiency, effectiveness and reliability of the total company's safety management system;
- to identify best practice, ideas and experience in the full range of the company being audited;
- to facilitate safety workshops in order to assist in a particular safety problem area;
- to provide information about performance management practices and developments; and
- to carry out fact-finding missions in companies, which may include presentation of results on performance audits.

Safety performance auditing covers a range of approaches in terms of scope, methodology and form of reporting, including:

- examination of documentary proof;
- visual observations of the workplace;
- evaluation of accident statistics; and
- interviewing of workers to substantiate findings.

The above may include reviewing and auditing a number of factors particular to a specific company, but should at least include:

- reviews of the compliance to a safety management programme by reference to results, standards or benchmarks;
- reviews of the safety systems and processes and of a company compared to international standards of good practice;
- reviews of the adequacy of safety performance measurement mechanisms;
- auditing of the accuracy and relevance of safety performance information; and

▓ an analysis of the contribution made by workers to the level of safety performance in the company and the relationship of safety achievements to other forms of performance reviews. *This is a key issue.*

For the purpose of safety auditing the safety management system can be divided up into three key elements that exist at different levels in an organisation, namely:
▓ management arrangements;
▓ risk control systems; and
▓ practical workplace precautions.

9.6.1 Management arrangements

The term 'arrangement' is used to cover all aspects, e.g.:
▓ policies/objectives;
▓ planning;
▓ organising;
▓ implementing;
▓ measuring performance;
▓ reviewing; and
▓ auditing.

These are managerial methods by which an organisation sets out to determine and provide adequate control systems.

9.6.2 Risk control systems

Risk control systems set out the way workplace precautions are implemented and maintained. These include baseline risk assessments, issue-based risk assessments and continuous risk assessments. Continuous risk assessment tools most often used include systems such as planned task observations, pre-use check lists, critical task analysis and standard operating procedures, to mention only a few of the more important tools.

9.6.3 Workplace precautions

Workplace precautions provide protection at the point of risk, and cover the following areas.
▓ *Physical conditions:* This includes safety hardware such as machine guarding, protective equipment, overhead protection, handrails, pressure relief valves and other safety devices.
▓ *Working practices:* This refers to the way things are done. In this regard cognisance must be given to the difference between the prescribed procedure and what actually happens in the workplace.
▓ *Control documents:* These documents include documentation such as permit to work forms, fireman's report, standard operating practices, planned task observations, pre-use checklists, etc.

9.6.4 Conclusion

The three elements discussed above support each other and cannot exist in isolation from one another. The interdependency of the levels identified for auditing can best be illustrated by comparing them to a pyramid, with workplace precautions forming the base of the pyramid.

Programme evaluation provides feedback on the performance of safety management programmes. It goes beyond simple performance measurement to assess performance in depth and judge the effectiveness of safety policy and programmes. Some companies have a systematised evaluation function requiring full involvement in safety programmes by the whole workforce and/or new policy proposals. These are to be submitted, with evaluations, only once the safety committee accepts them.

9.7   RISK ASSESSMENT

9.7.1 Introduction

The term 'risk assessment' is commonly used in various ways to describe some method of identifying, understanding and controlling risks. *Risk assessment* is a detailed systematic examination of any activity, location or operational system to identify risks, understand the likelihood and potential consequences of the risks and review the current or planned approaches to controlling the risks, adding new controls where required. Successful risk control can include outcomes such as improved safety, health, production, environment protection, community acceptance, etc.

From this it can be seen that risk assessment is by no means restricted to health and safety. Pure risk is said to be a risk that results only in loss, damage, disruption or injury with no potential for gain, profit or other advantage. Speculative risk on the other hand carries the potential of loss or gain. This chapter, however, will focus on the use of risk assessment to minimise the impact of pure risk.

There are many ways to do risk assessment. One of the most common and accepted approaches to access the health and safety risks present in a company is called *qualitative risk assessment*.

During the qualitative risk assessment process no absolute values are assigned to the hazards identified. There are no fixed rules about how to do a qualitative risk assessment. The assessment will depend entirely on the nature of the job and the type and extent of the hazards and risks.

The following general principles should be followed to ensure that a suitable and sufficient qualitative risk assessment would result.

- Assess all relevant risks and hazards systematically.
- Address what is actually happening on the job.
- Meaningfully involve a vertical slice of people from the organisation in the risk assessment process.
- The detail of a risk assessment should match the level of the potential risk.
- The risk assessment should take cognisance of the effect of existing safety measures.
- First prioritise the risks by means of a rough risk assessment and then use more sophisticated methods of risk assessment.
- Ensure that the outcomes of the risk assessment identify the groups and individual workers primarily at risk.

Quantitative risk assessment is part of regulatory requirements in many countries for approval of proposed chemical process facilities. In this application, quantitative risk assessment involves a detailed analysis of the chemical hazards that may pose a risk to the people who live and work around a proposed site.

For the purposes of this chapter nothing further will be said about quantitative risk assessment, as this type of risk assessment is not commonly used in South Africa for determining health and safety risks.

9.7.2 Hazard identification

The first step in the risk assessment process is the identification of hazards. This step requires the systematic listing of all hazards present in the workplace. It is during this phase that unskilled workers can contribute the most, as they are the ones present in the workplace daily.

Many risk assessment tools are available to assist in the hazard identification process. In general, these tools can be divided into two categories, namely top-down and bottom-up techniques. Typical examples of top-down techniques include checklists, fault-tree analysis and brainstorming, while the bottom-up techniques include HAZOP studies, failure mode effect analysis and structured what-if analysis.

9.7.3 Risk measurement

After the risks have been identified they should be prioritised in order to allow informed decisions to be made about them. Care should be taken not to make use of a measurement system that is overly complex or too sophisticated.

Various different approaches for the measurement of risk are available. The most common is a risk matrix approach. In using this approach the risk assessment team categorises the consequences of the hazard and its likelihood separately and then combines them in a matrix to determine the priority.

		LIKELIHOOD OF OCCURRENCE			
		Weekly	Monthly	Annually	1 in 10 years
Consequence	Fatality	1	2	3	4
	Permanent disability	2	3	4	5
	Hospitalisation	3	4	5	6
	Lost time	4	5	6	7

Figure 9.2 Risk matrix

A simple example is shown in figure 9.2. In this example a risk rating of 1 will indicate a very high risk and immediate attention will be required. At the same time a hazard resulting in a risk rating of 7 will be recorded but at this time no management action would be required.

This type of risk measurement does however have its limitations, as the exposure of the workforce does not get taken into account. Another, also very simple risk measurement tool utilised by some safety professionals is the vertical line graph assessment method.

This method entails having four vertical scales, as in figure 9.3, with the severity that the hazard can result in being displayed on the first vertical scale, with the worst case at the top and the least severe at the bottom. On the second vertical scale the frequency of the occurrence is plotted, with the highest frequency at the top and the lowest frequency at the bottom. The third vertical scale must indicate the percentage of the workforce being exposed to the hazard, with 100% at the top and 0% at the bottom. The fourth vertical scales indicate the priority risk ranking, with the highest at the top and the lowest at the bottom.

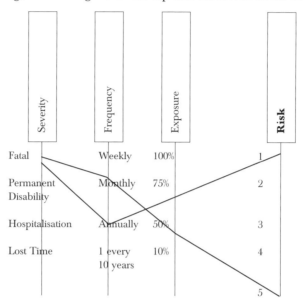

Figure 9.3 The vertical line graph

In figure 9.3, two hazards are being evaluated, both with a fatal severity. The one represented by the double line indicates that the frequency of this hazard occurring is monthly while the single line indicates a frequency of an annual occurrence. Based on this alone, it would appear that the hazard represented by the double line would require more urgent attention than the one represented by the single line. If the graph is completed by extending the lines from the frequency line through the point on the exposure line to the risk line, it becomes clear that the double line hazard has a much lower risk ranking than the single line. This is because the level of exposure of workers to the single line hazard is much higher than that for the double line hazard. Care must be taken with the selection of the magnitudes of the different vertical scales. A variation on this method is to allocate a subjective risk value factor to each of the three components – severity, frequency and exposure – and to multiply these factors by one another for each hazard. The resultant number will then represent relative risk ranking.

The results obtained during qualitative risk measurements should be interpreted as relative risk rankings and not as absolute risk values.

9.7.4 Reporting and recording

Risk assessment should not turn into a system of generating paper, as some safety systems have done in the past. It is however of critical importance to record what took place during risk assessment so as to have evidence of whether management is taking appropriate action to ensure the health and safety of the workers. This will ensure that the risk assessment is verifiable and repeatable, which is a requirement of a good risk assessment.

The recording should take place concurrently with the rest of the process. The identified risks and their respective risk rankings should be recorded in the company's master risk register (MRR). The MRR should also reflect existing and proposed safety measures and the extent to which they would control the risks.

9.7.5 Preventive and protective measures

In terms of the International Labour Organisation (ILO) convention on risk assessment, the process should include the determination of measures, including changes in the organisation of work and the design of safe systems of work, necessary to (in order of importance):
- eliminate risks identified;
- control the risk at the source; and
- minimise the risk to the individual.

And, in so far as the risk remains:
- provide for personal protective equipment; and
- institute a programme to monitor the risk to which employees may be exposed.

9.7.6 Cost implications

Safety measures always cost money and the employer must foot the bill. A perpetual conflict of interest exists between employer and employee as to the type and magnitude of safety measures that could be considered reasonably practical and reasonably necessary.

The principles of risk assessment indicate that the application of preventive measures must undergo a cost-benefit analysis to establish the extent of the preventive measures required to be implemented. In order to do a reasonable cost-benefit analysis it would be necessary to have an idea of the costs involved in accidents in the particular industry.

It is a well-known fact that flawed cost information can cause wrong management decisions to be taken. If management make a decision on information they believe to be true, but in reality the information is not a true reflection of the facts, it will most certainly lead to losses in the long run. Unreliable cost information is an open invitation to problems later.

Cost information is used in making a wide range of operational, management and strategic decisions. One of the only ways to be sure that the cost information used as a basis for decisions is accurate is to do an activity-based costing study.

The basic elements of the cost systems in general use today were isolated around the turn of the 20th century. During the last decade or so doubts have been expressed by managers and cost accountants alike about the suitability of this system of allocating costs in an ever-changing industrial and commercial environment. Incorrect cost information can put any organisation on a crisis course from which it may never recover. This may lead to problems the organisation can ill afford in today's competitive environment. It may focus line management on the wrong priorities and encourage them to solve the wrong problems.

In essence activity-based costing and the traditional costing system utilise a common setting, i.e. the allocation of overhead costs through a two-stage process. Overhead costs are pooled in both cases and then a series of cost rates is used to attach the accumulated overhead costs to the product lines. The difference exists in that activity-based costing is based upon the combining of overheads into a variety of activity-based cost centres that are directly coupled to outputs via a sequence of cost drivers.

Although there is a resemblance between the approach in traditional and activity-based costing systems, there is distinctly a greater degree of refinement in the way costs are assembled under the latter approach.

Activity-based costing principles are based on knowing which activities cause the application and usage of resources. The strength of activity-based costing is the ability it has to exactly delineate cost and non-financial data. This includes the illustration of the relationships between the two entities (see chapter 15 for more information on cost allocation).

The following ten elements have been identified as making up the cost of industrial accidents:

- compensation paid to the injured;
- loss of production;
- penalties due to delay in production;
- training and retraining costs of injured workers;
- investigation and enquiry costs;
- medical and hospitalisation costs;
- rehabilitation costs of accident site;
- costs of fines, statutory levies and penalties (criminal, civil and administrative);
- additional administration costs; and
- loss of tolerance and integrity with the workforce, customers and community.

Each of the above activity cost pools contributes in varying degrees to the costs of accidents. In some instances the costs of accidents are so enormous that companies are financially crippled and have to close down. Preventing accidents does not only make sense from a social perspective, it is essential for financial survival.

9.8   ACCIDENT INVESTIGATIONS

9.8.1 Introduction

Unreliable information as to the causes of accidents is a certain way to ensure disaster. Information pertaining to accidents may be used to make a wide range of operational, management and strategic decisions. One of the only ways to be sure that the accident information used as a basis for decisions is accurate is to do an in-depth study of the basic and root causes of each accident in an accident investigation.

9.8.2 The accident

Accidents mostly occur as a result of a combination of factors that must all be simultaneously or sequentially present. An unsafe act or situation does not give rise to an accident until someone is exposed to it and both physical and psychological factors, in combination with unsafe systems of work, trigger the accident. Combining environmental hazards and human factor hazards multiplies accident potential. The larger the number of hazards, the quicker the accident potential will increase.

When notification of an accident is received, you should establish at least the following:
- the location of the accident;
- the number of people involved;
- the extent to which the accident is affecting operations;
- the name, position and contact details of the person in charge;
- all the relevant details of the working place;
- the date and time of the accident;
- the initial identified causes of the accident;
- the severity of the accident;
- the progress made with the rescue operations; and
- the hospital details of people already sent there.

9.8.3 The investigation team

Where possible, the investigator should be free of the operational control of the supervisor(s) concerned, in order that the investigation may be objective. The investigation should be thorough and attempt to identify underlying causes. The tendency to blame an accident on an employee's 'carelessness' should be avoided because the term is too vague and usually hides problems that could be corrected if identified. It will generally not be possible to conduct a detailed investigation into every accident that occurs in a large company.

All accidents should be carefully scrutinised and a risk category allocated. Generally four risk categories are used to identify risk as follows:
1. high risk;
2. moderately high risk;
3. moderately low risk; and
4. low risk.

The risk category should be determined by taking into account the probability of the hazard resulting in accidents, the exposure of the workforce to the hazard and the potential consequence of the hazard.

Once the decision has been made that an accident will be investigated and by whom, the investigator(s) should travel to the accident site as quickly as reasonably practicable.

9.8.4 Preparations

All the risks that were taken that led up to an accident have to be evaluated during the accident investigation. The evaluation should result in identifying the inadequacies in the safety management programme and standard procedures, as well as shortcomings in compliance with standards.

When an accident occurs, the prime concern is for the injured, who should have immediate access to first aid and medical facilities. Unless the injured are well enough to be questioned at the scene, they should not be further upset with questions. The investigation process should begin promptly once those injured are properly taken care of.

The *in loco* inspection is a very important part of any accident investigation. The following is a suggested sequence of events when such an inspection is planned.

- Identify the stakeholders, i.e. the family(ies) of the injured, unions, government departments, etc.
- Notify the stakeholders of the accident.
- Prepare for the *in loco* inspection.
- Carry out the *in loco* inspection.

In preparation for the *in loco* inspection, the following steps should be followed.
- Arrange an appropriate date and time for the inspection. Remember that the sooner you get to the scene of the accident the more accurate the information obtained will be.
- Obtain all relevant documentation pertaining to the accident and familiarise yourself with the relevant portions of all the documents. This could include:
  - codes of practice
  - standards and designs
  - special instructions
  - layout plans and maps
  - procedures
  - statutory inspection reports.
- Arrange for specialised equipment that may be required at the accident scene, depending on the circumstances, such as:
  - light meters
  - sound meters
  - flow meters
  - thermometers
  - stop watches
  - cameras (infrared, digital or video)
  - dust level meters
  - tape measures or disto-mats
  - radiation level meters etc.
- Arrange for the appropriate protective clothing, a notebook and other personal items.

9.8.5 The *in loco* inspection

During the *in loco* inspection ensure that all those concerned with the accident (potential witnesses) are available for interviews and to give explanations of what may have happened. On arrival at the site, the accident scene must be secured to prevent important physical evidence from being disturbed or destroyed.

The investigator(s) must take charge of the accident site. Once the area is secured, the necessary sketches and photographs should be prepared. Each piece of evidence should be carefully labelled and accurate records should be kept of the time and place that it refers to and the person from whom the evidence was received.

Prior to entering the accident site, assess the safety of the party. Special care should be taken not to expose the investigation team and other members assisting them to unnecessary risks.

Inspect the scene for any hazards that could cause another accident. Keep everyone away from the immediate area so that the scene remains undisturbed until all the facts are collected. Personally interview everyone involved, i.e. the injured person(s), nearby employees and other personnel who may provide clues to the causes of the accident. Investigators must emphasise that their investigation is a fact-finding, not fault-finding mission, otherwise the workers and supervisors who have the information may conceal the information to protect themselves and their fellow employees.

The more information obtained the sooner after the accident the more accurate the investigation will be. With this in mind, all victims and potential witnesses should be identified and interviewed as soon as possible. Also interview those who were present prior to the accident and those who arrived at the accident scene shortly after the accident. The investigator should not try and remember what every witness had to say and also should not try and record everything that every witness says. He/she should make notes of the salient points mentioned by each witness and ensure that he/she knows who said what in the event of him/her wanting to clear up any uncertainty.

The reconstruction/re-enactment of the accident should only be allowed if it can be 100 per cent ensured that the accident will not be repeated.

9.8.6 The investigation

Before the investigation starts a number of arrangements must be made to ensure its successful completion. These arrangements will include at least the following activities.
- Confirm the date and venue of the investigation with all the interested and affected parties.
- Study the appropriate laws, regulations, codes of practice, special instructions and national codes.
- Compare company-specific standards with industry standards and guidelines.
- Evaluate previous similar accidents.
- Prepare specific questions for each witness.
- Arrange for specialist witnesses as appropriate, e.g.
  - medical experts
  - rope experts
  - explosives experts
  - rock engineering experts
  - chemical experts.
- Arrange for appropriate record-keeping facilities.
- Arrange for an interpreter, if required.
- Get another investigator to review the preparations.

The accident investigator must define the scope of the investigation, if this has not already been done by somebody else. The team leader of the accident investigation team must assign specific tasks to each member of the team. In the event of the team consisting of one person only, the investigator must ensure that he/she keeps track of requests made for information.

Always keep in mind that the purpose of an accident investigation is the determination of the causes of the accident and not to try and determine who was responsible for the accident. A separate statutory inquiry could and in some cases will be held to determine liability.

Accidents represent problems that must be solved through different investigation techniques. To solve a problem, the investigator may utilise any one of a number of formal problem-solving techniques. The most commonly used techniques for accident investigation are change analysis (CA) and job safety analysis (JSA).

Once all the data pertaining to the accident has been collected, it should be analysed in a structured way. The identification of any unsafe acts and conditions should be noted and the unsafe portion of the procedure and design should be isolated. Any lack of appropriate training and leadership should be identified as a minimum requirement.

With the above information available it should be possible for the investigation team to determine why the accident occurred.

9.8.7 Concluding the investigation

At the conclusion of the analysis phase of the accident investigation, a post-investigation briefing should be held to inform the management team of the outcome of the investigation. In order to prevent a similar accident from occurring, a summary report that includes recommended preventive actions must be prepared and distributed in accordance with the applicable instructions.

9.9 ACCIDENT STATISTICS

9.9.1 Background

Most people are afraid of the word 'statistics'. This fear is probably based on incorrect information. Statistics deal with numbers, but are not merely counting. The basis for statistics comprises figures, data and information utilised to explain complex situations in terms that are easy to understand.

The science of statistics is concerned with the comparison of numbers. Data is normally too complex to evaluate effectively in its raw form. Figures in statistics are meaningless unless they are compared with other related figures. Once all the data is repackaged and displayed in a statistical format, the data becomes information that can be intelligently applied.

When comparing statistical figures one should keep the following basic rules in mind.

- Refer to the same group of items.
- Ensure that the figures are of the same kind.
- Only compare figures if they are in the same unit(s).
- Compare only statistical figures that were measured with the same degree of accuracy (i.e. decimal point accuracy).

The use of statistics to convince people to act in a specific way is a very popular custom. In this section the focus will be on the methods in use that will produce useable accident statistics for the engineer who is serious about reducing accidents and who will not use statistics to hide accidents.

When dealing with accident statistics, it is important to ensure that the integrity of the accident statistics used is beyond reproach. In order to ensure this, a systematic approach is recommended. The process involves five steps that can briefly be summarised as follows.

- Review the sampling design.
- Conduct a preliminary data review.
- Select a statistical test.
- Verify the assumptions of the statistical test.
- Draw conclusions from the data.

9.9.2 Collecting accident statistics

It is not intended to explain the statistical theory of sampling here, but simply to give an introduction to different accident data-collection techniques. If for any reason the person wanting to analyse the data does not want to make use of all the available information on accidents, he/she needs to take a scientific sample of the available data. This sounds simple enough, until the statisticians start to talk. In reality a number of rules must be followed in order to ensure that the sample is of significance.

The sample must be of a certain minimum size to give reliable results. When selecting data to be included in the sample, each piece of information must have an equal chance of being included in the final sample. As an example, take 10 apples and assume that you want to randomly select 5. This sounds like a simple act until you realise that if you remove the first apple (which had a 1 in 10 chance of being selected), the next apple you select only has a 1 in 9 chance of being selected and the fifth apple a 1 in 6 chance. This results in not all the apples having an equal chance of being selected. It is for this reason that different sampling methodologies have been developed, each with its own unique characteristic.

Some of the more known random sampling techniques used are:
- simple random sampling;
- systematic random sampling;
- stratified sampling;
- multi-stage sampling;
- cluster sampling; and
- quota sampling.

Depending on the intended use of the sample and the statistical accuracy required, a different sampling technique would be utilised.

9.9.3 Analysing accident statistics

Once the sample has been taken and the user is satisfied that the data is valid and reliable, it must be analysed. The type and method of analysis will depend on the desired results.

There are a number of different types of tests that could be performed on the data collected. They are divided into main groupings, with different tests available to establish the same criteria. These tests are grouped as follows:
- tests for distribution assumption;
- tests for trends;
- tests for outliers;
- tests for dispersions;
- transformations tests; and
- tests to deal with values below detection limits.

The tests most often utilised to interpret accident data are those for trends.

The detailed discussion of the different tests lies outside the scope of this chapter. It is suggested that a statistician be approached to analyse the accident statistics, as the use of most of these tests requires above-average knowledge of and skill with statistics.

9.10 CONCLUSION

The object of this chapter has been to provide a general overview of safety management for engineers, technologists and scientists. After a brief introduction the theories and concepts of safety management were discussed. Our daily lives are controlled by means of laws and our workplace is no different. It is therefore appropriate that the next section in this chapter discusses the statutory implications of accidents.

A number of leading engineers and safety professionals agree that safety management and ethics are closely linked, which is the reason for the inclusion of a section on ethics in this chapter.

Systems safety and the safety audits, together with risk assessments, form the core of safety management and therefore also of this chapter. This was followed by a brief overview of accident investigations. The final part of this chapter consists of a brief discussion on the use of accident statistics.

The occurrence of industrial accidents and injuries has been with us since the first industrial accident of the Industrial Revolution. It is the challenge of the new millennium to change the acceptance of industrial accidents to a paradigm of zero tolerance towards accidents.

Self-assessment

9.1   Discuss the elements involved in an accident, indicating the effect that timing has on the occurrence of accidents.

9.2   Reflect on the legal implications an accident might have on companies and engineers respectively.

9.3   Compare the moral and ethical responsibilities of engineers.

9.4   Draw up a safety audit programme for an engineering company.

9.5   Discuss the difference between a quantitative and qualitative risk assessment.

9.6   Do a qualitative risk assessment of five hazards you have identified in an engineering environment using two different risk measurement systems and compare the resultant risk rankings obtained.

9.7   Evaluate and compare the difference between emergency response and business continuity management.

9.8   List ten cost drivers responsible for causing accidents that will reduce company profits and discuss the implications.

9.9   Apply the activities involved in an accident investigation on a recent industrial accident in your workplace and record the actions you would take during each step of the investigation.

References

Anderson, V.L. & McLean, R.A. 1974. *Design of Experiments: A Realistic Approach*. New York: Marcel Dekker.

Beedle, W. & Harrington, G. 1996. 'Accident investigation: The ergonomic eye.' *Occupational Health & Safety Canada*, 12 (5), September/October, pp. 36–8.

Bird, F.E. & Germain, G.L. 1996. *Loss Control Management: Practical Loss Control Leadership*. Revised edition. Loganville: Det Norske Veritas.

Bolton, F. 1996. 'Basic guide to accident investigation and loss control.' *American Industrial Hygiene Association Journal*, 57 (1), January, pp. 77–8.

Brannen, J. (ed.). 1992. *Mixing Methods: Qualitative and Quantitative Research*. Aldershot: Avebury.

Carroll, D. (ed.). 1990. *The State of Theory*. New York: Columbia University Press.

Caulcutt, R. 1983. *Statistics in Research and Development*. London: Chapman & Hall.

Corbett, M.M. & Buchanan, J.L. 1985. *The Quantum of Damages in Bodily and Fatal Injury Cases: General Principles*. Cape Town: Juta.

Curtis R. 1995. *Outdoor Action Program*. Princeton: Princeton University Press.

Davies, N.V. & Teasdale, P. 1994. *The Costs to the British Economy of Work Accidents and Work-related Ill Health*. London: HSE Books.

De Beer, C.S. 1991. *Pitfalls in the Research Process: Some Philosophical Perspectives*. Pretoria: HSRC.

Dehaas, D. 1996. 'The art of the interview.' *Occupational Health & Safety Canada*, 12 (5), September/October, pp. 72–9.

Einstein, A. 1961. *Relativity: The Special and General Theory*. New York: Wings Books.

Fewell, P. n.d. *The Cost Aspect of Safety*. Johannesburg: Gencor.

Firestone, C.M. 1996. 'Guide to accident investigation and loss control.' *Plant Engineering*, 'Product/Literature Supplement', March, p. 12.

Gregory, D. & Ward, H. 1974. *Statistics for Business Studies*. 2nd edition. London: McGraw-Hill.

Guarnieri M. 1992. 'Landmarks in the history of safety.' *Journal of Safety Research,* 23 (3).

Guy, D.M. 1981. *An Introduction to Statistical Sampling in Auditing*. New York: John Wiley.

Huysamen, G.K. 1983. *Beskrywende Statestiek vir die Sosiale Wetenskappe*. 2nd edition. Cape Town: Academica.

Huysamen, G.K. 1993. *Metodologie vir die Sosiale en Gedragswetenskappe*. Halfway House: Southern.

International Risk Control Africa. 1994. *Hazard and Operability Studies*. Training manual.

Janicak, C.A. 1994. 'Significant accident prediction.' *Professional Safety*, 39 (7), July, pp. 20–5.

Kemp, D.A. & Kemp, D.A. 1992. *The Quantum of Damages in Personal Injury and Fatal Accident Claims*. London: Sweet & Maxwell.

Kertesz, L. 1994. 'Control costs by investigating accidents.' *Business Insurance*, 28 (18), May, pp. 21–2.

Krause, T.R. & Russell, L.R. 1994. 'The behaviour-based approach to proactive accident investigation.' *Professional Safety*, 39 (3), March, pp. 22–6.

Kuhlman R.L. 1977. *Professional Accident Investigation*. Loganville: Institute Press.

Lehtonen, R. & Pahkinen, E.J. 1995. *Practical Methods for Design and Analysis of Complex Surveys*. Chichester: John Wiley.

Maisel, R. & Persell, C.H. 1996. *How Sampling Works*. California: Pine Forge Press.

Marx C. 1996. 'An activity-based costing approach to fall of ground accidents in a South African gold mine.' MBA dissertation, PU for CHE.

Minerisk Africa. 1998. *System Safety Accident Investigations*. Randburg: Miningtech CSIR.

Minerisk Australia. 1995. *Systems Safety Accident Investigation Skills Workshop*. Sydney: Minerisk.

National Occupational Safety Association (NOSA). 1997. *Incident Investigation.* Arcadia: NOSA.

Oxman, S.V. & Myers, G.K. 1994. 'Accident investigations can cut future losses.' *National Underwriter*, 98 (29), July, pp. 11, 28.

Petersen, D. 1994. 'Integrating safety into total quality management.' *Professional Safety*, 39 (6), June, pp. 28–30.

Powers, K. & Arnstein, C. 1995. 'Getting them back to work safely.' *Occupational Health & Safety*, 64 (2), February, pp. 42–5.

Render, B. & Stair, R.M. 1991. *Quantitative Analysis for Management.* Boston: Allyn & Bacon.

Safety & Health Practitioner. 1996. 'Incident and accident investigation.' 14 (3), March, p. 58.

Sayers, D. 1994. 'Accident investigations.' *Occupational Health & Safety Canada*, 10 (3), May/June, pp. S12–S15.

Senecal, P. & Burke, E. 1994. 'Root cause analysis: What took us so long?' *Occupational Hazards*, 56 (3), March, pp. 63–5.

Smith, S.L. 1995. 'Safety incentives: A program guide.' *Occupational Hazards*, 57 (6), June, pp. S5–S10.

Smith, S.L. 1996. 'Reaping the rewards of safety incentives.' *Occupational Hazards*, 58 (1), January, pp. 99–102.

Thomen, J.R. 1996. 'Root cause: Holy Grail or fatal trap?' *Professional Safety*, 41 (9), September, pp. 31–2.

University of South Australia. 1995. 'Procedure for accident and incident reporting and investigation.' Available at http://www.infoserv.unisa.edu.au/admininfo/proceed/ohs/accident.htm, accessed 30 September 1997.

Valsamakis, A.S., Vivian, R.W. & Du Toit, G.S. 1996. *The Theory and Principles of Risk Management.* Isando: Heinemann.

Vincoli, J.W. 1993. *Basic Guide to Safety Systems.* New York: Nostrand Rienhold.

Vincoli, J.W. 1994. *Basic Guide to Accident Investigation and Loss Control.* New York: Nostrand Rienhold.

Western Australian Department of Minerals and Energy. 1997. *Accident and Incident Investigation Manual.* 5th edition. Perth.

Wojcik, J. 1996. 'Avoiding a crisis of cumulative trauma.' *Business Insurance*, 30 (18), April, p. 39.

Woodward, J.L. 1995. 'Structuring accident investigations.' *Chemical Engineering*, 102, 6 June, pp. 88–9.

Woolsey, C. 1995. 'Accident sleuths can help improve safety.' *Business Insurance*, 29 (19), May, p. 41.

Workers Compensation Board of British Columbia, Worksafe Prevention Division. 1996. *Investigation of Accidents and Diseases: An Overview.* Vancouver.

10 Maintenance Management

Krige Visser

Study objectives

After studying this chapter, you should be able to:
- calculate reliability and maintainability functions
- perform a system breakdown for a technical system
- define a vision and key objectives for the maintenance department
- develop a maintenance plan for a technical system
- organise and structure the maintenance resources
- control maintenance performance
- apply the approaches of reliability-centred maintenance (RCM), business-centred maintenance (BCM) and total productive maintenance (TPM)

The aim of this chapter is to present introductory knowledge of maintenance management to the reader, with an emphasis on the necessary tools to manage the maintenance function of an enterprise.

10.1 INTRODUCTION

10.1.1 Maintenance management perspectives

All physical assets (also referred to as technical systems in this chapter) require some form of maintenance during their useful life. However, for a long time maintenance was regarded as an unimportant sub-function of production or operations. In the last decade or two, maintenance and maintenance management has become increasingly important in many companies due to the increasing complexity and size of many technical systems. The maintenance function influences the profitability of an organisation because the availability of a physical asset directly affects the outputs of the asset (products or services).

Maintenance management is a relatively new field, especially as an academic discipline. However, continuing research in maintenance and maintenance management is now providing the necessary framework of principles and procedures that can be applied practically by maintenance managers and supervisors in real world situations. The main objective of maintenance is to maximise the availability of a physical asset through increased reliability and maintainability, and to achieve an acceptable quality of the products or services provided by the asset.

10.1.2 The cost of maintenance

Many senior executives and managers are surprised at the total cost of maintenance in manufacturing or service enterprises. The maintenance costs vary greatly from one industry or organisation to another, and are typically between 3 and 50 per cent of the total production cost. The total cost of maintenance in South Africa was estimated by comparing output and manufacturing costs with international companies and by extrapolating from historical data. The estimated total maintenance cost for South Africa was approximately R6 800 million in 1995.

10.1.3 History of maintenance

Maintenance has always been necessary, but most assets that were built in the 1800s were non-complex and mostly operated as individual units. The home industries in Europe resulted in the owner doing the design, construction, operation and maintenance of the asset. Thus, maintenance was necessary but maintenance management was not. The Industrial Revolution brought bigger manufacturing systems that consisted of a variety of machines and equipment that were integrated into large units. Technical systems increased in size and complexity, and more sophisticated technologies were introduced. The nature of the workforce also changed and the need for proper management of the workload in maintenance became more important. The need for effective management of the maintenance function thus increased.

Before World War II, maintenance management, even in large companies, was not practiced effectively. In the 1950s, industrial development across the world increased rapidly and elements of preventive maintenance were introduced in many manufacturing systems. Predictive maintenance technologies were introduced to maximise the useful life of equipment. In the 1960s, reliability-centred maintenance (RCM) was developed to fulfil the need for safe and reliable aeroplanes. Total productive maintenance (TPM) developed in Japan to fulfil the need for greater operator and general worker participation in the maintenance function. In the 1970s, computerised maintenance information systems were introduced and these systems have improved continuously in the last decade.

10.1.4 World-class maintenance

Many manufacturing companies all over the world are realising that world-class manufacturing must be achieved to remain internationally competitive. Wireman (1990) states that: 'World-class manufacturing requires the elimination of complexity. It requires simplicity in design and manufacturing processes.' He also feels that organisations can only achieve world-class standards of manufacturing if maintenance management is put at the same level as other management positions. Campbell (1995) and Tarita (1998) define the fundamental building blocks to achieve a world-class maintenance organisation as illustrated in Figure 10.1

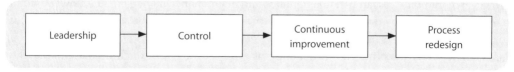

Figure 10.1 Building blocks of world-class maintenance

Leadership is concerned with establishing a maintenance strategy with a clear vision of what is to be achieved by the maintenance department. It also requires effective management of the maintenance resources with special consideration for the human resource. The next step towards world-class maintenance is to introduce adequate control mechanisms, such as setting quantitative objectives or standards, planning and scheduling maintenance tasks, selecting effective maintenance actions to improve reliability and availability, and introducing a computerised maintenance information system.

The next step in progressing towards world-class maintenance is to implement a well-established maintenance approach or strategy, such as business-centred maintenance (BCM), TPM, or RCM, or some combination of these, as described by Prahdan (1994) and Sherwin (2000). If a company achieves the first three steps outlined above, and the maintenance performance is still insufficient, process re-engineering (or business process re-engineering) should be considered. Part of the re-engineering process is to re-engineer maintenance through an extensive analysis of the maintenance transformation process and to radically change those elements of the process that do not provide adequate performance.

10.1.5 Input-output model for the maintenance system

The maintenance system can be regarded as a sub-system of the production or manufacturing system, which is a sub-system of the business enterprise system. The maintenance system can be represented by the well-known input-output model for a system, as indicated in figure 10.2.

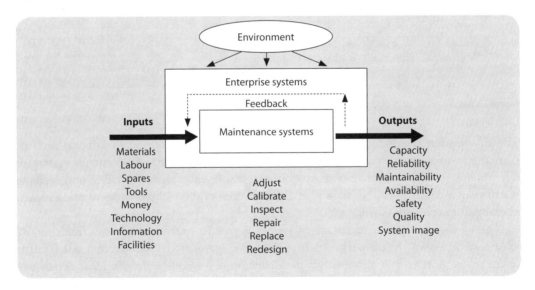

Figure 10.2 The input-output model for the maintenance system

The inputs to the maintenance process are the resources, typically workers, spares, materials, information, facilities and money. The outputs are various performance parameters that are related to the inputs. The outputs of the maintenance system are somewhat difficult to define, since many of the 'traditional' output parameters such as availability, reliability and safety are not the sole responsibility of the maintenance department. The production or operations department can also influence these parameters significantly. Changes to the inputs, the maintenance transformation process or the environment will cause changes in the outputs of the maintenance system or process. If the inputs to the maintenance system are reduced, e.g. the human resources or spares, and the outputs have to remain the same, some changes to the maintenance transformation process are required to maintain the same output. This model of the maintenance system therefore indicates the dynamic relationship among the inputs, process and outputs.

10.1.6 Framework for maintenance management

Any system must be managed and this management function can be partitioned into the sub-functions of planning, organising, staffing, leading and controlling. These functions also apply to the management of the maintenance system, and a general maintenance management framework as formulated by Visser (1997) is shown in figure 10.3.

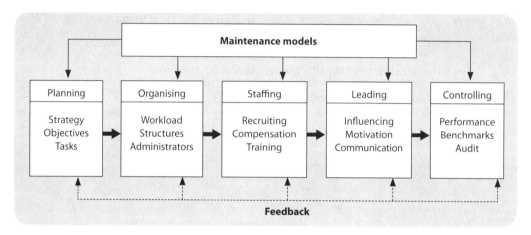

Figure 10.3 Maintenance management framework

The *planning* function in maintenance involves both strategic and tactical planning. Strategic planning focuses on the vision, mission and key objectives for the maintenance system or department. Tactical planning is more concerned with short-term objectives and leads to the development of a maintenance plan for each item of the technical system and detailed planning for each maintenance task.

The *organisation* of the maintenance system is mostly concerned with the matching of all the resources of maintenance (such as workers, spares, tools, information, etc.) to the maintenance workload. It also involves the definition of a resource structure that determines the geographical location of the resources and their function, composition, size and logistics of movement.

Maintenance *staffing* comprises the filling of positions in the maintenance organisation structure through identifying workforce requirements, recruiting, selecting, placing, promoting, compensating and training candidates and current jobholders. The maintenance manager is especially involved in recruiting the right people and providing training until a satisfactory performance level has been attained.

Leading in maintenance can be defined as the process of influencing maintenance workers so that they will contribute to the goals of the maintenance department and of the total enterprise. Both the maintenance manager and supervisors must apply their managerial skills in influencing and motivating maintenance workers in such a way that the creativity of each worker is used in the best interests of the total enterprise.

Controlling is viewed as the measurement and correction of maintenance performance to make sure that maintenance objectives and the plans devised to attain them are accomplished. The control systems must ensure that the maintenance resources are applied and properly directed towards the achievement of the maintenance objectives. Controlling and planning go together and there is a constant interchange between these two activities.

Various *models* are increasingly being used in engineering and engineering management to enhance the decision-making process. Many models, including mathematical models, can be used by maintenance practitioners to optimise the maintenance function. This includes reliability, maintainability, spares inventory management, queue behaviour and workload forecasting.

10.2 MODELS IN MAINTENANCE

10.2.1 The need for models

Modelling is the most common method used for analysing a system and supporting the decision-making capability during both the design and operational phase of the system life cycle. Modelling is the cheapest and fastest way of studying the effects of changes in key design variables on system performance. The purpose of modelling is to understand, predict, control and ultimately improve system behaviour. Maintenance models are used to enable the maintenance manager to make better decisions. Typical decisions for the maintenance manager or supervisor are as follows.

- What maintenance type or strategy should be used for a component or item?
- How frequently should preventive maintenance be performed?
- When should an asset be discarded and replaced?
- What spares and how many of each type should be kept in the store?

This decision-making process is simplified by using models, especially analytical or mathematical models. The complexity of the maintenance environment has led to the development of a variety of mathematical models for different aspects of the maintenance function. Some basic principles of availability, reliability and maintainability are discussed in the next sections. Detailed discussions on this specialised topic are provided by Lewis (1995) and Smith (2001).

10.2.2 Reliability

The *reliability* of an item can be defined as the probability that the item will function without failure for a specified time under specified conditions. Most reliability phenomena are understood in terms of the number of failures in a time interval or time until failure. Reliability is expressed mathematically by means of five related functions. These functions and symbols commonly used are indicated in table 10.1 on the next page.

Table 10.1 Functions used in reliability

Function	Symbol	Unit
Failure density	$f(t)$	
Reliability	$R(t)$	
Cumulative distribution function for failure (CDF)	$F(t)$	
Failure or hazard rate	$\lambda(t)$	time^{-1}
Mean time to failure	MTTF	*time*

Knowledge of the failure probability distribution, e.g. the function $f(t)$ or $\lambda(t)$, can be used to determine any of the other functions in table 10.1. Formulas for each function are given by Lewis (1995), but these functions can also be calculated for specific distributions by means of spreadsheets that have built-in probability distributions. Most reliability phenomena can be described by means of the exponential distribution that results from random failure behaviour of an item. If the failure rate of an item is constant, therefore independent of its age or operating history, the reliability is expressed by the exponential distribution. The failure rate $\lambda(t)$ is just a constant, λ, and the reliability functions are then given by:

$$f(t) = \lambda \cdot \exp(-\lambda t)$$
$$R(t) = \exp(-\lambda t)$$
$$MTTF = \frac{1}{\lambda}$$

However, not all reliability phenomena can be described by the constant failure rate model. Some complex assemblies or items exhibit early failures, while failures due to wear-out also occur. This leads to the well-known bathtub failure rate model that is shown in figure 10.4.

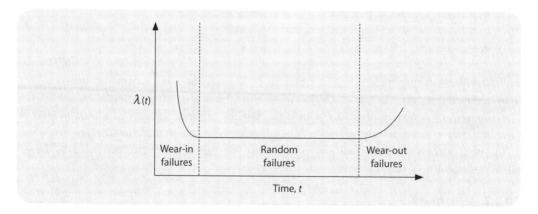

Figure 10.4 The bathtub curve

Probability distributions with a time dependent failure rate are needed to model the wear-in failures (a decreasing failure rate) and the wear-out failures (an increasing failure rate). The normal distribution can be used to model an increasing failure rate and the lognormal distribution can be used to model a decreasing failure rate. However, the well-known Weibull distribution is generally preferred to model an increasing, constant or decreasing failure rate.

The reliability functions for the two-parameter Weibull distribution are:

$$\lambda(t) = \frac{m}{\theta}\left(\frac{t}{\theta}\right)^{m-1}$$

$$f(t) = \frac{m}{\theta}\left(\frac{t}{\theta}\right)^{m-1} \cdot \exp\left[-\left(\frac{t}{\theta}\right)^{m}\right]$$

$$R(t) = \exp\left[-\left(\frac{t}{\theta}\right)^{m}\right]$$

In these equations the two parameters m and θ must both be positive. The parameter m is known as the *shape parameter* and θ is known as the *characteristic life*. The mean time to failure, *MTTF*, for the Weibull distribution cannot be calculated by means of a simple equation, but the techniques are described in the literature (refer to Lewis, 1995). The value of the parameter m determines the nature of the reliability functions of the Weibull distribution. If $m < 1$ a decreasing failure rate is modelled, for $m = 1$ the constant failure rate is modelled, and for $m > 1$, an increasing failure rate is modelled.

Example 10.1 Reliability calculation for the Weibull distribution

The failure characteristics of a switch in an electrical distribution board have a Weibull distribution. The value of the shape parameter, m, is 3,5 and the characteristic life, θ, is 60 days. Calculate the reliability of the switch after 40 days and after 80 days.

Solution:

a) Reliability after 40 days: $R(40) = \exp\left[-\left(\frac{40}{60}\right)^{3,5}\right] = 0,785 \rightarrow$

b) Reliability after 80 days: $R(80) = \exp\left[-\left(\frac{80}{60}\right)^{3,5}\right] = 0,065 \rightarrow$

Graphical illustrations of the failure density, reliability and failure rate functions for the exponential, normal and Weibull distributions are given by Pintelon *et al.* (1995).

Knowledge of the failure characteristics of an item is of utmost importance in selecting an appropriate maintenance policy or action. Some data is available for certain classes of components (electrical, electronic, mechanical, etc.), but data should also be collected for a specific component or item to build some failure history. Curve fitting techniques to determine the parameters of the Weibull (m and θ), normal or exponential distributions from failure data are described by Smith (2001) and Lewis (1995).

10.2.3 System reliability

The components of a complex system are functionally related to each other through a *series* relationship, a *parallel* relationship or a *combination of these*, as illustrated in figure 10.5 on the next page.

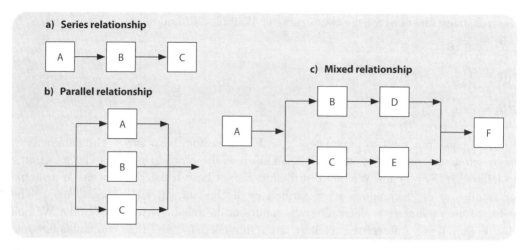

Figure 10.5 Component relationships for system reliability

The following equations can be used to calculate the reliability of the total system, $R_S(t)$, for each of the configurations given in figure 10.5:

a) $R_S(t) = R_A(t) \cdot R_B(t) \cdot R_C(t)$

b) $R_S(t) = 1 - [1 - R_A(t)] \cdot [1 - R_B(t)] \cdot [1 - R_C(t)]$

c) $R_S(t) = R_A \cdot \{1 - (1 - R_B \cdot R_D) \cdot (1 - R_C \cdot R_E)\} \cdot R_F$

In evaluating combined series-parallel configurations, the reliability of sub-assemblies is evaluated first, and the total system reliability is then evaluated by combining the sub-assemblies. A reliability block diagram (RBD) indicating a breakdown from system to components is usually constructed before system reliability can be determined (see Aslaksen and Belcher (1992) for examples).

10.2.4 Maintainability

The *maintainability* of an item can be defined as the probability that an item can be returned to a specified functional level within a specified time period. Maintainability therefore addresses issues such as diagnostics, standardisation, accessibility, interchangeability, mounting, labelling, etc. A system should be designed in such a way that that it can be maintained easily without large investments in time, money and resources (personnel, materials, tools and facilities). Maintainability can also be expressed mathematically by means of five interrelated functions, similar to those defined for reliability. These functions are given in table 10.2.

Table 10.2 Functions used in maintainability

Function	Symbol	Unit
Repair density	$m(t)$	
Maintainability	$M(t)$	
Cumulative distribution function (CDF) for repair	$G(t)$	
Repair rate	$v(t)$	$time^{-1}$
Mean time to repair	MTTR	*time*

The same distributions that are used for reliability can also be used for modelling repair times. If a constant repair rate, $v(t) = v$, is assumed, the exponential distribution applies and the following equations can be used:

$m(t) = v \cdot \exp(-vt)$

$M(t) = 1 - \exp(-vt)$

$MTTR = \dfrac{1}{v}$

The repair distribution (exponential, Weibull, etc.), as well as the values of the parameters of the distribution, can be obtained for simple maintenance tasks by measuring the time taken to perform the task several times and applying a curve fitting technique to the CDF curve. Maintainability is mostly expressed in terms of the mean time to repair ($MTTR$).

Example 10.2 Maintainability calculations

The repair rate for the replacement of the bearings of an electric motor is constant at 0,05 min⁻¹. Determine the maintainability of the bearing and the MTTR.

Solution:

Constant repair rate $v = 0,05$ min^{-1}

Maintainability $M = \exp(-vt) = \exp(-0,05 \cdot 90) = 0,011$ →

Mean time to repair $MTTR = \dfrac{1}{v} = \dfrac{1}{0,05} = 20$ min →

10.2.5 Availability

The *availability* of an item can be defined as the probability that the item will, when used under specified conditions, operate satisfactorily and effectively. Various measures of availability are used in industry, but the most general measure is the operational availability, A_o, which is given by the equation:

$A_0 = \dfrac{\text{uptime}}{\text{uptime} + \text{downtime}}$

The physical assets of an enterprise are acquired to provide availability and therefore a certain capacity of production or service. Availability is an extremely important attribute of any system and comprises the reliability and maintainability of the system. If an item has a constant failure rate, $\lambda(t) = \lambda$, as well as a constant repair rate, $v(t) = v$, the availability of the item is given by the following equation:

$A_0 = \dfrac{v}{v + \lambda}$

Since reliability and maintainability are mostly fixed by the design of the technical system, the availability is also fixed. The design and systems engineer of a new system has a great responsibility to design for reliability and maintainability to ensure that the availability of the system is adequate.

10.3 MAINTENANCE PLANNING

10.3.1 Background

The managerial function of *planning* is mostly concerned with the selection of a vision, mission and objectives, and the actions to achieve these objectives. These actions involve much decision making, especially in choosing among a number of options. At the highest level, planning is required to bridge the gap between the vision of the business enterprise and the current situation.

Maintenance planning is therefore concerned with establishing feasible objectives for the maintenance department. These objectives must be in harmony with the production objectives and the overall objectives of the manufacturing enterprise. It is crucial to define the objectives for the maintenance system in terms of performance parameters, which are fully or at least partially under the control of the maintenance department. Maintenance planning also involves the development of a suitable maintenance strategy, which defines the actions that are required to achieve the selected maintenance objectives. The formulation of a maintenance strategy involves aspects such as the selection of a maintenance procedure (corrective, preventive, design-out, etc.) and schedule for each maintenance significant item (MSI) of the total technical system.

As in other managerial functions of the enterprise, maintenance planning can also be divided into long-term planning and short-term planning. Long-term maintenance planning (between two and five years ahead) is dependent on long-term sales forecasts and long-term production forecasts or requirements. Maintenance planners, working with other managers in the company, plan the actions and schedules that are required to achieve certain objectives in the future.

Short-term maintenance planning is more concerned with the annual objectives and actions of the maintenance department. It involves the effective application of the maintenance resources to cope with the average maintenance workload. It also involves detailed work planning and scheduling of all maintenance work, especially major maintenance events such as large overhauls and turnarounds (shutdowns).

10.3.2 Strategic planning in maintenance

The modern enterprise is concerned with both *internal activities* and the *external environment*. The external environment consists of the economic, political, legal, social and natural environments. Many of these will also impact on the maintenance department and must therefore be incorporated into the strategic planning of the maintenance manager.

Pearce and Robinson (1994) view strategy at the enterprise level as a company's large-scale, future-oriented plans for interacting with the competitive environment in order to achieve company objectives. Strategies for the production and operations department or the maintenance department are at the functional level and support the overall strategies at the business and corporate levels. The strategic management process for the maintenance department can be represented by the flow diagram in figure 10.6 on the next page.

The first step of the strategic management process is to define a vision and mission for the maintenance department. The next step is commonly referred to as a SWOT (strengths, weaknesses, opportunities and threats) analysis. The output of this step is then used to formulate both long-term and short-term objectives for the maintenance department. Strategies (action plans) are then formulated to achieve the objectives and are then implemented in the maintenance department. The total process is repeated periodically, e.g. every three or five years.

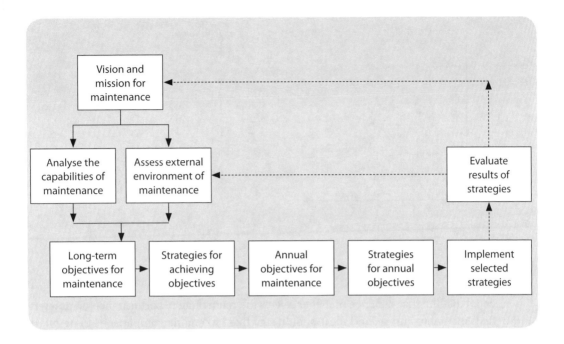

Figure 10.6 Strategic management process for maintenanc

10.3.3 Maintenance performance measurement

The *overall performance* of any business enterprise should be measured and maximised. Performance objectives for the enterprise are defined as part of strategic planning and the overall performance is allocated to the sub-systems of the enterprise. A variety of maintenance performance parameters can be used by maintenance management to implement the methodology outlined above, e.g. as provided by Wireman (1990), Suzuki (1992) and De Groote (1995).

An overall maintenance performance indicator, total maintenance effectiveness (TME), can be defined as the sum of a number of weighted maintenance performance indicators. The number of indicators should be between five and ten. TME is therefore given by the following equation:

$$TME = \sum_{j=1}^{n} W_j \cdot U_j$$

In this equation, W_j is the weight factor for parameter j and U_j represents a benefit or utility function. A benefit function must be defined for each parameter that is used and examples of linear benefit functions are given in figure 10.7. The benefit or utility function transforms the value of a parameter to a normalised benefit or utility value, typically 0–1 or 0–100.

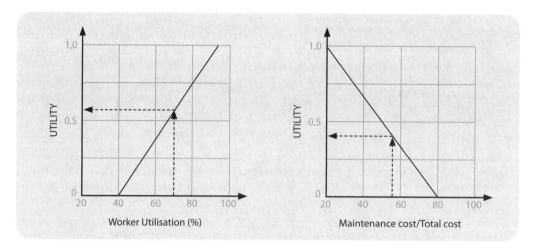

Figure 10.7 Typical benefit functions for maintenance performance

It is essential that the maintenance manager selects the maintenance performance parameters and defines the utility curves, and these should stay fixed to enable comparisons to be made from one year to another. The total maintenance effectiveness, calculated as shown above, is a value between 0 and 1, or 0 and 100. The value of TME should be calculated and monitored at least monthly and some corrective action should be taken by the maintenance manager if performance is not adequate.

10.3.4 Tactical maintenance planning

Medium-to-short-term maintenance planning involves three main activities, namely to define an appropriate short-term overall objective, to formulate a short-term maintenance strategy and to formulate a detailed maintenance plan comprising maintenance tasks and schedules for these tasks. All maintenance interventions of a technical system can be classified as either corrective maintenance, preventive maintenance or improvement/modification maintenance. These top-level categories can be sub-divided into next-level classes, as indicated in figure 10.8.

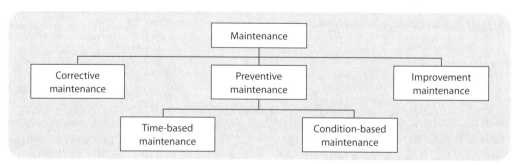

Figure 10.8 Maintenance policies

Among the main strategic decisions that need to be taken in maintenance is the level of maintenance that is required to achieve the maintenance objectives within the budget constraints. If modification/improvement maintenance is viewed separately, the main actions/

policies for maintenance are preventive and corrective maintenance. The maintenance manager must therefore decide how much preventive maintenance is required to provide adequate availability and safety, and optimise the cost of maintenance.

- *Corrective maintenance* is the maintenance that is performed after failure of an item has occurred. It is usually applied to items and components that exhibit random failures, if the failure does not have a serious consequence.
- *Preventive maintenance* is the maintenance that is performed before a failure occurs to preserve an item or a system in its original (or satisfactory) condition.
- *Time-based maintenance* is preventive maintenance that is performed at regular time periods or at a certain load, e.g. number of kilometers travelled or number of cycles. Time-based maintenance is only effective if the component or item has a dominant wear-out mode and therefore exhibits a normal distribution, or the scale parameter of the Weibull distribution is larger that 1 (preferably greater than 4).
- *Condition-based maintenance* is preventive maintenance that is based on knowledge of the condition of the item from routine or continuous monitoring.

Improvement maintenance can be classified into two categories. The first category is the maintenance that is performed to improve the reliability or maintainability, and therefore also the availability of the system. This is usually accomplished by replacing components or higher-level assemblies with more reliable ones, or assemblies that can be maintained more easily. A second category is the maintenance that is performed to improve the technical performance of the system, e.g. to increase the output of the system.

10.3.5 The system breakdown structure

A maintenance strategy can only be formulated by reducing the complexity of the technical system. A system should therefore be partitioned into sub-systems, sub-sub-systems, assemblies, sub-assemblies, units, items, components, and so on. An example of the breakdown of a technical system, a metallurgical plant (system), is illustrated in figure 10.9. The system is broken down into its major units (sub-systems), each of which performs a major function of the total system. The units are partitioned into assemblies which are further sub-divided into sub-assemblies, items and eventually into the lowest level, e.g. components. Maintenance is normally performed at the component level. When a component, e.g. a gear, fails or is unable to perform its intended function, it has to be replaced or repaired. Some component failures will affect the next-higher-level assembly or item (e.g. the gearbox) and the severity of the failure (consequence) will determine whether the total item should be replaced. Item replacement can only be justified if it is much quicker to replace the item and do repairs in the workshop than to repair or replace the faulty component *in situ*.

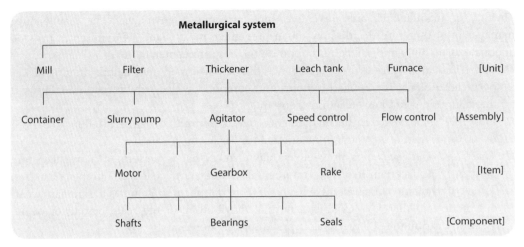

Figure 10.9 Hierarchical breakdown of a metallurgical plant

10.3.6 The maintenance life plan

The system breakdown structure leads to the next step in developing a maintenance plan, namely to define an appropriate maintenance type or strategy for each maintenance significant item or component of the system. Typical maintenance types were mentioned in the previous section, e.g. condition-based maintenance (CBM), time-based maintenance (TBM) or failure-based maintenance (FBM), also referred to as operate to failure (OTF). A typical maintenance plan for a lawn mower is indicated in table 10.3.

Table 10.3 A typical maintenance life plan for a lawn mower

Item	Maintenance type	On-/Off-line	Frequency	Secondary action
Start rope	CBM (visual inspection)	On	2 weeks	Replace if worn
Cooling fins	CBM (visual inspection)	On	2 weeks	Replace if damaged
Fuel tank	CBM (inspection)	On	1 week	Replace if damaged or leaking
Fuel filter	TBM (replace)	Off	18 weeks	
Air filter	TBM (replace)	Off	9 weeks	
Ignition coil	OTF (replace)	Off	–	
Spark plug	TBM (replace)	Off	28 weeks	
Cutting blade	TBM (replace)	Off	18 weeks	
Wheels	CBM (visual inspection)	On	Daily	Repair/replace if damaged
Collection bin	CBM (visual inspection)	On	2 weeks	Repair/replace if damaged

Having defined all the preventive maintenance tasks that have to be performed for all the components and items of the system, the next step involves the formulation of maintenance schedules. Tasks that have a similar frequency are grouped together. For a large technical system, the formulation of all maintenance tasks and schedules is a tedious exercise and a computerised maintenance systems can to be used perform this planning and scheduling function.

10.4 ORGANISING THE MAINTENANCE RESOURCES

10.4.1 Background

The *organisation* of human resources implies the formulation of a formalised intentional structure of roles or positions. All the resources (i.e. workers, spares, tools, information, facilities and technology) that are used in the maintenance transformation process must be organised effectively. Inadequate organisation will result in a low utilisation of the maintenance workers; excessive downtime due to unavailability of spares, tools and data; excessive overtime, etc.

The maintenance organisation system can only be defined if the maintenance workload is known fairly accurately. The workload for preventive maintenance is obtained from the maintenance plan, as described in section 10.3. The corrective maintenance workload is probabilistic and can only be estimated by using historical data for failures and repairs. The total maintenance workload is then determined by adding the preventive workload, corrective workload and any improvement work together. This should be done for all maintenance trades or disciplines in the company, e.g. electrical, mechanical, plumbing, instrumentation, etc.

The primary task of maintenance organisation is to apply the maintenance resources to the maintenance work that has been identified through the maintenance planning function. Maintenance organisation is not static and should react according to the dynamics in the enterprise. Modification will be necessary to respond to changing requirements that are prompted by a change in the environment of the enterprise. Changes to the production system, which in turn reacts to sales and demand, will also affect the maintenance organisation system.

10.4.2 Elements of organising in maintenance

As already defined, the purpose and objective of maintenance organising is to divide and structure the maintenance work into small units that can be managed properly. The design of an organisational system should make use of the well-established basic elements of organising.

- **A division of the work:** This involves the partitioning of the maintenance work into smaller parts, for example inspect, check oil level, adjust, calibrate.
- **Division of labour and specialisation:** This involves the allocation of maintenance tasks to work teams and specific individuals, matching the work to the specialisation of the person or team.
- **Authority and responsibility:** Management should define the sub-groups (teams) within the maintenance division, indicating the lines of authority, using organisational charts, job descriptions, etc.
- **Unity of direction:** All activities in the maintenance department should be directed at achieving specific maintenance objectives.

▓ **Span of control:** This refers to the number of workers reporting directly to a manager. In maintenance the span of control is usually 8–12, but modern trends indicate that 15–20 maintenance workers can report to a maintenance foreman of supervisor.

10.4.3 The maintenance resources

Two of the most important decisions for the maintenance manager are the staffing of the maintenance organisation and the geographical division of the resources. Both of these aspects mainly involve the human resource. The following resources are relevant in the maintenance environment:

▓ people;
▓ spare parts;
▓ tools;
▓ information;
▓ facilities;
▓ technology; and
▓ cleaning materials.

The following methods are commonly used to staff the maintenance organisation system.

▓ **Totally in-house:** The maintenance workers are all on the payroll of the enterprise.
▓ **Combined in-house/contract:** In-house workers perform most of the maintenance tasks, but some specialised maintenance tasks are outsourced.
▓ **Contract maintenance:** Most tasks are outsourced and maintenance supervisors are used to supervise work performed by contract maintenance workers.
▓ **Complete outsourcing:** All maintenance work is outsourced and contractors report to a production or operations manager.

Three main options exist for geographical division of maintenance resources.

▓ **Centralised maintenance:** All staff members report to a central location for work assignment.
▓ **Area maintenance:** This involves small maintenance teams and workshops that are distributed geographically over the company areas.
▓ **Combination of centralised and area maintenance:** A main group of maintenance workers is kept centralised, while some small work groups are decentralised.

10.4.4 The maintenance resource structure

The characteristics of some of the maintenance resources are discussed below. Kelly (1997a) gives a more detailed discussion of the maintenance resources.

Human resources

For a large technical system, the organisation of the human resources is quite a challenge for the maintenance manager. The compensation for the maintenance workers can be up to 70 per cent of the total maintenance cost and it is clear that low utilisation of the maintenance workforce will lead to high maintenance costs. The specific nature of the technical system will determine what maintenance specialisation is required. Most systems require mechanical fitters, electricians and instrument technicians. Most companies also have a number of buildings to maintain and therefore need some building artisans (painters, masons, plumbers and carpenters). Modern process plants also have sophisticated process computers, and computer technicians are therefore required.

Spare parts

The increasing complexity of modern manufacturing systems has resulted in the need for more sophisticated spare parts inventory management. The purpose of keeping spare parts in stock is to reduce the unavailability of the system or parts of the system due to an unexpected failure of a component or item. The organisation of the spare parts therefore tries to obtain a balance between the cost of keeping spares in stock and the cost of unavailability of the manufacturing or service system. Many software programs are available today to manage spare parts inventories. However, slow-moving spares still present a problem for the maintenance manager and only practical experience can eventually lead to an optimum inventory policy for these spares. Guidelines on managing the inventory of slow-moving spares are given by Kelly (1997b).

Maintenance tools

The tools used by maintenance workers can have a large impact in achieving a high maintainability. Many technical systems require special tools for maintenance and if these tools are not readily available for the artisan while doing a maintenance task, time will be lost and the worker utilisation will decrease. Inadequate tools can also cause subsequent problems if nuts, bolts and screws are damaged. A system should be developed to issue tools from and return them to a centralised location.

Information

Maintenance information is any document, user manual, service manual, catalogue, specification, procedure or drawing needed to perform maintenance work. Information has become a critical resource in maintenance and computerised maintenance information systems are now used in most companies to manage information more effectively and to keep track of system configuration changes.

10.4.5 The maintenance workload

The maintenance organisational structure is formulated to cope with the expected *maintenance workload*. The amount of maintenance work is not fixed with time and can have a large probabilistic component if mainly corrective maintenance is performed. Maintenance was classified into three main categories in paragraph 10.3, namely:
- corrective maintenance;
- preventive maintenance; and
- modification or improvement work.

A company can sub-divide these main categories into sub-categories to manage the workload more effectively. Corrective maintenance can be divided into *emergency* work and *deferred* work. Deferred work is that maintenance that need not be done immediately, usually because there is a standby item to take over the function of the failed item, or a section of a plant (or system) is not required to operate for some length of time. Preventive maintenance

217

can be divided into inspections (usually on-line), minor preventive maintenance (usually off-line) and shutdown work, which involves major overhaul of sections of the technical system, which must be done off-line.

Modification or improvement work is not classified as maintenance work, but the maintenance department is frequently requested to assist in doing this work and it must therefore be taken into account and planned. It is essential to separate normal maintenance work from modification work for accounting purposes, so that correct costs can be allocated and presented to management.

One way of establishing the workload characteristics of a system is to plot the time spent on each category of maintenance work for a time period, typically one month. This is indicated in figure 10.10.

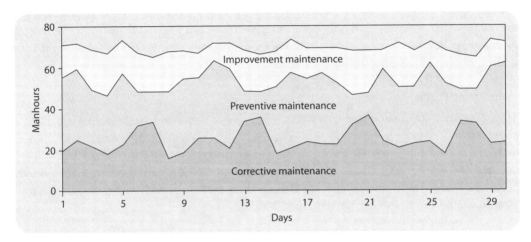

Figure 10.10 Maintenance workload for one month

The variation in maintenance workload can be seen in figure 10.10. The peaks in the corrective maintenance workload occur about every six or seven days, suggesting that the system was operated for five days and was then switched off for two days. This is typical of some plants or systems that operate for eight hours a day, five days per week. Weekends are then used for all maintenance that has to be performed off-line. Preventive maintenance work can be used to smooth the maintenance workload.

10.4.6 Administrative structure

The structure of an organisation (or department or division) can be expressed in a number of ways, but the formal organisation chart (or administrative chart) is still one of the most widely used models to express the formal organisation structure. The organisation chart has been criticised in the past because of its rigidity and because it only shows the formal communication channels, but it is really an essential aid for a manager to understand formal authority relationships. A typical organisation chart for the maintenance department of a power station is illustrated in figure 10.11. The main activities of the maintenance department as outlined in figure 10.11 are mechanical maintenance (turbines and boilers), electrical maintenance and instrumentation maintenance.

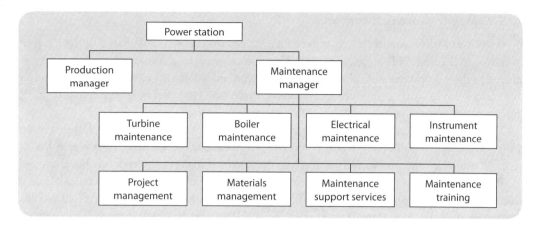

Figure 10.11 Example of organisation structure for a power station

10.5 CONTROLLING MAINTENANCE PERFORMANCE

10.5.1 Background

Control is essential to achieve management objectives and is one of the main functions of management. It is therefore also necessary to understand and describe the main control systems needed for the maintenance function and the maintenance system control must be directed at ensuring that maintenance resources and their application are monitored and properly directed towards achievement of the maintenance objective. A basic model for controlling a system is given in figure 10.12. This model can be applied for the control of physical parameters such as temperature or speed, or for the control of management parameters such as cost or worker utilisation.

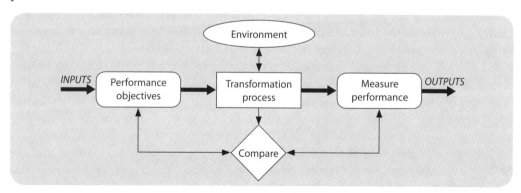

Figure 10.12 Basic model for the control of a system

The main elements of control for an organisational system can be derived from figure 10.12, namely performance objectives for the organisational unit (e.g. maintenance department), measurement of the performance of the department, comparison of actual performance with the objectives and some corrective action if the discrepancy is too large. In the maintenance environment, typical corrective measures could be to reduce or increase certain resources, to change the maintenance type (CBM, TBM, FBM), or to reduce or increase intervals for time-based maintenance or condition-based maintenance.

10.5.2 Maintenance cost control

Budgets are used to obtain the objectives for cost control. The maintenance budget is part of the overall budget of an organisation. The components of the maintenance budget are similar to the components of other budgets in the organisation and involve an estimation of the cost of the resources (labour, spares, tools, materials, etc.) that will be needed to meet the expected maintenance workload.

The maintenance budget can be constructed by means of the bottom-up technique, using the organisation structure as a basis. If a record of the maintenance costs over a number of years is available, the amount of corrective maintenance that is required can be estimated for the following time period. This information is not available for a new technical system and it is therefore important to collect cost information accurately during the first few months of operation of a new system. The corrective workload and resultant costs will not remain constant as a system ages and they can be expected to increase as long-life items reach the end of their system life.

Preventive maintenance work can be estimated using the techniques discussed in paragraph 10.3. The forecast workload for preventive maintenance should be checked against actual workload at the end of a time period and adjusted for the next time period if required.

10.5.3 Reliability and availability control

The main purpose of preventive maintenance is to improve the reliability and availability of the items, units and ultimately the total technical system. Proper monitoring of the reliability and availability at various hierarchical levels should therefore be performed in order to monitor whether the preventive maintenance programme is really effective. Supervisors and maintenance workers must provide the required information on the cause of failure by means of suitable checklists that can also be incorporated in the work orders. A prerequisite for the control of reliability is adequate records of failures, failure causes, time to repair, etc. Many companies do not keep adequate records and the reliability or mean time to failure of items and components can therefore not be compared with design values.

Several software programs are available for extracting the parameters of failure distributions from failure data, but regression analysis tools in spreadsheets can also be used. Most maintenance management information systems can also be used to perform the necessary calculations. If the reliability of a complex unit is to be compared with design values, some modelling or simulation might be required.

The Pareto technique is also quite useful to identify the components (or items) that are the major contributors towards failures of units and systems. The Pareto principle predicts that about 20 per cent of the items of the system will be responsible for about 80 per cent of the total cost of maintenance (or cause 80 per cent of the total downtime). The maintenance effort should be concentrated on the 20 per cent if the reliability of the total system is to be controlled properly. (See Herbaty (1990) for a detailed discussion of the Pareto technique.)

10.6 MAINTENANCE APPROACHES AND STRATEGIES

10.6.1 Overview

Currently there are three well-documented maintenance approaches or strategies that are used in manufacturing and service companies:

- business-centred maintenance (BCM);
- reliability-centred maintenance (RCM); and
- total productive maintenance (TPM).

Harris (1996) says: 'it seems to me that currently there are three distinguishable, but inter-related, maintenance strategies, viz. business-centred TDBU (*à la* Kelly), people-centred TPM (*à la* Nakajima) and asset-centred RCM (*à la* Nowlan and Heap).' Various maintenance practitioners have also developed and described variants of RCM and TPM, as well as combinations of these documented approaches, which are discussed briefly in the following paragraphs. Various books can be consulted for a detailed discussion on all three approaches (see Kelly (1997a), Campbell (1995) and Moubray (1990).

10.6.2 Business-centred maintenance (BCM)

Kelly (1997a) developed a maintenance approach in the 1980s that he termed *business-centred maintenance*, or *BCM*. This approach is also discussed briefly by Sherwin (2000). The BCM approach was developed in response to the need for a more cost-effective approach towards maintenance, but with a high priority for safety. This is a generic approach that can be applied in most industries or for most manufacturing/production systems or service systems, e.g. chemical process plants, assembly plants, power stations, fleet-type systems (buses, trains) or communication networks. The full BCM approach is summarised in figure 10.13.

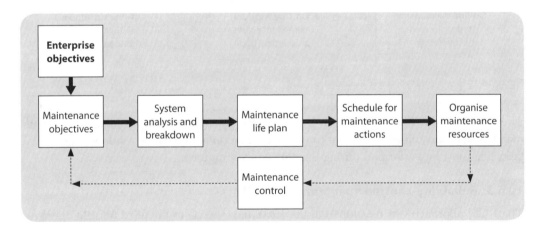

Figure 10.13 The business-centred maintenance approach

The starting point of the BCM approach is a review of the objectives of the overall enterprise, which are translated into objectives for the maintenance system or department. A thorough understanding of the operation of the system is then used as an input for the top-down-bottom-up (TDBU) analysis, in which a life plan is developed for the components, items and units of the overall system. The life plan for a unit defines which maintenance procedures (e.g. condition-based maintenance, fixed-time maintenance or operate to failure) are effective and a maintenance schedule can thus be formulated for each unit. The maintenance life plan is then used to determine the workload for preventive maintenance and historical information can be used to estimate the workload for corrective maintenance. The BCM strategy also addresses the organisation and control functions of maintenance management.

The TDBU analysis is an integral part of the BCM approach for maintenance decision making, and it includes the following three main steps, as outlined by Kelly (1989).
1. Analyse and understand the characteristics of the system operation.
2. Establish a maintenance life plan for each unit of the system.
3. Establish a maintenance schedule for the system.

A flow diagram that outlines the TDBU approach is shown in figure 10.14.

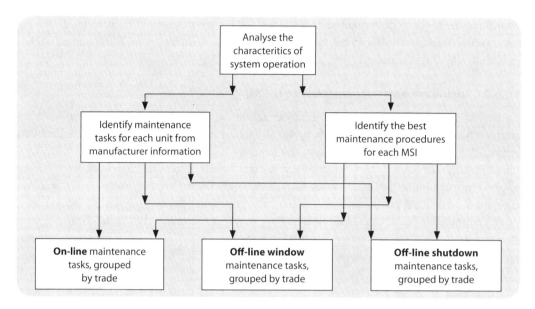

Figure 10.14 Flow diagram of the TDBU analysis

The output of the TDBU process is a list of all maintenance tasks, grouped into on-line maintenance tasks, off-line tasks that can be performed in maintenance windows, and off-line tasks that must be performed during a maintenance shutdown.

10.6.3   Reliability-centred maintenance (RCM)

The RCM approach was developed in the 1960s in response to the need for safe operation of new-generation aeroplanes that utilised new technologies such as fly-by-wire and jet engines. RCM is an asset-centred approach that has as its main objective the realisation of the inherent reliability of a technical system. This is accomplished through a sophisticated decision algorithm that takes into account the failure pattern of a component, as well as the consequence of each probable failure mode of the component. The RCM process is summarised by the flow diagram in figure 10.15.

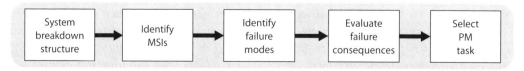

Figure 10.15 The reliability-centred maintenance approach

222

As in BCM, the first step in the RCM process also involves a proper system breakdown to an appropriate level of items and components. On large systems with thousands of components it is impossible to develop failure modes for all components, therefore only maintenance significant items (MSIs) are selected for further analysis. Likely failure modes for these items are then identified through the definition of functions for the items, the specific functional failure (how the function can be lost) and then identifying the root causes of these functional failures. Each failure mode has a consequence for the organisation and in RCM these consequences are classified as *safety, operational, non-operational* and *hidden failure consequences*. The RCM approach focuses on the formulation of maintenance tasks that are effective in dealing successfully with the consequences of a specific failure mode. The severity of the consequence of a failure determines the maintenance effort that is required to try and prevent the failure, or to reduce the consequence of the failure. The output of the RCM process is also a maintenance life plan for each component or item of a technical system. It involves a list of preventive tasks (scheduled discard tasks, scheduled restoration tasks or on-condition tasks) for the total system. Scheduled failure finding tasks are also identified for items that have hidden failure consequences. The RCM approach is documented in detail by Nowlan and Heap (1974) and Moubray (1990).

10.6.4 Total productive maintenance (TPM)

The TPM approach was developed in the 1970s in response to the need for higher overall equipment effectiveness in Japanese industries, specifically the assembly line industries. This is accomplished through the elimination of all losses and by adopting a zero defect, zero loss and zero failure approach. TPM can be viewed as a people-centred approach towards maintenance and is a philosophy for all workers of the enterprise. Many books have been published on TPM, but the text by Suzuki (1992) is a comprehensive treatise on the subject. TPM is defined by the Japanese Institute of Plant Maintenance (JIPM) in the following five phases (refer to Suzuki, 1992).

- Build a corporate constitution that will maximise the effectiveness of the production systems.
- Using a shop floor approach, build an organisation that prevents every type of loss for the life of the production system.
- Involve all departments in implementing TPM, including development, sales and administration.
- Involve everyone, from top management to shop-floor workers.
- Conduct zero loss activity through overlapping small-group activities.

Some of the main features of TPM are:
- the maximising of equipment effectiveness through the elimination of all machine losses;
- creating a sense of ownership in the operators of the systems through a programme of training and development; and
- the promotion of continuous improvement through small-group activities involving all departments of the enterprise.

Self-assessment

The following six performance parameters, with weights as well as minimum and maximum values, have been selected to monitor the performance of a maintenance system at a process plant that is operated continuously. The actual measured value in one particular month is given. Determine the total maintenance effectiveness (TME) for the maintenance system and what could be the reason for the low value of OEE in comparison with the availability.

Parameter	Weight	Range	Measured value
Availability (%)	0,20	80–90	85
Overall equipment effectiveness (OEE) (%)	0,10	50–80	63
Workforce utilisation (%)	0,20	50–80	72
Maintenance cost/Total cost (%)	0,20	50–10	23
PM cost/Total maintenance cost (%)	0,15	0–30	24
Work orders completed on time (%)	0,15	20–80	61

References

Aslaksen, E. & Belcher, R. 1992. *Systems Engineering*. Sydney: Prentice-Hall.

Campbell, J.D. 1995. *Uptime: Strategies for Excellence in Maintenance Management*. Portland: Productivity Press.

De Groote, P. 1995. 'Maintenance performance analysis: A practical approach.' *Journal of Quality in Maintenance Engineering*, 1 (2), pp. 4–24.

Harris, J. 1996. 'Editor's foreword.' *Maintenance*, 11 (1), January/February.

Herbaty, F. 1990. *Maintenance Management Handbook: Cost Effective Practices*. Park Ridge: Noyes.

Kelly, A. 1989. 'Maintenance and its Management.' Conference communication.

Kelly, A. 1997a. *Maintenance Strategy*. Oxford: Butterworth-Heinemann.

Kelly, A. 1997b. *Maintenance Organisation and Systems*. Oxford: Butterworth-Heinemann.

Lewis, E.E. 1995. *Introduction to Reliability Engineering*. 2nd edition. Singapore: John Wiley.

Moubray, J. 1990. *Reliability-Centred Maintenance*. Oxford: Butterworth-Heinemann.

Nowlan, L.S. & Heap, H.F. 1974. *Reliability-Centered Maintenance*. National Technical Information Service, US Department of Commerce.

Pearce, J.A. & Robinson, R.B. 1994. *Strategic Management: Formulation, Implementation and Control*. 5th edition. Burr Ridge: Irwin.

Pintelon, L., Gelders, L. & Van Puyvelde, F. 1995. *Maintenance Management*. Leuven: Acco.

Prahdan, S. 1994. 'Maintenance strategies for greater availability.' *Hydrocarbon Processing*, 73 (1), January, pp. 39–44.

Sherwin, D. 2000. 'A review of overall models for maintenance management.' *Journal of Quality in Maintenance Engineering*, 6 (3), pp. 138–64.

Smith, D.J. 2001. *Reliability, Maintainability and Risk*. 6th edition. Oxford: Butterworth-Heinemann

Suzuki, T. 1992. *TPM in Process Industries*. Portland: Productivity Press.

Tarita, A. 1998. 'RCM lite.' Paper presented at the Maintenance Engineering Conference of Zimbabwe (MECZIM), Harare, Zimbabwe, March.

Visser, J.K. 1997. 'A conceptual framework for understanding and teaching maintenance.' *Mechanical Transactions*, ME22 (3 & 4), pp. 61–9.

Wireman, T. 1990. *World Class Maintenance Management*. New York: Industrial Press.

11 Marketing for Technical People

Darryl Aberdein, Wilhelm Nel and Danie Petzer

Study objectives

After studying this chapter, you should be able to:
- describe the function of marketing
- explain what customer satisfaction is
- address consumer needs
- explain the relationship between product development and the market
- identify where in the product life cycle a product is
- differentiate among consumer adoption rates
- describe the implications of product classification
- describe the consumer decision-making process
- explain the marketing mix

The aim of this chapter is to introduce the reader to principles of marketing and explain the role of technical people in the management process.

11.1 INTRODUCTION

Marketing is concerned with the exchange of goods, services, technologies, business systems, information, concepts and ideas between buyers and sellers (Morris, 1992:4). Marketing management is the process of planning and executing the conception, pricing, promotion and distribution of goods, services and ideas to create exchanges with target groups that satisfy customer and organisational objectives (American Marketing Association, 1985).

Technical people are mostly involved in the marketing of industrial products and intellectual property rather than the marketing of end-user products and services. This is because industrial products are fairly complex. Industrial buyers are therefore technically trained, more sophisticated and usually require more technical information before making purchasing decisions than end users. The marketing of industrial products is also referred to as business-to-business marketing or organisational marketing. Most of the marketing principles discussed in this chapter apply to both business-to-consumer marketing and business-to-business marketing.

It is usually a difficult task to successfully introduce new technology and products to the market. (Note that most consumer products, e.g. a car, may have many types of technologies embedded in them, e.g. in the example of a car, engine technology, braking technology, information and communication technology, and materials technology.) Sometimes seemingly superior products that contain leading technologies and designs fail to gain significant market share while seemingly less-superior products may gain most of the market share. Some new technologies may also take much longer (e.g. decades) than others to be commercialised and diffuse (spread) through society. A good example is the fuel cell vehicle, which is superior from an environmental and efficiency perspective to vehicles driven by the internal combustion engine. In spite of the fact that the fuel cell principle was invented in 1839, only a few hundred fuel cell buses and cars are operating world-wide. One of the reasons why hydrogen fuel cell cars are taking so long to diffuse through society is because they rely on the co-diffusion of hydrogen generation and distribution networks that require a lot of money to implement (Nel, 2004:287). Would you (as end user) buy a product such as a cellular phone or fuel cell vehicle if the complementary network of base stations and refuelling stations were not available? Would you (as entrepreneur) build a hydrogen refuelling station if there were not many vehicles yet that would provide you with selling opportunities? This is essentially the 'chicken and egg' problem that is associated with the introduction of some new products that rely on complementary networks for their diffusion.

In this chapter you will be introduced to a number of marketing principles, the role of technical people in the marketing process and some of the issues related to the marketing of technology.

11.2 THE ROLE OF MARKETING

Society as a whole has needs and wants that it is continually striving to satisfy. People need food, clothing, transport, entertainment, and other products and services daily. Organisations need various raw materials, components, technical know-how and equipment that they transform or use in the transformation or value-adding process to produce products and services. Marketing aims to influence the demand for specific products or services. Marketing and production therefore have to combine forces to provide customers with the products or services they need. Marketing can be defined in a variety of ways, but in the words of management guru Peter Drucker (1991), 'Marketing is so basic that it cannot be considered a separate function. It is the whole business seen from the customers' point of view.' This implies that everybody in an organisation has to be involved in marketing directly or indirectly. If this is not the case then marketers will promise customers one thing but production will deliver something else.

It is very important to understand that any successful product or service must satisfy a large group of customers' needs. One of the roles of the engineer is to design and develop products that will satisfy the customer. In today's competitive business environment, businesses have to:

- increase customer satisfaction;
- ensure the production of quality products;
- reduce operating costs; and
- shorten lead times (the time it takes to develop and introduce a new product to the market).

By constantly working at the above, success, profitability, corporate survival and growth may be achieved by an organisation. The above illustrates the integrated effort that is required within an organisation. This is illustrated in figure 11.1 on the next page.

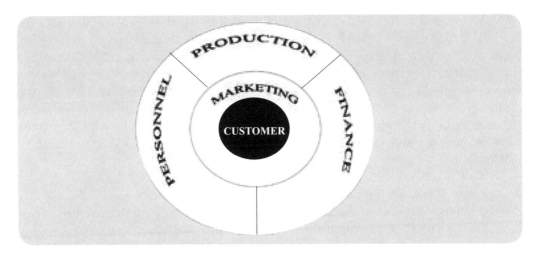

Figure 11.1 Marketing as integrator between various functional areas in the organisation and the customer

It would be incorrect to deduce from figure 11.1 that only people that are employed by the marketing department interact with customers. The role of technical people in the marketing process is further discussed in the next section.

11.3 THE ROLE OF TECHNICAL PEOPLE IN THE MARKETING PROCESS

Engineering managers, engineers and other technical staff often engage with organisational customers and end users. Sometimes they have to collaborate with marketing staff when customers require detailed technical information. A few examples of how technical people typically engage with customers are as follows.

- A workshop manager at an automobile repair shop listens to clients and ensures that problems are fixed.
- A lecturer discusses course content with industry role-players when designing a new course to ensure that industry-specific training needs are addressed.
- An engineer discusses the technical requirements of a product or process with a customer during the design process to ensure that it will meet the specific requirements of the customer.
- A project manager communicates with various project stakeholders to ensure that their inputs are obtained and incorporated in the project plan.
- A technical consultant provides organisational customers with advice and guidance.
- A sales engineer from a big aircraft manufacturer meets with clients to ensure that all the technical questions that they may have on life cycle issues such as maintenance and the provision of spare parts are answered.

Many companies never sell products or services to end users. They sell services, technical know-how, designs, patents and other intellectual property to organisational clients. Examples of such organisations are firms of consulting engineers, fabless (factory less) companies (organisations that design products but out-contract their production), process designers, value engineers and various types of contractors that may offer specialised services.

Many technical people believe that consumers will favour those products that offer the best quality, performance or innovative features. That may be because such people are often responsible for delivering such things. This approach to marketing is called the product concept orientation to the marketplace (Kotler, 1994:15). Such an orientation may lead to a focus on the product rather than on customers' needs. It is therefore important to determine the features, level of performance and quality that is desired by different (types of) customers. Determining this with current marketing research techniques seems to be difficult at times. Kloss, an inventor and entrepreneur, describes the limitations of marketing research as follows:

> The kinds of products that people might want are not limited to what people have said they want or what people, when you knock on their door, say that they will want. In the first case it's too late if people express the desire for what they want (Burgelman *et al., 2004:58*).

Designing and developing successful products and services is therefore a complex process that requires continuous learning and assessment of changes in the market and the technological environment. The next section provides more information on how to satisfy the customer.

11.4 CUSTOMER SATISFACTION AND PROFITABLE CUSTOMER RELATIONSHIPS

Marketing focuses on building profitable relationships with customers. Companies have realised that losing a customer is much more than losing a single sales transaction. It means that the company is losing a flow of purchases over the lifetime of the customer. It is thus important for a company to focus on creating customer lifetime value. A company captures value by creating loyal customers and retaining them over time and also by growing the portion of purchases that a customer makes from the company. Doing this leads to customer equity (Kotler & Armstrong, 2006:19). An automobile repair shop will thus become more profitable over time if it is able to retain and grow customers. A retained customer will service all the motor vehicles he/she will own over his/her lifetime at the repair shop. For the repair shop, this means a constant revenue stream over time from the retained customer.

Customer focus is the most important determinant of business success. Drucker (1991) explains it as follows.

> There is only one valid definition of business purpose ... to create a customer. It is the customer who determines what the business is. What the business thinks it produces is not of first importance, especially not to the future of the business and to its success. In fact, what the customer thinks he/she is buying, what he/she considers 'value' is decisive. The customer decides what the business is, what it produces and whether it will prosper.

Satisfied customers are loyal customers and this is exactly what businesses need. It is estimated that it costs five times more to gain new customers than to keep those you already have. So what does it take to produce and deliver customer satisfaction and retain customer loyalty? The *value exchange model* needs to be considered in this regard.

11.4.1 Value exchange model

Gabbott (2004:3 & 30) recognises the importance of the creation of value in marketing and proposes a value exchange model (VEM) for marketing. The model comprises two components. One component represents the 'customer' and the other the 'provider' or 'company'.

The model proposes that a customer makes certain investments in order to achieve *desired* benefits and/or outcomes. These investments include buying the product, consuming it and disposing of it. Providers or companies, on the other hand, also make investments to achieve certain *designed benefits and/or outcomes. Companies invest in materials, labour, infrastructure and systems to meet customer demands.*

A customer with desired outcomes links up with a provider or company with its *designed* outcomes. For example, an airline company requiring aircraft that are economical and easy to operate links up with an aircraft manufacturer like Boeing that provides a range of different aircraft models aimed at meeting the demands of airline company customers. At this point a value-facilitating exchange takes place. From this interaction the customer and company realise certain outcomes. These outcomes are sometimes different from what either the customer or provider expects. The difference between what was desired or designed and the actual outcome represents the *value* created for the customer and provider concerned. For example, the more successful the Boeing aircraft are in meeting the airline company's desired outcomes, the more value the airline company obtains from the exchange. For a company to survive in the long term, it should create value for its customers and know what drives this value (Gabbott, 2004:38–60). But how does an organisation create value?

Value is created in an organisation or industry along the value chain. Every company consists of a chain of activities from design, production, marketing and distribution to product support. All the functions – personnel, finance, production, IT, etc. – within the company play a role and form part of a bigger picture. In recent years, companies have grown to understand the impact of all the role-players in the business arena, and they are now looking closely at the working relationships that they have with suppliers and distributors in order to improve customer service. Figure 11.2 highlights the main players in a typical value chain.

Figure 11.2 The business value chain

The focus is on developing customer orientation in all functions and partners in the value chain. By implication, anybody who adds value to the product or service the consumer uses forms part of the value chain. The contrary is also true. Those who don't add value should be removed from the chain, as they are likely only to increase costs for the customer.

Total quality marketing (TQM) focuses on the continuous improvement of products and services provided by an organisation, utilising people from all levels. A company's marketing strategies will be ineffective if they are only entrusted to the marketing department. Impressive marketing strategies cannot compensate for deficient products or services. For example, a customer who cannot understand the operating instructions for a product will mentally downgrade the company producing it.

There is a connection between product quality, customer satisfaction and company profitability. A quality product or service meets or exceeds customers' expectations. For example, a high-priced car may deliver higher performance quality than a lower-priced one, but both can be said to deliver conformance quality if their respective target markets get what is expected. It is therefore important that technical people:

- correctly assess customers' needs when engaging with them;
- communicate these needs and expectations to colleagues;
- ensure that orders are correct and delivered on time;
- keep in touch with the customer; and
- convey customer suggestions to colleagues.

Measures do exist to determine the impact of customer satisfaction, customer retention and customer loyalty on profitability and are referred to as market performance metrics (Best, 2005:4).

Customer focus, satisfaction and retention are not only the responsibility of the marketing department.

11.5 MARKETING PRINCIPLES

Marketing management involves identifying attractive market opportunities, identifying specific customers (target marketing) and developing a marketing mix to satisfy their needs. The *marketing mix* refers to the tools a marketer can use to influence target customers. These are known as the 4 Ps of marketing: *product, place, promotion* and *price.* A change in any one of these areas will have an impact on the others. To fully understand these principles, let's first consider market segmentation.

The bases of *market segmentation* are geographical (where the customer lives), psychographical (personality, lifestyle) and demographical (age, sex, occupation, income, etc.). Any of these areas could be used to segment prospective customers. Target marketing involves selecting one of the segments that have been identified. Think of the example of a company that wants to enter the transport industry. Possible customer groups are school children, young adults needing transport to work, and high-income executives who travel the globe. All of these markets have different needs and one company cannot always address all of them. Targeting one segment will focus business attention on providing the most suitable products and services for the customers in that market segment. The right combination or mix of the 4 Ps should be used when targeting a specific market segment. The product must be appropriate to the segment targeted and the price of the product should be affordable to that segment. The product or service should also be easily available to that segment. The marketing message should also be directed at that segment. An additional 3 Ps apply to the marketing of services. They are *people, processes* and *physical evidence.* 'People' refers to those individuals who are involved in the delivery of the service, while 'processes' refers to the procedures and systems used to deliver the service. 'Physical evidence' refers to the service environment, the appearance of staff and company equipment (Bennett, 2002:183).

Product is the first and most important element of the marketing mix and deserves further discussion. The *core benefit* of the product is the essential benefit the customer buys. Products can be classified according to features such as durability and whether they are consumer or industrial goods. *Consumer goods* can again be further classified according to the shopping habits of the consumer, e.g. convenience, speciality and unsought products. *Industrial goods* can be further classified according to the production process, e.g. materials, capital equipment, supplies and services.

It is important to understand the type of product the company is producing, as this will impact on marketing decisions. A consumer product must be available off the shelf or the buyer will simply take the next best option. For example, if one particular brand of washing powder is unavailable due to production problems, the consumer will simply purchase another brand. However, in the case of a specialised product, like a satellite, customers will be prepared to wait several months or even years for it.

Companies cannot be all things to all people. A good understanding of portfolio analysis will enable a company to set objectives and allocate limited resources effectively. This is called the *product mix*. The following illustrates the product mix of Samsung.

SAMSUNG	HOME ENTERTAINMENT	HOME APPLIANCES	OFFICE EQUIPMENT
Product mix	Television Video tape recorder Hi-fi	Washing machine Tumble drier Dishwasher	Photocopier Fax Computer

Notice how the company focuses on different products in different market segments. By understanding the product mix, a company can expand its business in several ways. Firstly, it can *add new product lines*. Secondly, it can *lengthen product lines* (Samsung can add ovens to its home appliance range). Thirdly, more product *variants* can be added to each product (Opel can introduce a diesel option in the Corsa bakkie range) and, finally, it can pursue *more product lines* that will create a strong reputation in a specific field (HP Compaq computers can add a printer range). Products in a product mix, and more specifically in product lines, are grouped together not only according to a market segment focus. Buying decision making is often the reason behind products sharing the same product line.

Positioning refers to the way consumers perceive a product or a service. Positioning is the place that a product or service occupies in consumers' minds. For example, Rolex watches are associated with prestige and wealth. *Product differentiation* is used to gain competitive advantage. Differentiation is a set of differences used to distinguish one company's offering from that of its competitors. The generic business strategy of differentiation and its link with organisational-level technology strategy is discussed in chapter 17 ('Business and Technology Strategy').

The concept of *branding* is linked to that of differentiation. A brand is a name, term, sign, symbol or design, or a combination of them, intended to identify the goods or services of one seller or group of sellers and to differentiate them from those of competitors (Kotler, 1994:444).

11.6 PRODUCT LIFE CYCLES

The concept of product life cycle is important because it indicates distinct stages in the sales history of a product or service. Corresponding to these changes are distinct opportunities and problems linked to marketing strategy and profit potential. A company can make better plans if it knows the stage a product is in, or heading towards. The stages are *introduction, growth, maturity* and *decline*, as illustrated in figure 11.3.

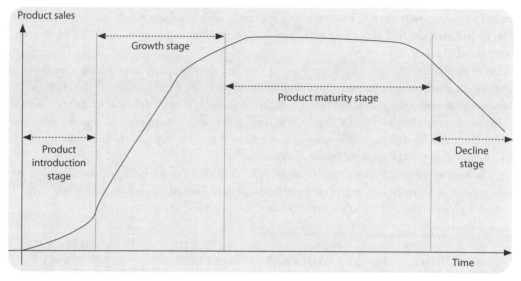

Figure 11.3 Stages in a product's life cycle

The life cycle model implies that:
- products and services have limited lives;
- products cannot sustain market growth indefinitely;
- organisations that define themselves in terms of a specific product will fade as the product matures and declines;
- different stages of the product life cycle present different challenges to the seller;
- profits and cash flow rise and fall during these stages; and
- different strategies are needed at each stage of the life cycle.

The duration of a life cycle differs from product to product. For example, a fashion item or a product in a fast-changing industry might have a short life cycle of only a few months compared to that of the much longer life cycle of a mining truck, which can be measured in years. There are specific generic characteristics for each stage of a product's life cycle. These are listed in table 11.1.

Table 11.1 Stages of the product life cycle

Introduction	Growth	Maturity	Decline
Slow sales increase	Fast sales increase	Sales slow down	Poor sales
Few customers	Better prices	Market saturation	Competitors move in
Limited production	More customers	Competition grows	Profit declines
High promotion cost	Higher production	Heavy promotions	Lack of interest
High prices	Better distribution	Less profit	
	Few competitors		

As the product reaches the end of its life span, some inspired engineering intervention or repositioning can be adopted to keep it alive when it reaches the maturity phase. For example, Volkswagen SA reintroduced the Golf as the City Golf. Developers should take note of strategies available in the different stages of the product life cycle. In the *growth stage*

232

the focus should be on product quality and new models of the same range. *Maturity* stage strategies include the improvement of the durability and efficiency of the product, e.g. the introduction of electric motors to lawn mowers to increase cutting speed. Motor manufacturers give their cars 'facelifts' by adding safety features and other functional changes to give the model a longer life. In the *decline stage*, weak product elimination is important. A reduction in research and development work and capital equipment investment is advisable.

Given the intense competition in most markets, companies that fail to develop new products and services expose themselves to great risk. Their existing products are vulnerable to changing consumer needs and tastes, new technologies and increased competition. New product development is risky. It is estimated that 80 per cent of new products fail, 75 per cent of which occur at the launch stage. Studies have indicated that successful new product launches share the following characteristics:

- unique and superior products (good quality, new features, greater value);
- a well-defined product concept prior to development;
- technology and marketing synergy, and quality of execution in all stages;
- in-depth understanding of customers' needs;
- introduction ahead of the competition;
- cross-functional teamwork;
- sufficient spending on the announcement and launch of the product; and
- top management support.

The marketing of new products is further discussed in the next section.

11.7 MARKETING NEW PRODUCTS

Marketing new technology products often involves a significantly different approach to that which is used for more familiar products and services. For a start, users do not have a strongly developed frame of reference in which they can position the product and as a result will not be able to tell you how they will use the product or how they could benefit from it. Secondly, users are justifiably sceptical about new technologies, as these technologies are often full of glitches or are not reliable. Thus, the innovator is faced with the challenges of determining the real user needs when users are unable to enunciate them properly, and then convincing users to adopt the product or service when that decision could involve significant risk.

Finally, users are looking for a 'whole product',[1] which is often difficult for a single company to provide. For example, it not enough to develop an electric car if there is not a network of recharging points available to users. The ability to provide, or at least influence, the provision of these externalities should form part of the decision as to which new products should form part of a product portfolio.

Many of the answers to these challenges are to found in so-called Diffusion Theory (Rogers, 1995), which describes how users interact with and adopt innovations. Geoffrey Moore (Moore, 1991, 2000) has expanded on diffusion theory and given it a practical slant. He refers to the marketing problem that innovators face as 'Crossing the Chasm'. In order to market new technologies, one first needs to understand the profiles of potential users. The key to overcoming the difficulties in marketing new technologies is to use the traits of each of the different types of users to the innovator's advantage.

11.7.1 Traits of potential users

Adopters of new technologies have long been classified in terms of what is called the Technology Lifecycle Adoption Curve. This is shown in figure 11.4.

The first adopters of a technology in this model are called *innovators* (not to be confused with the definition of an innovator that you will find in chapter 18). Innovators are a small but important group of users who are prepared to use new technology with all its associated glitches and problems. Indeed, innovators always want to be first and really enjoy experimenting with new products. Because they continually try out new products, innovators usually have a well-developed frame of reference and can often understand the potential product's benefits long before others can.

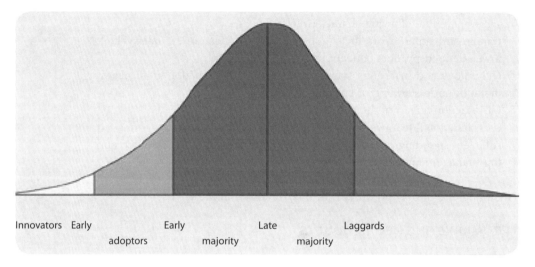

Figure 11.4 Technology Lifecycle Adoption Curve

While innovators represent a very small market and are often not prepared to pay much for new products, they represent a useful source of user feedback and provide invaluable help with product definition. Innovators may even be prepared to assist with bug testing and fixing.

The next category, that of the *early adopters*, does represent a real commercial market. The early adopters will buy a product for its uses and benefits, as opposed to the innovators, who simply just want to experiment. What distinguishes the early adopters from other commercial users is that they perceive the product as a means of gaining some commercial or personal advantage. They could be up-and-coming executives trying to establish a name and reputation, or a small company that is ambitious in its goals. Early adopters are calculated risk takers and are prepared to live with small glitches as long as the product can support their goals. They are often not very price sensitive and may even be prepared to help fund some of the development work.

Unfortunately, early adopters represent a very small part of the total market and are often mistakenly seen as part of the mainstream market, which is the preserve of the *early majority* and *late majority*. The early and late majorities represent the largest segments of the market. Their distinguishing characteristic is their conservatism. It is extremely unlikely that they will try out a new product without first getting positive references from several other mainstream customers. They are also unlikely to adopt unless they can perceive that the whole product is mostly in place. This could include aspects such as spares, manuals, repair centres in all major centres, 24-hour support, country-wide coverage (e.g. for cell phones), etc.

The *laggard* segment is relatively small and unimportant. Users in the laggard segment will only adopt new technologies when there is no other choice, e.g. when it is no longer possible to repair the old technology.

The problem lies in moving from the early adopters into the mainstream markets. Mainstream users have very limited contact with and often distrust the advice of early adaptors and innovators. This disconnect is one of the biggest reasons for marketing failure in new products, and is what Moore terms the 'Chasm'. At the same time, the problem is compounded by the reality that it is unlikely that innovating companies will have the resources to provide a whole product in all segments at product launch.

11.7.2 Strategies for entering new technology markets

Faced with these problems, the solution for entering new technology markets is to segment the market into small (tiny, if necessary) segments of connected users in which at least one visionary user can be identified. As whole-product needs differ markedly across segments, careful segmentation can make their provision in each small segment achievable with limited resources. Domination of a small segment provides confidence to other subsequent segments, and enables the whole product to be built in affordable chunks.

The most difficult decision is to select these initial segments. The following criteria should be used.

- The segment should be small or unimportant enough that the other players in the industry leave it largely 'undefended'. The last thing that one wants is to get into a battle with the incumbents (i.e. those providers of the existing technology or product).
- The segment should be small enough for you to dominate (and thus establish credibility for future segments) in a reasonable period. Domination in this context normally means a market share of 50 per cent or more.
- It should be possible to identify one or more visionary clients who are relatively influential in the segment. These clients should have a compelling need for your product, which will make selling the product to them easier.
- The whole product for the segment (i.e. the core product plus other things like support, maintenance, user groups, etc.) should be within your, or a strategic partner's, ability to provide.
- The segment should be connected to other market segments to which you will market. The terms 'connected' means that the users in the following market segment communicate with, and trust the opinions of, users in the prior market segments.

Using this strategy, the various segments are approached in a bowling-pin approach. Each successive market segment is approached using the credibility gained in prior markets. At the same time, 'pieces' are added to the product offering, building out the whole product and the network until the technology becomes the *de facto* standard or dominant design and the technology diffuses almost of its own accord.

Example 11.1 'Crossing the Chasm': Call centre technology

A South African company was one of the first in the world to provide PC-based call centre technology in the mid 1980s. This meant that the calls were recorded on a hard disk and not on magnetic tapes, as was common at the time.

After trying unsuccessfully to sell the new technology to the large banks, it was decided to sell to the emergency services segment first, where the need to reliably record and manipulate emergency calls was most pressing. A local fire chief of a small satellite town proved to be a visionary client. The various users in the emergency services segment are very well connected with each other, and in a short while the majority of emergency calls in South Africa were recorded on hard disk technology.

Once the credibility of the technology had been established with the emergency services segment, the police and insurance industry segments could reference other mainstream users whom they could trust, and were quick to adopt the same technology. It was only then that the banking industry, through referencing the insurance industry, followed.

An important point to note is that product features and functionality (i.e. defining what constitutes a whole product for each segment) is an important part of the marketing strategy. It is therefore not surprising that marketing strategy inputs are an essential part of the product development process.

11.7.3 Positioning a new technology

When marketing new technology, the marketer often has to assist the potential user to define how the specific product can be of benefit to him/her, and how the new product differs from existing solutions. Initially, early adopters (see section 11.7.1) can be very helpful in defining how a new technology can be used in practice and what the benefits to the user are. The marketer will take this information and use it to 'package' a marketing proposition for each segment. This is often called an 'elevator pitch', referring to the brief time one might have to explain a particular offering to a prospective client.

The elevator pitch should position the product in the user's mind and give a clear indication of the benefits to the client. The elevator pitch should at least contain the following components (Moore, 2000:148):

- who the typical users of the product are;
- why they are dissatisfied with the current market alternatives;
- a description of the new product category/type;
- a description of the key problems that it addresses; and
- a description of why it is different from the current market offerings.

The elevator pitch should be no longer than one page in written form and would typically be the core of a product brochure or web page.

11.7.4 Channel strategies

Distribution channels (or simply 'channels', as they have become known) refer to both the organisational and physical aspects of transferring products from the factory gate to the user. The physical distribution strategy is normally relatively easy to determine, as it is standard across product types. In contrast, the selection of the organisational aspects of a distribution channel for a new product is a far more complex and important decision. The existing channels are often unsuited for new product types and in some cases the channels need to be set up from scratch – which is a costly and time-consuming activity!

When determining a channel strategy, you have to understand how the channel will support your overall objectives (Moore, 2000:164). The following considerations are important.

- Is the channel a *demand creator* or *demand fulfiller?* Certain channels (e.g. see the retail channel) are essentially demand fulfillers, in that they display a product on a retail shelf and passively wait for a customer to pick it up. Other channels (e.g. direct sales) are demand creators and attempt to make a sale to a client that might not have occurred otherwise.
- What is the role of the channel in providing the whole product? Certain products require extended product items such as spares, maintenance and advice. Should the channel provide these services or are they just an unnecessary expense?
- What is the potential of the channel to scale to high volumes? Retail channels are optimised for high volumes and low cost, while direct selling offers little in the way of economies of scale.

Example 11.2 Channel strategy

The first generation consumer radios used thermionic valves as part of the amplifier and control circuits. Thermionic valves are like miniature television tubes with delicate filaments encased in fragile glass. These valves needed occasional replacement.

The distribution channel that developed for the early valve radios required that radio shops carry a range of spare valves and employ a person on site who was trained in replacing defective valves.

When transistors were introduced, this level of expertise proved to be unnecessarily expensive, as transistors were inherently more reliable than valves. Distribution shifted to cheaper alternatives such as department stores.

Initially, when introducing a new product to market, one would probably place an emphasis on customer intimacy and demand creation in order to move across the chasm. Understanding customer needs and communicating the benefits in each chosen segment are an important part of the strategy and can rarely be achieved through third-party distributors or through impersonal channels such as Internet sales. This often means that you will use a two-phased distribution strategy: a lower-volume model with high customer interaction at first, and then move to a higher-volume, lower-cost model once the product has established credibility in the market.

The choices for a sales channel are essentially as follows (some hybrid strategies have been omitted).

- **Direct sales:** This is traditionally the most successful channel in high-technology marketing. At its heart is a salesperson who interacts intensively with the customer, understanding his/her needs and drawing up a proposal that leads (it is hoped) to an order. The direct sales model is expensive and the costs need to be built into the price. It works best where the offering is reasonably comprehensive and of a relatively high value. This model was once very popular (e.g. selling mainframe computers), but is less so these days due to the rapid commoditisation of high-tech products that quickly brings prices down below the point that this model can work. It is often the preferred model for introducing new technology, due to the close contact with customers and because it is focused on a demand creation rather than demand fulfilment.

- **Retail sales:** This channel is aimed at high volumes, and existing brands will always get priority on shelf space. It is also largely aimed at demand fulfilment rather than demand creation. For these reasons, it is normally not a suitable channel, at least initially, for new technology products. Typically, products sold through retail channels are lower priced – in the order of R20 000 as a typical maximum (think of fridges, audio equipment and personal computers as examples).

- **Value-added resellers:** Often simply called VARs, they bridge the gap between direct sales and retail sales. They can provide some level of support and infrastructure, normally with a low overhead structure, and are therefore useful when some additional support and consulting are required. Some VARs are small and have only a local footprint and for this reason do not scale well as the demand grows. Other 'super-VARs' operate nationally and provide the scale required, but are often not able to service the smaller customers, due to their higher overhead costs. VARs are often essentially consulting firms that leverage the product sales to generate more consulting income – sometimes at the expense of selling products.

- **OEM distribution** – The original equipment manufacturer/marketer (OEM) is part of a growing phenomenon of what is more correctly called outsourced design and manufacture (ODM), where OEM companies with a global marketing reach and system integration capabilities source the design and manufacture of complete sub-systems. In the PC industry, companies such as HP and Dell sell PCs built up from hard drives, motherboards and DVDs from often-unknown companies. Because the OEM market mostly requires sub-systems to fit established standards and interfaces within the system architecture, it is unlikely that this channel will suit new, disruptive technologies that do not comply with the existing system architectural standards.

- **Internet:** The Internet as a sales channel is still relatively new but is undoubtedly set to evolve. For the most part, however, it is best suited to large-volume, low-customer-interaction sales. Along with mail order and similar sales channels, it shares the advantage of having no limit with respect to physical shelf space. The Internet affords the ability for customers to search at virtually zero cost for speciality items. Despite this, the Internet's lack of customer interaction makes it a poor choice, at least initially, for marketing high-tech products. The Internet does provide an excellent medium for physically distributing certain products (e.g. software, knowledge, pre-paid vouchers), as well as for some extended product components such as manuals and user support.

11.7.5 Forecasting diffusion potential

When planning the marketing of an innovation, one needs to consider what the potential size of the market is, and how fast the market will grow to this level. A number of different models are available to the technology marketer to make these predictions.

Some of the *statistical* models (Tidd *et al.*, 2001:188) use observed data points at the start of a diffusion curve and extrapolate these to make forecasts. The predictions can be inaccurate, as they often ignore important factors that influence diffusion.

Example 11.3 Technology trend forecasting

There has been much speculation as to when the electric car will supersede the petrol-powered car. One approach to this would be to look at the various trends of the component technologies and forecast when they will all be available at sufficient performance levels and reasonable prices to form a new dominant design.

The dominant design for cars likely consists of a price point of less than R200 000, able to travel 500 km on one tank/charge, can refill/recharge within 10 minutes, can carry 5 people and has a 30 litre luggage capacity. One woule then examine the various technology trends (e.g. battery size, weight and cost) to forecast when this dominant design could likely be achieved.

One would also then look at the availability of complementary products (service stations, recharge points) and make similar predictions as to when they might become available in sufficient numbers.

Lastly, one would look at environmental trends, political and social pressures, and other external factors that might support the new technology and work against the old technology (e.g. fuel reserves, environmental treaties and legislation).

Trend models look at the underlying technologies and markets for signs that they will be able to combine to form a viable dominant design (see example 11.3). Along with this, one would consider external factors such as the availability of the necessary infrastructure to support the diffusion of the technology. This is especially important where there are what economists call *network externalities* present. Network externalities simply mean that the value of a particular product to a user is dependant on an external network being present. For example, a car is almost valueless if there is no road network, while a fax machine is valueless unless there are many other fax machines connected to the network.

While trends and statistical forecasting may be useful in certain circumstances, the challenge facing innovators is to be able to make marketing predictions at a very early stage of the technological and product life cycle. For this, sociological rather than statistical methodologies are often used. The factors most closely correlated with diffusion potential are the *relative advantage of the product, the degree to which it supports existing norms and values, the lack of complexity, the trialability and the observability of the product.*

The most important attribute of an innovation affecting diffusion is the *relative advantage* the innovation has over competing or substitute technologies. This is the degree the product is perceived to be better than competing or substitute technologies. Relative advantage has two components: performance and price. Both of these tend to change over the life cycle of the technology, with performance increasing and price decreasing.

Another attribute affecting the diffusion of an innovation is the degree that the innovation is *compatible with the values, norms and experience of the users*. Innovations that are compatible with the values and norms of the community have a far better chance of diffusion in that community. Innovations that are linked to other innovations already accepted in the marketplace also have a better chance of diffusing. For example, basing Internet banking access on the ATM card and password and having similar user interface screens enhance the diffusion of Internet banking.

The third attribute affecting diffusion is *complexity*. Complexity is the degree the product is perceived to be difficult to use or understand. Complexity generally retards diffusion. The diffusion of personal computers has been enhanced by innovations such as 'plug and play' and graphic user interfaces, all of which reduce the complexity to the user.

The fourth attribute affecting diffusion of an innovation is the *trialability* of the innovation. Trialability is the extent to which the innovation can be tried out in a low-risk situation. Trialability is more important for early adopters, as they are not generally able to observe others using the product. Late adopters are less likely to want to try out an innovation, as they will use the experience of early adopters as a surrogate for their own trials.

The last attribute affecting diffusion of an innovation is the *observability* of the innovation. Observability is the degree to which the results of the innovation are visible to others.

Example 11.4 Observability of automated check-ins

One of the airlines at Johannesburg International Airport's local departures terminal placed its new automated check-in terminal next to the manual check-in queue. In this way, frustrated passengers in the manual queue could observe users of the new facility receiving their boarding passes quickly and easily.

In the next section the thinking and decision-making process that takes place in the buyer's brain before, during and after purchasing is discussed.

11.8 THE CONSUMER DECISION-MAKING PROCESS

The consumer adoption process focuses on the *mental process* through which an individual passes from first hearing about the product until he/she finally buys it. It is the process whereby customers learn about new products, try them, then adopt or reject them. Marketers must therefore look beyond the various influences on buyers and develop an understanding of how consumers make their buying decisions. Buying behaviour can vary from complex, high-involvement decision making to habitual, low-involvement decision making. In *complex decision making,* the buyer is highly involved. The product is expensive and not bought frequently (for example, motor cars). The marketer must understand that information is vitally important at this stage, to inform the buyer of what to look at.

In *dissonance-reducing* buying behaviour, the buyer is still very much involved in the process, but fails to see the differences between the items on offer. He/she will shop around and decide quickly because of the lack of perceived brand differentiation. The buyer will be influenced by a good price or purchase convenience. Communication must be aimed at making the buyer feel good about his/her purchase.

Variety-seeking buying behaviour is characterised by low consumer involvement and a lot of brand switching. Buying biscuits is a typical example. Here the product must dominate shelf space and out-of-stock situations must be avoided.

Habitual buying behaviour also sees low involvement from the consumer. There are no real perceived brand differences. A typical example of such a product is salt. Advertising repetition creates product or brand familiarity and customers buy because they are familiar with the name. Visual symbols and imagery are important and marketers should focus only on one or two key product points. Most consumers pass through five stages when buying something.

Figure 11.5 The customer purchasing decision-making process

Figure 11.5 illustrates that the buying process starts long before the actual purchase and has consequences long after the product has been bought. *Need recognition* can be triggered by a variety of stimuli. For example, passing a bakery and smelling freshly baked bread may stimulate a person's appetite. Marketers must also develop strategies that trigger consumer interest. For example, the Camel Trophy sparks interest in the recreational vehicle market. In the *information-gathering stage*, people are more aware of information that relates to their needs. People wanting to buy a new cell phone will spot special offers much more easily than someone who doesn't want or already has one.

In *evaluating* alternatives, buyers see every product as having a bundle of attributes. Each attribute is seen as having some utility function. If the marketer can focus on the attributes that will satisfy the major utility functions of the product, the product is sold. The *purchase decision* can be influenced by the attitudes of others and by unanticipated situational factors. A buyer's preference for a particular brand will increase if someone he/she likes favours the same brand. Situational factors can include a rude sales person or the loss of income.

Post-purchase behaviour is the level of satisfaction or dissatisfaction the buyer experiences after the sale has taken place. The marketers' job continues into this phase. Satisfied customers say good things about products and services that are invaluable to a company. Communication with the user after the sale has been made is important to confirm the belief that the right purchase was made. Companies that follow up on sales and those who provide accessible channels for customer complaints have experienced fewer returns or cancellations.

11.9 CONCLUSION

An understanding of basic marketing principles can be beneficial to engineers. Companies can no longer afford functional approaches to product development. Customer service is the primary element of successful businesses. To attract and retain customers, companies and their personnel must adopt sound marketing strategies.

Self-assessment

Please note: Suggested solutions are provided to questions that are marked by an asterisk (*). You will find the suggested solutions at the back of the book.

11.1 Describe the role of marketing in an organisation.

11.2 Describe the value chain of which an organisation forms a part.

11.3 List the focus points of total quality marketing.

11.4 Define:
- i) marketing mix
- ii) market segmentation
- iii) market positioning
- iv) product differentiation
- v) diffusion process
- vi) adoption process
- vii) marketing management
- viii) branding
- ix) network externalities.

11.5 Products go through four distinct stages during their lives. The four stages are introduction, growth, maturity and decline. This is called the product life cycle. Describe the generic characteristics of each of these four stages.

11.6 Describe the various issues that should be considered by an engineer when developing new products.

11.7 Explain the relationship between product development and the market.

11.8 Describe the consumer decision-making process.

11.9 Define 'market segmentation' and explain how each of the following markets could be segmented:
- i) the market for motor vehicles
- ii) computers.

11.10* You are the owner of a quarry that produces building stone (aggregate). Identify at least two market segments within the building stone market. Explain how you would combine the different marketing instruments to form various marketing offerings that will address the needs of the different segments.

11.11 Describe the problems that the marketers of new technological products face. How can these be overcome?

11.12 List the factors/attributes that affect the diffusion of new technologies and innovations. Give a brief description of each factor or attribute.

Mini-projects

Note: Mini-projects should be answered in the format of a report. You will find guidelines at the back of the book.

A. **Product life cycle:** Select one to a maximum of five products or services that your employer (or a company that you are familiar with) offers. Use the product life cycle model to analyse these products. Explain in what stage of its life each product is and what the consequences for the company are in terms of research and development spending, marketing, phasing out, etc.

B. New business opportunities often involve a company moving into new markets with new technologies. What risks does such a company face and how would it manage those risks?

C. **Marketing strategy:** Interview a sales engineer, marketing manager or sales representative of a company. Ask this person to tell you more about:
   - one of his/her company's products;
   - the market and how it can be segmented;
   - the segment(s) that this company is targeting; and
   - how this company uses the marketing mix to formulate a market offering that addresses the needs of the market segment.

   Record your findings in a report.

D. **Minerals marketing:** Minerals such as platinum, gold, diamonds, coal and zircon are marketed in slightly different ways. Interview at least one person that is involved with the marketing of one type of mineral and record your findings in a report. You may choose any number of minerals for this project, but preferably focus on one only. Try to answer the following questions in your report.
   - How is the mineral promoted?
   - Describe the organisations that may be involved in the promotion of the mineral and how they promote it.
   - Describe the main use(s) of the mineral. How is the market for this specific mineral segmented?
   - Is the mineral sold on a market such as the London Metals Exchange or by means of contractual agreements between mines and purchasers?
   - Describe whether branding of the mineral is done.
   - What percentage of the local production of this mineral is locally processed into consumer or industrial products?
   - What can be done to promote further beneficiation of the mineral in South Africa?

E. A number of car manufacturers have built concept fuel cell cars, buses and motor cycles and are continuing to spend lots of money to further improve them. By the end of 2004 there were about 500 such vehicles in the world and it may still take a while before a fuel cell car will become a viable alternative to internal-combustion-driven automobiles for a large percentage of customers. Apply your knowledge of the requirements that a new product should meet and the factors that assist the diffusion of a new technology in the marketplace to specify some characteristics that such a fuel cell car will need to compete with traditional (internal combustion engine) cars. You may include supporting infrastructure such as fuel distribution in your analysis.

References

American Marketing Association. 1985. *Strategic Marketing and Management*. New York: John Wiley.

Bennett, J.A. 2002. 'Marketing Management.' In *Business Management: A Value Chain Approach*. Pretoria: Van Schaik.

Best, R.J. 2005. *Market-Based Management: Strategies for Growing Customer Value and Profitability*. 4th edition. Upper Saddle River: Pearson Prentice-Hall.

Burgelman, R.A., Christensen, C.M. & Wheelwright, S.C. 2004. *Strategic Management of Technology and Innovation*. 4th edition. Boston:McGraw-Hill.

Drucker, P. 1991. *The Practice of Management*. Oxford: Butterworth-Heinemann.

Gabbot, M. (ed.). 2004. *Introduction to Marketing: A Value Exchange Process*. French's Forest: Pearson Prentice-Hall.

Kotler, R. 1994. *Marketing Management*. Englewood Cliffs: Prentice-Hall.

Kotler, P. & Armstrong, G. 2006. *Principles of Marketing*. 11th edition. Upper Saddle River: Pearson Prentice-Hall.

Moore, G.A. 1991. *Crossing the Chasm: Marketing and Selling Technology Products to Mainstream Customers*. New York: Harper Business.

Moore, G.A. 2000. *Crossing the Chasm: Marketing and Selling Technology Products to Mainstream Customers*. 2nd edition. Oxford: Capstone.

Morris, M.H. 1992. *Industrial and Organizational Marketing*. 2nd edition. New York: Macmillan.

Nel, W.P. 2004 'The diffusion of fuel cell vehicles and its impact on the demand for platinum group metals: Research framework and initial results.' Paper presented at the International Platinum Conference, 'Platinum Adding Value.' Sun City, South Africa, 3–7 October. SAIMM Symposium Series S38, pp. 287–328.

Peters, T. 1984. *In Search of Excellence*. New York: Warner.

Rogers, E.M. 1995. *Diffusion of Innovations*. New York: Free Press.

Tidd, J. Bessant, J. & Pavitt, K. 2001. *Managing Innovation: Integrating Technological, Market and Organisational Change*. Chichester: John Wiley.

Weihrich, H. & Koontz, H. 1993. *Management: A Global Perspective*. New York: McGraw-Hill.

(Footnotes)

[1] The whole product refers to the core product plus the augmented product, which encompasses things like support, manuals, maintenance, complementary services, etc.

12 The Engineer, User of Information and Communication Systems

Wilhelm Nel

Study objectives

After studying this chapter, you should be able to:

- explain the information needs of engineers, scientists and technologists
- describe and list the information technology needed by an engineer involved in design, development, research, manufacturing and management
- explain what is meant by a database and list its benefits
- explain how information systems can assist engineers to make better decisions
- explain why information is an important organisational resource
- explain how information technology can assist engineers and engineering managers to improve products and services, process efficiencies and competitiveness
- list various sources of information
- describe some of the opportunities the Internet offers businesses

The aim of this chapter is to provide an overview of the use of information and communication technology by engineers in the various activities they may be involved in. The use of information technology to improve decision making, products and services, process efficiencies and competitiveness is emphasised.

12.1 INTRODUCTION

New information and knowledge can change our perceptions, views and lives. They influence the way managers plan and organise. They are the basis of innovation, growth and the advancement of humans and countries. In their book *Not by Genes Alone*, Boyd and Richerson explain that humans are the only species that can accumulate knowledge over long periods of time and transfer it so that the next generation can improve on it. This is one of the main reasons why modern humans are such a successful species. As far back as 4500 BCE, some civilisations kept records of receipts, disbursements, inventories, loans, purchases, sales, leases and contracts on clay tablets. Modern technology has brought about many more choices for recording, storing, and processing data in order to improve decision making. The earliest use of computers was for data processing, which reduced clerical costs and the volume of paperwork. As technology advanced, organisations began to use computers to process information relating to production, marketing, inventory control and other business functions (see figure 12.1 on the next page). Organisations do not only apply computers in areas where they can save money on administration, but also where they improve methods of operation and provide a competitive edge through the improvement of products, services and relationships with clients. Some of these are given in table 12.1 and example 12.1.

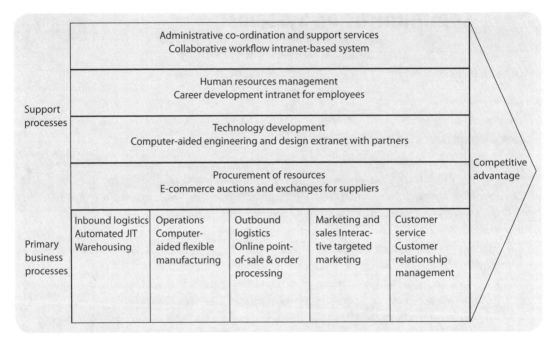

Figure 12.1 Information systems in the value chain of an organisation (O'Brien, 2002:50)

Table 12.1 Some examples of how information and communication technologies (ICT) are influencing the workplace (Nel, 2005:291)

Corporate communication/diffusion (spreading) of information (e.g. intranet, Internet, e-mail, bulletin boards, video conferencing)
Changes in marketing (call centre technology, Internet advertising, database marketing, PC–TV integration, wireless-enabled marketing)
Decision making (management information systems, data mining)
Business-to-business communication and e-commerce (EDI, ExtraNet)
Streamlining of the value chain (cutting the middle man out – buyer and seller are in direct contact – lower transaction costs)
Mass customisation: Providing personalised experiences to customers. Companies that provide computers (e.g. Dell), jeans (e.g. Levi), cars, bread and so forth according to customer specifications. Although such products may seem ordinary they have the information provided by a specific client embedded in them.
Telecommuting – workers at home or at remote centers using ICT to communicate with co-workers and supervisors
Virtual office – companies taking salespeople and executives out of their offices and sending them to work in cars, hotels and customers' offices. They communicate with the office and their co-workers via laptop computers (Smither et al., 1996:54).

Information and communication technology and systems assist organisations to maximise the use of information – which is considered to be an important factor in deciding the competitiveness of countries.

Example 12.1

Mitsubishi uses microcomputers in its vehicles to control body roll during cornering, making swerving for an emergency at high speed easier and safer. Electronic suspension also eliminates the tendency of a car's nose to dive when brakes are applied hard or its tail to squat with sudden acceleration. The smooth ride of the car gives the company a sales advantage (Hussain, 1992:1).

Engineers, scientists and technologists use information to make management decisions and in other activities such as research, design, development, production and maintenance. More specific examples of the use of information systems by engineers are listed in table 12.2 and provided in examples 12.1 and 12.2.

Table 12.2 Examples of the use of computers and networks in various engineering disciplines

Engineering discipline	Applications of information and communication systems
Aeronautical engineering	The design, wind tunnel testing and production of aircraft
Chemical engineering	Design and control of plants
Civil engineering	Stress analysis
Electrical (including electronic) engineering	Circuit design and analysis, power distribution and control, 2G/3G network planning optimisation software
Mining engineering	Mine design, mine optimisation, production scheduling

Example 12.2 The effect of computers on structural engineering

'The computer enables engineers to make more calculations more quickly than was conceivable with either the slide rule or the calculator; hence the computer can be programmed to attack problems in structural analysis that would never have been attempted in the pre-computer days. Now, the computer not only can perform millions of simple, repetitive calculations automatically in reasonable amounts of time but also can be used to analyse structures that engineers of the slide rule era found too complex' (Petroski,1985:193).

Decision making in an environment where multiple factors impact on a situation can be very complex. A vast amount of data may be available, but models that can transfer it into useful, quality information may be lacking. Engineering and management decisions are therefore sometimes taken under imperfect conditions. The lack of perfect knowledge or information does not mean that one has to feel stranded. Engineers and scientists often have to make assumptions in order to progress. Safety factors can be built into designs when perfect knowledge about the materials that are used is unavailable. Scientists often make assumptions or hypotheses, then find evidence to either support them or prove them wrong. The cycle of creating theories, developing and applying practical methods and tools based on the theories, leading to new insights that improve the theories, is the primary growth engine of science and technology. Many organisations are adapting methods of experimenting and learning and call themselves learning organisations (Senge, 1994:51). This explains why some management theorists are starting to view companies as knowledge-compiling and knowledge-processing systems, where the creation of new knowledge becomes both a means and an end.

In an environment where information is no longer the preserve of the select few, it has become more important for management to act timeously on the opportunities that new information and knowledge offer. Complicated multi-disciplinary engineering projects often call for knowledge that is both deep and broad, and it is therefore important that companies actively seek to increase their knowledge base outside the company as well.

12.2 DATA, INFORMATION AND KNOWLEDGE

Engineers, scientists and technologists work with lots of measurements, facts, figures, codes and names. All of these are called data. Data is processed through models before it is used for decision making. In a useful format it is called information. Processed data should therefore meet certain criteria before it becomes useful input in decision-making processes. The following are some of the criteria that information should meet. It must be:

- relevant to the decision that must be made;
- timely;
- cost-effective;
- accurate.

Example 12.3

Poor representation of data could have resulted in the explosion of the Challenger space shuttle in 1986. Although NASA had all the facts, the (wrong) decision was made to launch the shuttle because these facts were not represented in a useful format – i.e. they did not provide the information upon which proper decisions could be made (*Management Review*, 1998:24).

Information is the source from which knowledge originates and insight originates from knowledge. Insight is what is needed for sound decision making: '... knowledge is the primary resource for individuals and for the economy overall. Land, labour, and capital – the economist's traditional factors of production – do not disappear, but they become secondary' (Drucker, 1985).

A significant percentage of the workforce is engaged in the production of information or employed by organisations that manufacture or sell information products. Accountants, clerks, engineers, lawyers, computer programmers, systems analysts, managers, physicians, librarians and auditors are all information workers. Even workers directly employed in operations are involved in information functions and require information aids (Burch & Grudnitski, 1989:9).

	Creates and collects information	
Knowledge worker	Acquires information resources	**Knowledge industry**
	Organises information	
	Stores information	
	Analyses information	
	Presents or uses information	
	Distributes information	
	Makes decisions based on information	
	Maintains information systems	

Figure 12.2 Jobs performed by a knowledge worker (Hussain, 1992:341)

12.3 INFORMATION NEEDS AND RESPONSIBILITIES

Information is the fuel that drives organisations. A major function of an engineering manager is to convert information into action through the process of decision making. It is important to generate only the information that is necessary for effective decisions. The effectiveness of the manager's action depends on how complete, relevant and reliable the information is. Other users of information inside an organisation also need relevant information in time, in the correct form and at an acceptable cost. But who is responsible for supplying these users with information and what should be done to make such information available? Computer professionals, management and users share this responsibility. The correct technology in the form of hardware, applications software and networks should be put in place and regularly updated as technology and users' needs change. End users, like functional managers, may choose equipment, design and develop systems, handle their own computer operations and are therefore faced with the problems of how to:

- select and acquire hardware and software;
- provide enough computer power;
- make technology as accessible and easy to use as possible; and
- ensure the privacy, security and backup of data (Hussain, 1992:4).

In a large organisation, information managers and employees in the information systems or computer services department may support end users. The information manager and his/her staff may be responsible for buying computers, seeing to their maintenance, installing networks, facilitating communication, data processing, systems design and implementing information systems. Users usually have the responsibility of deciding what data needs to be recorded and in what format. Transactions must be recorded and the information must be made available to decision makers in the most suitable form.

The following questions can be used to identify the information needs of decision makers.

- What types of decisions do you make regularly?
- What information do you need to make these decisions?
- What information do you get regularly?
- What information would you like to get that you are not getting now?

- What information would you want daily? Weekly? Monthly? Yearly?
- What data analysis programs would you like to use?
- How much information is needed?
- How, when and by whom will the information be used?
- In what form is information needed?
- What information is necessary for planning and controlling operations at different organisational levels?
- What information is needed to allocate resources?
- What information is needed to evaluate performance?
- What information is needed to improve products and services?
- What information is needed to improve customer relations?

12.4 INFORMATION SYSTEM BUILDING BLOCKS

Since the introduction of the first computer in 1944, computer technology has progressed at an astounding rate. Today's computers are significantly faster, smaller, more reliable and cheaper than earlier versions. A computer on its own has little value for users if they do not have complimentary products to use with it, e.g. applications to run on it or networks that connect it with other network devices. Software cannot be written if we do not have a model on which it can be based. Input, models, technology and databases therefore provide the building blocks for a computer-based information system. What technology is needed by an engineer who wants to make a drawing at home and then send it to somebody? Several options are available – some low-tech and others high-tech. One option may be to use a drawing board, paper and the postal service. Another method may be to use computer-aided drafting/design (CAD) software and send the finished drawing (stored as a file) to a colleague electronically. For this, the engineer will need a personal computer, CAD software, e-mail software, a modem and the services of an Internet service provider. The computer hardware will consist of a central processing unit (CPU), a hard disk drive, a keyboard, a mouse and a monitor. CAD software runs on the computer and provides the engineer with the means to draw various geometric shapes at specified positions and sizes. It helps the engineer to make a drawing of much higher precision than may have been possible using a pencil. Drawing takes less time than using mechanical methods, assuming the engineer is skilled in using the software. The modem connects the computer with the telephone line or cellular phone and the Internet service provider's connecting software will instruct the modem to dial the access number from where the electronic message will be transferred through another network such as the Internet. On the receiver's side, similar hardware and software will be needed to receive and read the electronic file. The affordability of not only the desktop computer but also the software that runs on it has resulted in a dramatic shift of information processing power away from the data-processing department to the desks of the engineer and project manager.

12.4.1 Hardware

Hardware resources include processors (minicomputers, microcomputers, mainframes), terminals (hard copy, cathode ray tubes (CRTs), graphic devices, storage devices (tape drives, disk drives, diskette drives) and communications networks.

12.4.2 Models and software

Some of the models used in businesses to change or process data into information on which various types of decisions can be based are as follows.

- **Book-keeping model:** This model sets up a procedure for classifying, recording and reporting an organisation's financial transactions (see chapter 14).
- **Cost-volume-profit model:** This model helps management to make decisions about selling price, sales volume, product mix, and so on (see chapter 15).
- **Budget and performance analysis model** (see chapter 14).
- **Project network model, e.g. PERT (programme evaluation and review technique):** This is an example of a network model used to plan and control projects (see chapter 13).

Various types of software exist. The operating system of a computer consists of an integrated system of programs that supervise the various components of the computer. Software languages (e.g. FORTRAN, Pascal, Basic, C, C++ and COBOL) allow engineers, technologists and scientists to write their own programs. Query languages (e.g. SQL) are used in conjunction with a database management system. Application and service software are directed to the job that needs to be done or the problem that needs to be solved. Many types of software packages can be bought off the shelf, from office applications like word processing, spreadsheets and database management, to specialist software such as CAD.

12.4.3 Telecommunications and networking

A network consist of nodes (e.g. a printer, a fax machine, a personal computer and a mainframe computer) that are linked by wire, cable, microwave, laser, optical fibre or satellite. Telecommunication networks link computers at places far away from each other and provide the means to move massive volumes of data in seconds. The advantage of telecommunications is that of compressing time and space, e.g. employers of international companies do not have to get on an aeroplane each time they want to meet, but can use video conferencing technologies instead.

Digital links, offered by an integrated services digital network (ISDN), for example, allow the integration of all forms of information communication (e.g. voice, fax, video, graphics, text) within a network. Some people refer to information and communication technology (ICT), due to the convergence of many information and communication technologies. Companies use networks for various purposes, e.g. efficiency can be increased if customers are linked to the order entry system of a supplier. Electronic data interchange (EDI) enables business activities and transactions such as ordering, billing and paying to be done between companies electronically. A provider company linked to customers has a competitive advantage because customers find it easier to place orders.

A local area network (LAN) is a private network that links various devices such as processors, printers and auxiliary storage within a company. It is usually installed in a single building. A wide area network (WAN) covers larger areas and therefore uses a combination of terrestrial transmitters and satellites.

Successful companies are those that consistently create new knowledge, disseminate it throughout the organisation, and quickly embody it in new technologies and products. In a brain-based economy, the organisations that emerge as winners will be those committed to spreading the flow of knowledge so that their people can quickly gather it and then do something interesting with it quickly (Harari, 1997:35). An intranet can help with the dissemination of knowledge through an organisation.

12.4.4 Databases and database management systems

Many users may require information simultaneously in a large, complex organisation. The database as the key building block of an organisation's information system should supply this information to managers, accountants, salespeople, manufacturing personnel, programmers and so on. There should be a strong correlation between the processing and decision-making needs of an organisation, and the structure and composition of its database. Computers are excellent devices for data storage and retrieval because of their speed and storage capacity. Entities for which data is stored may include employees, customers, an inventory item, an event, a software project or a profit centre. Each of these entities has certain attributes that database users may wish to record. For an inventory item, one probably wants to keep track of attributes such as the inventory number, description, size, price, unit of measure and quantity on hand. Table 12.3 is an example of how objects may be described. To the entity 'employee', attributes such as education, training and date of last promotion may be added (Burch & Grudnitski, 1989:397).

Table 12.3 Data attribute descriptors (Burch & Grudnitski, 1989:398)

Entity	Data attribute	Sample data
Customer	Customer number	509836
	Customer name	XYZ Ltd
	Credit limit	R450 000,00
Employee	Employee number	678936
	Employee name	Joe Soap
	Department	Engineering
	Hourly pay rate	R87,50
	Job classification	Fitter
Inventory item	Item number	AB112
	Size	16513FS
	Description	Motor tyre
	Price	R287,59

A database contains data only. Database management system (DBMS) software allows users to create, maintain and manipulate data records. The data in a database should be organised in a way that makes it possible to access specific information. In many computer applications, data in a database file is organised in records, with fields holding related data items. When the data is arranged in this way, the database is known as a relational database. Databases and a database management system are beneficial to organisations because:

- data is organised into a logical structure that allows multiple relationships to be defined between data entities;
- a DBMS usually contains a high-level query language that allows data to be obtained from the database without having to write an application program;
- data inconsistencies are less likely due to the way the database is designed; and

■ they provide ease of application development and management (Burch & Grudnitski, 1989:403).

12.5 USES, ADVANTAGES AND DISADVANTAGES OF INFORMATION SYSTEMS

Example 12.4 Software helps engineers innovate.

Engineers stumped by difficult design or manufacturing problems once turned to their smartest associates for guidance. Now they can rely on their computers and a program called IM-Phenomenon. The software consists of a database containing thousands of physical principles, natural laws, and effects, and .... A search method employs artificial intelligence to quickly examine this huge database and pull up several entries applying to the problem at hand (*Machine Design*, 1998).

12.5.1 Managing complexity

Fortunately for most motorists, they do not have to think or even know too much about the inner workings of a car in order to use it as a means of transport. They only have to know how to execute various procedures like starting, stopping, accelerating and so on. The same is true when it comes to using software applications. We do not have to think about all the complex issues that are dealt with inside the programs, but only how to use them correctly. However, it is important that we know the limitations and correct procedures for utilising software or a car.

12.5.2 Information systems make processes more efficient

It is easy to see that information technology contributes to greater efficiency when it is used to minimise uncreative, routine procedures and to reduce duplication. Word processing, spreadsheet software, CAD software, document delivery procedures and some expert system applications serve as examples. However, in some areas it may be difficult to determine the increase in efficiency as people tend to use information systems more once they are in place. The correct way to introduce information systems is not to duplicate traditional systems but to re-evaluate what is done (sometimes called business process re-engineering) (Bawden & Blakeman, 1990:127).

12.5.3 Faster product design and development

Suppliers need to be fast on their feet and supply superior products more quickly and effectively than ever before if they wish to stay in business. Customers demand more variety, more quickly and at a better price. The competitive environment in which suppliers of products and services operate has given rise to several rapid developments in product supply management (see example 12.5 on the next page). IT systems and software applications are playing an increasingly prominent role in these processes, e.g. electronic pre-assembly techniques assist manufacturers in designing cars for quicker assembly.

Example 12.5 Increasing model variety and decreasing volume per model in the automobile industry

The Chevrolet Impala was the best-selling automobile in the mid-1960s in the United States. Approximately 1,5 million units were sold per year of the platform on which it was based. In 1991 the best-selling automobile in the US was the Honda Accord, which sold about 400 000 units per year. Although the automobile market was significantly bigger in 1991 compared to the mid-1960s in the USA, the volume per model has dropped by a factor of about 4. In 1991 about 600 different automobile models were offered for sale in the US (Wheelwright & Clark, 1992:3).

12.5.4 Using IT to gain competitive advantage

Information technology (IT) is just one of hundreds of different types of technologies that could provide a competitive advantage to an organisation if it has access to technology that others do not have or if it uses existing technology in a new, innovative way to address the needs of customers. The difference between IT and most other technologies, e.g. biotechnology, however, is that IT is used in all types of organisational value chains, whether big or small (see figure 12.1).

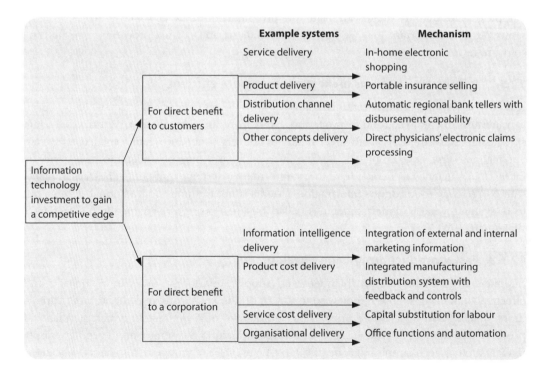

Figure 12.3 The information system as a weapon (Ives & Learmouth, 1984)

In chapter 17 the various competitive forces identified by Michael Porter are discussed: the power of buyers, the power of suppliers, the threat of new entrants and the threat of substitute products. Information technology can be used to influence all of these competitive forces. Information systems can control the power of buyers by increasing the cost of switching from one supplier to another, e.g. when American Hospital Supply placed on-line ordering entry terminals in hospitals, purchasing personnel in hospitals became familiar with the system and did not want to switch to alternative suppliers. Information systems can, however, also empower consumers and citizens. The Internet enables engineers to shop around by querying the databases of components offered by several suppliers until they arrive at the best prices (Schulteis & Sumner, 1992:12, 13) and allows the average citizen to disseminate information by means of blogs (a form of electronic personal journal with no editorial control and censorship beyond that of the owner of the blog).

12.5.5 Use computer software and systems with care

'When one automates a mess, one gets an automated mess' (Source unknown).

Just as bridges, buildings and excavations fail due to poor design, so do software applications. This may have relatively small or very serious consequences, ranging from the loss of data to the loss of lives. Software engineers, like all engineers, must design and develop software applications according to certain standards. However, it is just as important that the users of the application use it correctly for the purpose it was designed to serve. Users should also be aware of the limitations of each application and interpret outputs correctly.

Like the steam engine, the steel mill, the dynamo, the computer is an opportunity to be exploited, an immensely powerful extension of man's ingenuity and power in the service of his will. But it is also a source ... of concern. If we put the wrong things into it, if we select the wrong problems or state the right problems incorrectly, we will get unsatisfactory solutions. Perhaps the easiest way to put it is that in using the computer, man will get the answers he deserves to get (Morison, 1966:79).

Computer systems should also be secured against viruses to prevent the loss of data. High-tech criminals can steal substantial amounts of money and information if a system is not properly secured. Other security hazards include power and communication failures, fires, sabotage and natural disasters.

12.6 IT IN DESIGN AND PRODUCT DEVELOPMENT

Example 12.6

Aircraft manufacturer Boeing used three-dimensional interactive software and networked teams to create an electronic mock-up of the 777, virtually eliminating drawings and physical models, shaving off nearly a third of design time (*Time*, 1997).

Most engineering design is created with the aid of computers. Computer-aided design (also called computer-aided engineering, CAE) allows most design tasks to be completed quicker than by manual methods. CAE is essential for the design of very complex products, such as large integrated circuits (O'Connor & Galvin, 1997:72).

> **Example 12.7**
>
> ADAMS/Pre-aids engineering teams in creating virtual prototypes of a vehicle and enables them to test dynamic characteristics such as breaking and steering prior to building a physical prototype (*Automotive Engineering International*, 1998:40).

12.7 IT IN PRODUCTION AND MANUFACTURING

Manufacturers use information systems to co-ordinate and control a variety of operations. Many manufacturing companies have implemented material requirement planning (MRP) and manufacturing resource planning (MRPII) to control manufacturing operations.

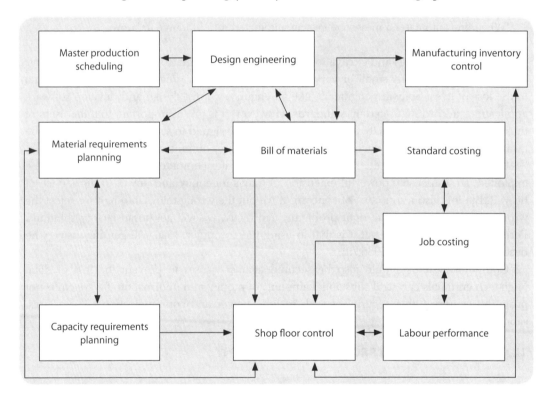

Figure 12.4 Schematic of an information system that supports the manufacturing function

12.8 ICT IN ENGINEERING MANAGEMENT

12.8.1 ICT in project management

Project management software is widely used to assist project managers with the development of project schedules and with project administration (Burke, 2003:321). The use of such software can be learned in a short period of time if one has a basic understanding of project management methodology and techniques (see chapter 13 for more information on project management). Project team members make extensive use of Information and Communiction Technologies (ICTs) such as e-mail, video conferencing and cellular phones because frequent communication is a necessity in such an environment where much co-ordination between different project contributors is required.

12.8.2 Management decision making

Information systems have changed the decision-making process of managers because of the former's ability to process and deliver information on which decision making is based. A manager may, for example, receive information on production quotas, orders, inventory levels, sales targets, accounts payable and many other factors, depending on his/her position in the organisation. Middle managers who are responsible for control will receive summarised data on a section's performance. Top management will receive consolidated information that will help them to answer 'What if?' questions. Some 30–80 per cent of the decisions at operational and middle management levels are programmable.

12.8.3 Classes of decisions

A distinction can be made between completely structured decisions and unstructured ones. The process of making a completely structured decision is algorithmic, whereas the process of making an unstructured decision is heuristic. Decision making is a stepwise process comprising three main phases:
- intelligence;
- analysis and design; and
- selection or choice.

12.8.4 Management information systems

A management information system (MIS) can be defined in terms of its application to the different classes of decisions. It makes structured decisions, e.g. a computer program can process inventory transactions and automatically reorder optimal replenishments. An MIS supports the process of making unstructured or semi-structured decisions by performing some of the phases of the decision-making process and providing supporting information for other phases, e.g. computer programs can report production cost over-runs, thus performing the intelligence phase mentioned in section 12.8.2, above. They can calculate the impact of various alternatives for solving the problems in the design phase, leaving a manager to complete the design phase and choose the best alternative.

Two components of the MIS can be identified:
- structured decision systems (SDS) that make the structured decisions; and
- decision support systems (DSS) that support unstructured and semi-structured decisions.

The primary purpose of a DSS is to provide the manager with the necessary information to make intelligent decisions. A DSS is a specialised MIS that supports a manager's skills at all stages of decision making – identifying the problem, choosing the relevant data, selecting the approach to be used in making the decision and evaluating alternative courses of action.

For example, a salesperson passes an order for a product that must be delivered in four months, time to the production department, where a computerised order-processing system reports that the order cannot be filled in the specified time. Instead of rejecting the order, the production manager uses a DSS to evaluate the consequences of refusing the order. A query language with a DBMS is used to check the mark-up on the product required and to determine whether the customer is a regular who places orders of high monetary value. The answers supplied by the system will help the production manager to determine losses if the order is rejected. He/She will also be able to study alternatives like delaying the order of a less important customer (Hussain, 1992:297).

12.8.5 Computer-assisted decision making

The objective of artificial intelligence is to enable computers to process information in the same way as a human being. This involves developing the capability of the computer to work in the same way a human brain would process information, draw analogies and solve puzzles. In other words, it involves making the computer 'think'. Since the computer is faster, more reliable and more objective, it should be able to make more effective decisions. Some programs have been developed to play chess, backgammon, make generalisations, draw inferences and make analogies. Although actual learning and true creativity are currently beyond the capability of computers, some people believe that they will be accomplished one day.

Expert systems, a form of artificial intelligence, can also assist decision makers. Such a computer program captures the knowledge of human experts in a knowledge base and is used to solve problems that would normally require human expertise. Decisions made by expert systems cannot be solved by conventional algorithms and programming. Similar to a decision support system, expert systems store and retrieve data, are interactive, manipulate data and develop models. However, they also diagnose problems, recommend alternative solutions, offer reasons for their diagnoses and recommendations, and even 'learn' from experience by adding information developed in solving similar problems to the current knowledge base. These exciting systems guide users through problems with a set of logical and orderly questions about a situation and draw conclusions based on the answers provided.

12.9 ENGINEERS AND THE INTERNET

No-one can ignore the opportunities that the Internet offers businesses as an additional distribution channel. The Internet enables the smallest business with a website to establish an international presence, as potential buyers may visit the site from all over the world. Examples exist of businesses that are not only using the Internet for marketing, but have developed a completely new business model based on it.

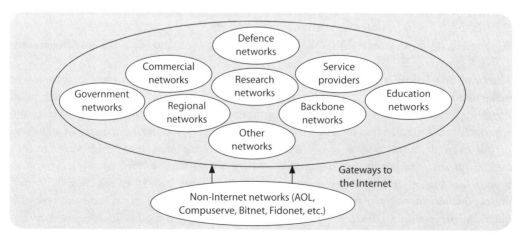

Figure 12.5 The Internet: A network of networks (Angell & Hessop, 1995: 2)

12.9.1 A new way of doing business

Amazon.com is one of the outstanding examples of how the Internet can be used for doing business. A large percentage of companies are already using the Internet for various purposes. Dell sells computers via the Net and General Electric procures goods over the Net. A number of companies use the Internet to distribute digital products such as music, e-books and software at virtually no (distribution) cost (see example 12.8).

Example 12.8 The Napster saga and the digital distribution of music

Although the Napster saga was brought to a close in a court of law in February 2001, it has challenged the long-held dominance of a small number of big record companies. Napster enabled millions of PC users to swap digitally encoded MP3 music files across the Internet for free by means of the file-sharing service that it provided. At the time it was stopped, Napster had 63 million members, which it gained over a relatively short period of time. The Napster phenomenon had shown that there is enormous consumer demand for online access to music. Digitally distributed music is very convenient (Cha & Rajgopal, 2004:378).

12.9.2 The Internet as source of business and technical information

The World Wide Web (WWW) provides access to multimedia and complex documents and databases. The Web is one of the most effective methods of providing information because of its visual impact and advanced features. Web application programs can access many of the Internet services. The Web can be used as a complete presentation medium for a company's corporate information or for information on all its products and services (Herbig & Hole, 1997:96).

Client	Client requests file by directing his/her request at a specific URL (e.g.www.unisa.ac.za)	Server
	Server sends file to client	

Figure 12.6 The document transfer process

12.9.3 The role of the Internet in research

The goal of the World Wide Web project, which was started in 1989 by Tim Berners-Lee at the CERN high-energy physics laboratory, was to find a way to share research and ideas with other employees and researchers scattered around the world. This goal will probably only be completely accomplished when all research papers and dissertations are available on the Net. Some universities already require that dissertations be submitted in a format for publication on the Internet.

12.9.4 Attractive characteristics of the Web

One attraction of the Web is its size. The Internet consists of thousands of networks that communicate with one another through the Internet Protocol (IP). The more connections there are, the more powerful and useful the network becomes. The larger it becomes, the more information is on offer and, therefore, the greater the attraction to surf the Net. The volume of people surfing the Net determines the size of a potential market or market segment.

A range of services can be offered via the Web – from providing potential customers with product information, to online shopping and education.

Companies already linked to the Internet receive the advantages of high-speed telecommunications and continuously evolving technology while learning invaluable lessons about the management of networked organisations. We are just beginning to understand the impact of networked communication on our daily lives and our way of doing business (Cronin, 1995:7).

Marketing executives are convinced that the Internet is a very exciting channel available for marketing products and services. The Internet is proving to be an outstanding marketing tool and has been called the most effective promotional tool of the century. It has become a place where any firm can open a website in an electronic mall and make its products available to millions of potential customers in a short period of time. The Web provides two of the most important aspects of modern marketing philosophies: the ability to target selected groups of buyers and to open interactive dialogue. The Web, unlike traditional marketing channels, is not a one-way communication between a vendor and a buyer. It provides a more intimate means of communication. This has the potential of getting new customers and keeping them interested (Herbig & Hole, 1997:95, 96). The challenge is to get your target market to visit your website and to design the website so that it will facilitate trading and dialogue.

12.9.5 Managing an Internet business

Management have to make several decisions about establishing a Web presence, e.g.:
- whether to buy and maintain a Web server or hire the space from a service provider;
- the type of information to be made available on the website;
- the range of services to be provided to clients;
- whether to design documents in-house or hire a company that specialises in website design; and
- how to secure the Web server.

Management cannot make these decisions without having a clear idea of what they want to achieve with their Web presence and how that fits in with the overall strategy of the business. Mary Cronin (1995:ix) states that developing a successful corporate strategy for the Internet and the World Wide Web involves more than keeping up with technical developments.

The greatest advantage comes from matching Internet capabilities with key opportunities for adding value to core functions like marketing, sales, customer support and information management.

Self-assessment

12.1   Describe the information you need daily, weekly and monthly.

12.2   Define data, information, software, hardware, network and Internet.

12.3   Explain why an engineer is described as a knowledge worker.

12.4   Explain why and how software applications and computer systems are useful to engineers.

12.5   List some issues engineers should be aware of when they use information systems.

12.6   Describe the information you would need to make a good decision in the following situations:

   a)   purchasing a personal computer, car or bicycle

   b)   deciding which career to follow.

12.7   List questions that could be helpful for identifying the information needs of a user.

12.8   Describe the purpose of a decision support system.

12.9   Explain how the Internet is changing the way business is done.

12.10 What is the main function of a database management system?

12.11 Briefly explain how information technology and information systems can be used to improve services, competitiveness and efficiency.

Mini-projects

A.   Describe what information you need to be effective in your current position. Explain where you get this information and whose responsibility it is to provide you with it. What type of information do you provide to other decision makers in your organisation? Why do you have to provide them with this information? In what format do you provide information to others inside and outside your organisation?

B.   You want to establish and manage a small business from home. You plan to build models that will simulate various clients' business processes and then offer them advice on how to optimise these. You want to design and establish your own website to promote your business' services. What hardware and software will you need to offer these services to your customers? Establish the cost of purchasing the hardware and software and acquiring the various services.

C.   Amazon.com is one of the good examples of how the Internet can be used to do business. Study the company and describe how its management uses the Internet to do business. Compare it with a traditional bookshop in terms of method of operation, market reach and the way information is made available to customers. Give an overview of all the hardware and software that Amazon.com has in place to do business.

References

Ahituv, N. & Neumann, S. 1986. *Principles of Information Systems Management*. Iowa: Brown.

Angell, D. & Hessop, B.D. 1995. *The Internet Business Companion: Growing Your Business in the Electronic Age*. Reading, Mass.: Addison-Wesley.

Automotive Engineering International. 1998. 'Vehicle simulation tool designed for virtual prototyping.' March.

Bawden, D. & Blakeman, K. 1990. *IT Strategies for Information Management*. London: Butterworths.

Boyd, R. & Richerson, P.J. 2004. *Not by Genes Alone: How Culture Transformed Human Evolution*. Chicago: University of Chicago Press.

Burch, J.G. & Grudnitski, G. 1989. *Information Systems Theory and Practice*. 5th edition. New York: John Wiley.

Burke, R. 2003. *Project Management: Planning and Control Techniques*. 4th edition. Cape Town: Burke.

Cha, B. & Rajgopal, K. 2004. 'Digital distribution and the music industry in 2001.' In Burgelman, R.A., Christensen, C.M. & Wheelwright, S.C. *Strategic Management of Technology and Innovation*. Boston: McGraw-Hill, pp. 378–98.

Cronin, M.J. 1995. *Doing More Business on the Internet: How the Electronic Highway Is Transforming American Companies*. New York: Van Nostrand Reinhold.

Drucker, P.F. 1985. *Innovation and Entrepreneurship: Practice and Principles*. London: Heinemann.

Duffy, N.M. & Assad, M.G. 1980. *Information Management: An Executive Approach*. Oxford: Oxford University Press.

Engineering News. 1999. 'The sky is the limit for Internet business.' 21 May, p. 14.

Harari, O. 1997. 'Flood your organisation with knowledge.' *Management Review*, November, p. 35.

Herbig, P. & Hole, B. 1997. 'Internet the marketing challenge of the twentieth century.' *Internet Research*, 7 (2), pp. 95, 96.

Hussain, D.S. 1992. *Information Management: Organization Management and Control of Computer Processing*. New York: Prentice-Hall.

Ives, B. & Learmouth, G. 1984. 'The information system as a competitive weapon.' *Communications of the ACM*, 27 (2), pp. 1193–201.

Knott-Craig, A. 2005. 'The future of Telecoms in South Africa.' Talk by CEO of Vodacom at Unisa's SBL, 5 May.

Maasdorp, E.F.De V. & Van Vuuren, J.J. 1998. 'Information management.' In Marx, S., Van Rooyen, D.C., Bosch, J.K. & Reynders, H.J.J. (eds). *Business Management*. 2nd edition. Pretoria: Van Schaik.

Machine Design. 1998. 'Software helps engineers innovate.' 5 February, p. 23.

Management Review. 1998. 'They had the facts.' October, p. 24.Morison, E.E. 1966. *Men, Machines and Modern Times*. Cambridge, Mass.: MIT Press.

Nel, W.P. 2005. 'Technology and OD.' In Botha, E. & Meyer, M. (eds). *Organisational Development and Transformation*. 2nd edition. Durban: Butterworths.

O'Brien, J.A. 2003. *Management Information Systems: Managing Information Technology in the E-Business Enterprise*. 5th edition. Boston: McGraw-Hill Irwin.

O'Connor, J. & Galvin, E. 1997. *Marketing and Information Technology: The Strategic Application of IT in Marketing*. London: Pitman.

Owen, D. & Kruse, G. 1997. 'Follow the customer.' *Manufacturing Engineer*, April, p. 65.

PC Magazine. 1997. 'Food for thought about Internet advertising.' 18 November, p. 30.

Petroski, H. 1985. *To Engineer Is Human*. London: St Martin's Press.

Salmon, W.I. 1994. *Structures and Abstractions: An Introduction to Computer Science*. 2nd edition. Burr Ridge: Irwin.

Schulteis, R. & Sumner, M. 1992. *Management Information Systems*. 2nd edition. Boston: Irwin.

13 Principles of Project Management

Ad Sparrius

Study objectives

After studying this chapter, you should be able to:
- develop a life cycle description for a project
- compile a project brief
- develop a work breakdown structure
- allocate responsibility for each task from the work breakdown structure

The aim of this chapter is to obtain a good understanding of major project management issues.

13.1 INTRODUCTION

The Project Management Institute (PMI) is an international association of project managers. It is based in America, but also has a South African chapter. PMI has set itself the objective of professionalising project management world-wide. As an interim goal, it has defined and published *A Guide to the Project Management Body of Knowledge* (2004). This chapter will follow the outline of this title, although it goes very much beyond it.

13.2 WHAT IS A PROJECT?

A project is a unique and complex process consisting of interrelated tasks performed by various contributors to create a specific result within a well-defined schedule and a limited budget. This definition explicitly identifies the three constraints of a project: objective, schedule and budget. The objective is always too ambitious; and there is never enough time or money. Each project needs to be planned, measured on a continuous basis, and its success defined in terms of these three constraints.

Remember that each project is unique – it has never happened before and will never happen again. We cannot take the plan for a previous project and simply reuse it.

13.2.1 The project brief

Any project is performed to satisfy a customer's requirements. If there is no customer, there is no project. This is a crucial point of departure for project management. The objective, budget and schedule are aimed at creating a deliverable for the customer. But the deliverable itself is only a means to an end; the end is customer satisfaction, and the objective, budget and schedule merely form the roadmap to get to the deliverable.

Customer requirements should always be stated in terms of those business processes applicable to the customer. For example, the needs of a military customer might be stated in the form of military missions – destroy a given target (say a building) 1 000 km away within the next two hours. For a transportation customer, the requirement might be: transport two tons of cargo packed in standard ISO containers from point A to point B within three days.

Customer requirements are stated in the project brief. This does not specify the solution; it merely states the problem. There is a major difference between a statement of the problem and a specification of its solution. (The project brief may also define an opportunity to be exploited.) For example, the best solution for the military customer could be a surface-to-surface missile system, not an aircraft. If the project brief specified the solution, then all opportunities for trade-offs and optimisation are pre-empted. And in any event, does the customer have the skills to perform such trade-offs?

The project brief also defines the constraints of the project. For example, only so much money is available and the solution must be ready within three months. The project stakeholders may also be identified. Who is the customer? The customer is that individual or small group of individuals who sanctions the project brief, who determines whether the project has satisfied the customer requirements and who pays.

13.3 PROJECT STAKEHOLDERS

The way a project is defined means that only the customer will determine whether or not the project is successful. The systems approach has significantly influenced modern management thinking. The main contribution of system thinking applied to management has been its focus on an organisation's environment and the relationships between that organisation and its environment.

System thinking leads into the idea of marketing and public relations as mechanisms for getting to grips with the environment. The concept of strategy as a framework for the relationship between an organisation and its environment is a natural consequence. Strategy is not a long-term plan, but the conceptual framework on which such a plan might be based. Although some question the usefulness of a strategy, there is little doubt that a preoccupation with relationships with the environment will continue. Can system thinking be applied to project management? Is it relevant? Project management should be equally concerned with its environment and its relationships with it.

A fundamentally important concept is that of a project stakeholder. A stakeholder is someone with a real or imagined interest in the project or its outcome. For example, several years ago, the Botswana government started a project to dredge the Boro channel in the Okavango Delta for water supply to downstream users at Orapa. The project was dead on arrival – halted by Greenpeace. People will not meekly ask permission to become project stakeholders; they'll just do it. Nobody grants rights to stakeholders, as they are inherent in a democratic society.

Each project has its own unique set of stakeholders. Stakeholders may include internal or external individuals and organisations. They may be actively involved in the project, for instance the project team, as well as suppliers and contractors. They may also include those whose interests will be affected by the project, for instance owners, customers, financiers, concerned individuals and society at large.

There is a symbiotic relationship between a project and its stakeholders. A project cannot exist without its stakeholders and, conversely, stakeholders rely, to some extent, on the project for their existence. Is the customer a stakeholder? Of course, but with the important condition that the customer also pays for the project.

13.3.1 Stakeholders and project success

Traditionally, project success has meant satisfying the customer's requirements on specification, on schedule and on budget. A successful project satisfies the expectations of all its stakeholders. The project manager should thus identify all stakeholders, determine their objectives and expectations, and incorporate these into the project plan as far as is reasonably possible. It may be necessary to change the original project objectives, but a better project will almost always materialise. Unfortunately, stakeholders do not initially identify themselves and disclose their expectations. The identification of stakeholders and their expectations should be one of the major activities during the project's concept phase. For example, a public participation forum might need to be established to canvass stakeholders. Remember that an initial lack of interest does not mean that no stakeholders exist or that they agree with the project. Silence does not always mean consent.

Can all stakeholders' requirements be satisfied or are project objectives and stakeholder expectations in conflict? Since the competition is also a stakeholder who wishes to destroy the project, it is safe to say that it will seldom be possible to completely satisfy all stakeholders. Irreconcilable conflicts will probably exist. The resulting project problems depend on the relative power of the stakeholders and their willingness to exert that power. None of the stakeholders' stakes is undesirable in isolation, but their aggregate may leave the project manager with little freedom in decision making. Demands may be in conflict and most may affect the project's cost and schedule constraints.

Is the customer a mere stakeholder or does he/she have more rights? The customer should probably be first among equals and, if a trade-off needs to be made, the customer's objectives should receive priority. This is a highly contentious matter.

13.3.2 Project management implications of stakeholders

Project management in the presence of stakeholders will inevitably become more 'political'. Project managers will have to take the needs of their various stakeholders into account and attempt to balance their interests. Fortunately, project quality (the degree to which the project satisfies its stakeholders) will almost automatically increase. By forcing the project to justify itself in the glare of public scrutiny, stakeholders will undoubtedly increase the need for meticulous homework, but quality will improve immeasurably.

If a project cannot convince sceptical stakeholders to support it, it probably does not deserve support anyway. This places the onus on the customer to initiate public debate with potential stakeholders long before the project is formally launched.

13.3.3 Stakeholder management

A discussion on stakeholder management should be based on two points.
- A stakeholder does not need permission to be a stakeholder. It is through free choice that someone becomes a stakeholder.

■ The mindset and values that motivate and justify a stakeholder's behaviour may be radically different from those for the project.

How do stakeholders feel? Consider the following hypothetical situation: To your immense joy you have finally managed to buy your own house. You have had to stretch yourself financially to get a bond, but you've made it! Congratulations! Then Eskom decides, in its infinite wisdom, to build a nuclear power station down the road. The value of your house plummets – you can't give it away. Overnight you will become a stakeholder, and you will use all the tricks in the book to prevent the nuclear power station from being built. After all, as a tax-paying citizen, you have rights too!

Are stakeholders detrimental to a project?

Is this always the case? It has already been argued that stakeholder satisfaction defines project success. Stakeholder involvement is not inevitably disruptive, but can be constructive by significantly improving the project quality at virtually no extra cost.

Public participation

Stakeholder management starts with stakeholder engagement. Engaging stakeholders in the project's decision-making process allows them to become part of it. Project deliberations should be open to and influenced by stakeholder concerns. This may mean that the project's need-identification and concept phases will have to be reopened. This could delay the project and, at worst, completely derail it. Although this sounds dangerous, especially on a fast-track project, it is much better to face such problems earlier rather than later. Ignoring problems will not make them go away. Stakeholder engagement is commonly known as public participation and includes:

■ informing the public of the project;
■ obtaining the public's consent for the project;
■ allowing the public to make decisions, e.g. by selecting a solution from a range of alternatives; and
■ inviting the public to initiate projects.

Public participation is not a single event – it is an open-ended and protracted process. Setting agendas unilaterally and expecting that earlier decisions will simply be rubber-stamped is patronising and will inevitably be counterproductive. Don't try to tell stakeholders what their needs are. Initial mistrust might require a neutral facilitator until the details of the public participation process have been agreed upon by all.

13.4 THE PROJECT LIFE CYCLE

Projects have beginnings, middle periods and endings. This may seem self-evident, but it is not trivial if you are a project manager. Where you are in the project life cycle determines what you should be doing and what alternatives are open to you (Frame, 1995).

The project life cycle is fundamentally important to project management and dominates the selection and implementation of the overall project strategy of a project. Most project management problems are caused by not understanding the project life cycle.

13.4.1 The iron law of a life cycle

A project's limited duration implies a distinct project life cycle. All life cycles, without exception, are subject to the iron law of a life cycle: problems downstream are symptoms of neglect upstream. In other words, the problems you experience today are symptoms of things that you neglected earlier. The corollary of the iron law is: upstream problems can only be solved upstream. In other words, issues such as 'Who is the customer?' and 'What are the customer's needs?' should be solved up-front. If you neglect some issues, and only find out about them later, the only solution is to reset the clock and attempt to deal with them immediately. Since the project will have progressed considerably, the constraints on solving the issue are significant and it may turn out that the problem is now impossible to solve. For example, if the project manager of a new product development project neglected to define the project's customers, he/she may find it impossible to resolve this just before market introduction. It is well worth developing a life cycle that emphasises which things need to be done when.

13.4.2 The project life cycle

According to the PMI, the four generic phases of the project life cycle are: *concept, definition* (also called the *design phase*), *execution or implementation*, and *close-out* (also called the *handover phase*). Collectively, these project phases are known as the project life cycle and each requires tight management control. Each phase has a particular intent as far as the nature of work is concerned (see figure 13.1). The project is launched with a project brief, which is a statement of the problem experienced by the customer. Remember, the project brief does not specify the solution to the problem; it merely states the problem and defines the constraints to the solution. In the concept phase, this problem statement is transformed into the solution specification. In the definition phase, the solution specification is transformed into a detailed plan to implement it. The plan is executed in the implementation phase.

Concept phase	Definition phase	Implementation phase	Close-out phase

- **Concept phase:** Clarify customer requirements as defined in the project brief; identify all stakeholders and their expectations; finalise the project objectives; define key performance indicators to quantitatively measure project performance and completion; generate and evaluate alternative solutions; determine the best solution and assess its risk; develop a detailed plan for the definition phase and a preliminary plan for the rest of the life; and prepare a provisional baseline for the decision-making milestone meeting.

- **Definition phase:** Develop the work breakdown structure for the imple mentation phase; determine which contractor will perform each task and obtain cost and schedule estimates; develop a detailed plan for the implementation phase and a preliminary plan for the rest of the life; and prepare a provisional baseline for the next decision-making milestone meeting.

- **Implementation phase:** Contract out or contract in to perform each task of the implementation phase work breakdown structure; monitor and control project performance; and launch corrective action whenever needed.

■ **Close-out phase:** Acceptance test all deliverables; transfer the life cycle management responsibility to the owner; close project accounts; reassign the project team; and conduct a project post mortem.

Figure 13.1 The phases of a project and the issues to be resolved in each phase

Different industries traditionally use different names for very similar project phases. For example, in construction, one might name the phases: feasibility, preliminary design, detailed design and construction, and commissioning. These phases map directly onto the phases in figure 13.1. Remember to focus on the intent rather than on the name of each phase.

13.4.3 Decision-making milestones

How does the customer keep control over the project? The project usually builds up a momentum of its own, and it is very difficult to redirect or stop it. The solution is to demarcate the project phases with decision-making milestones where the customer must make a crucial decision (see figure 13.2).

At each milestone, two decisions need to be made by the customer. The words 'authenticate' and 'authorise' are used to describe the situation. Authenticate means that the customer accepts the recommendations of the previous project phase, and authorise means that the customer permits the next phase to start and for money to be spent. These two aspects of each decision-making milestone are always present.

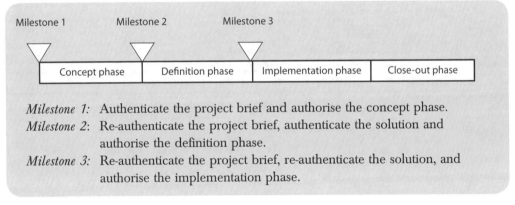

Milestone 1: Authenticate the project brief and authorise the concept phase.
Milestone 2: Re-authenticate the project brief, authenticate the solution and authorise the definition phase.
Milestone 3: Re-authenticate the project brief, re-authenticate the solution, and authorise the implementation phase.

Figure 13.2 Decision-making milestones

The decision at each milestone has one of three possible outcomes: stop; or continue; or repeat the previous phase, subject to constraints or hold.

At Milestone 1, the customer reviews the proposed project brief and authenticates that it truly reflects the problem. Authorisation is given for the concept phase to start, based on the detailed plan for this phase. This detailed plan for the concept phase contains all the following: a work breakdown structure, a responsibility assignment, a list of deliverables, a list of receivables, a budget and a schedule. At Milestone 2, the proposed specification of the solution is available. The proposed solution is authenticated. For example, does it fit in with the customer's strategy and does it have priority? Authorisation is given for the definition phase to start, based on a detailed plan for this phase. A preliminary plan for the rest of the life cycle is also made available – not for authorisation, but merely for informed consent. For instance, should the operators of the plant need new skills, then that should be made visible. The customer should never be able to blame the project for keeping important issues invisible. At Milestone 3, the implementation phase plan is ready. The plan of what the customer wants is authenticated. Are the necessary resources available? If so, the implementation phase is authorised to start. Again, a preliminary plan for the rest of the life cycle should be made available – not for authorisation, but merely for informed consent.

These decision-making milestones enable the customer to retain control over the project. The nature of the decision at each milestone is quite different, but logical. At Milestone 1 the problem statement is endorsed. At Milestone 2 the solution specification is authenticated. At Milestone 3 the detailed implementation plan is approved. The tasks in the various phases of the project life cycle are aimed at generating the information on which the customer can base these decisions.

13.4.4 Baselines

What does it actually mean when a customer authorises a plan to proceed? Essentially, when a customer authorises the next phase, the plan for that phase is *baselined*, i.e. the plan is placed under formal change control. This means that any change to the plan needs to be approved by all the parties concerned: the customer, the project team and all affected contractors. The baseline acts as a point of departure for both performance measurement and for change control. Each decision-making milestone thus corresponds with a baseline (see figure 13.3).

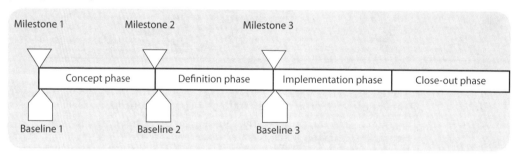

Figure 13.3 The integrated project life cycle

13.4.5 The project life cycle as an integrated package deal

The project life cycle comprises activity phases, decision-making milestones and baselines. Seeing the project life cycle as merely some phases is nonsense. What is the objective of each phase? Clearly each phase is aimed at generating information on which the customer can base the next decision. If the information already exists, there is no need for that phase.

13.4.6 Phases and processes

Some project managers use the phases: feasibility, planning, implementation and close-out. However, planning is an end-to-end process that occurs throughout the project, albeit it will vary. It is important to distinguish a life cycle phase from a process that occurs throughout the life cycle. The terms used here are the most descriptive for the objectives of each phase.

13.4.7 Risk management

It is fairly evident that the risk at Milestone 1 is much higher than the risk at Milestone 3. In this context, 'risk' means the certainty of the project outcome – not the amount of money at risk. The higher the uncertainty, the higher the risk. The project life cycle is a sophisticated risk-management strategy – the first tool in the risk manager's toolbox.

The project life cycle model is at heart a risk-management model. In an ideal world, the project would start with the implementation phase. An *uncertainty* is a lack of information that prevents the start of implementation. Risk is directly related to uncertainty – more uncertainty means more risk. As a little money is invested, uncertainty will diminish and risk will consequently decrease. One should only invest substantial money when project risk has been made acceptably low.

The typical uncertainties in a project are defined in figure 13.4. The logical sequence in which those uncertainties need to be resolved is also shown – first things first. It is pointless worrying about whether the solution is acceptable if the nature of the problem is uncertain. The milestone at which each uncertainty is eliminated is also shown.

Typical uncertainty	Sequence	Milestone
Who is the customer? What are the customer's needs? What are the constraints on the project?	1	1
What is the best solution to the problem? Is that solution feasible? Practical? What will its environmental impact be? Will the stakeholders accept it?	2	2
What will the implementation phase cost? Duration? Risk? Is the project still feasible?	3	3

Figure 13.4 Uncertainties in a project

The management construct to eliminate an uncertainty is shown in figure 13.5.

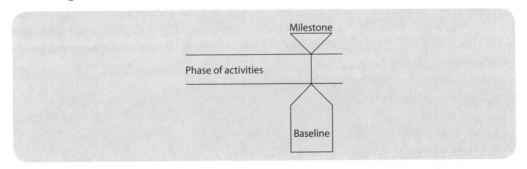

Figure 13.5 The management construct to eliminate an uncertainty

The phase generates the information that populates the baseline, and the baseline is authenticated and authorised by the customer at the milestone decision. After authentication and authorisation, there is no further argument – the uncertainty has been eliminated. If the decision turns out to be incorrect, then it can always be rectified, since the baseline is under change control.

The sequence of the uncertainties determines the sequence and nature of the milestone decisions and baselines. The project life cycle is then merely the concatenation of the management constructs. Since all projects have the uncertainties defined in figure 13.4, the project life cycle is as discussed earlier. However, let us assume that powerful stakeholders may question the environmental impact of the solution. Then it will be worthwhile to insert an additional milestone into the project life cycle where those stakeholders sign off on the environmental acceptability of the solution, say Milestone 2A.

Since each project is unique, including in its uncertainties, each project will have its own life cycle. Nevertheless, most projects follow the traditional life cycle. *It is impossible to plan a project without first determining the most appropriate life cycle.* Determine the project life cycle as follows.

- Explicitly identify all uncertainties.
- In which logical sequence should those uncertainties be eliminated? This defines the decision-making milestones. Name each phase in terminology familiar to the customer.
- What information is needed to eliminate each uncertainty? Which tasks will be needed to obtain the information to eliminate each uncertainty? This determines the nature of the tasks in each phase and is the start of the work breakdown structure for each phase.

The decision about whether the project has progressed sufficiently to schedule the next decision-making milestone is ultimately determined by the risk propensity of the customer. For example, if the customer is willing to accept huge risks, Milestone 3 could be scheduled earlier.

13.4.8 A slight contradiction

There is a minor contradiction in the life cycle model outlined above. Project work needs to happen before the project actually exists. The project brief and a plan for the concept phase must be created before there is a project.

13.4.9 Tailoring

The project life cycle model needs to be carefully tailored and customised for each particular situation. For example, Boeing uses three phases in an aircraft development project (Simpson *et al.*, 1988): project definition, cost definition and production. Seven milestone decisions are used in the project definition phase:

1. authorisation of start of project definition phase
2. market requirements review
3. project review
4. preparation for initial airline contacts
5. approval of market analysis
6. configuration selection
7. authorisation of start of cost definition phase.

13.5 INTEGRATIVE MANAGEMENT

Do all projects justify project management? Many more issues are associated with project management than its definition implies.

13.5.1 The challenges of integrative management

The main issue of project management is integrative management. There are several interpretations of integrative management, each one equally valid.

- The project manager should integrate the project's schedule, cost and objective, while satisfying the customer's stated and implied requirements, e.g. what is the implication of a three-month delay on a project's cost if the objective were held constant?
- The project manager should integrate the contributions of each participant in the project, e.g. a problem experienced by contributor A may best be solved by contributor B.
- The project manager should integrate into the project plan all those life cycle issues that may only materialise much later, e.g. the disposal of toxic waste should be a major design concern from the start.
- The project manager should integrate the project's results into the customer's organisation, e.g. in the construction of a new power station, the project should include the development of new operating procedures and the appropriate training of operators. In development circles, this issue is usually known as 'capacity building'.
- The project manager should integrate all stakeholders' concerns into the project plan as far as is reasonably possible.

Every organisation faces integrative management issues, e.g. specialist activities in various sections need to be carefully co-ordinated. However, in a project, these integration issues are much more severe. Often the contributors to the project are scattered throughout various organisations. Integrative management is the key issue of a project, but how is this achieved?

13.5.2 Integrative management mechanisms

Integrative mechanisms that can be adopted in any organisation include:

- hierarchical referral to the 'boss'. This is the most powerful and the default mechanism. Its main disadvantage is limited access to that boss. In a project, there may be no single individual who acts as boss to all contributors and the customer may not wish to fulfil this role;
- rules and procedures. The disadvantage of this mechanism is that rules and procedures are often inflexible. It is difficult to use a standard operating procedure for a project that is unique;
- planning, including scheduling and budgeting. If the plan were to actually materialise, then planning might be quite adequate. However, since Mother Nature does not read plans, they are seldom precisely achieved;
- direct contact between contributors via ad hoc or regularly scheduled meetings;
- a full-time liaison person, such as a project co-ordinator. However, little or no authority is usually delegated to a co-ordinator;
- an integrator person, a project manager, to whom authority has been delegated, that approximately matches accountability; and
- additional people who may be assigned to the project manager to create a project office.

Each integrative mechanism is progressively more powerful – but also more complicated and expensive – than the previous one. Integrative mechanisms that are lower on the list can be added to, not substituted for, those higher on the list. For example, it is pointless to appoint a project manager if the organisation does not use the approach of planning, measuring against that plan and then holding people accountable for its achievement. Only once this approach is in place and turns out to be not powerful enough will it be useful to use another instrument lower on the list.

13.5.3 The project manager as the single point of integrative management

The project manager is primarily an integrative management mechanism to manage a unique and complex project. The project manager should be a single point of integrative responsibility who makes extensive use of all other integrating mechanisms on the list (Archibald, 1976). The first characteristic associated with project management is that of integrative management. The second equally important issue is that the project manager is accountable for the tasks of all contributors (i.e. is responsible for all integrative management for the project) without having direct line authority over them.

13.5.4 The project manager needs integrative management tools

If the project manager's primary role is to be the single point of integrative management, then integrative tools are needed. Not surprisingly, information technology is fundamentally important in developing these integrative tools. The main integrative management tools are:

- traditional project planning tools such as work breakdown structures, statements of work, linear responsibility charts, scheduling and budgeting; and
- traditional project control tools such as earned value management.

Project management is a powerful method for achieving integration. Planning, with or without software tools, is another. It would be unwise to implement project management without empowering the project manager with all other available integrative mechanisms.

13.6 THE WORK BREAKDOWN STRUCTURE AS A CONSOLIDATED MANAGEMENT FRAMEWORK

Integrative planning based on the work breakdown structure (WBS) should be the project manager's main management instrument. A WBS is a system-oriented breakdown of all work activities required for the execution of a project phase. The WBS is a task inventory that forms a consolidated management framework for:

- defining all work tasks needed to accomplish that project phase;
- assigning responsibility for performing each task;
- defining the deliverables and receivables for each task;
- planning the cost and time required for each task;
- measuring the cost, schedule and technical performance while performing each task; and
- reporting and summarising the progress of each task of the entire project.

Figure 13.6 on the next page shows a simple WBS for a camping trip. Note the following.

- The WBS should be action oriented. Start each WBS task with a verb to clearly define the activity to be accomplished.

- The WBS should be exhaustive. All tasks required to accomplish the project should appear on the WBS.
- The tasks should be mutually exclusive (not overlap on work content).
- The WBS tasks are not listed in any specific sequence. Task schedules and budgets can be prepared after the task inventory has been completed.
- The tasks on this WBS may look obvious. If you cannot design a WBS, you know too little about the project to be able to accomplish it or even to manage it!

Project brief: Try out the new 4 × 4 off-road vehicle by means of a one-week camping trip in the Etosha game reserve in Namibia. The entire family will go on the trip (husband, wife, toddler and baby), together with three other couples and their children. The trip has been scheduled for the second week in August. Additional costs over and above staying at home include fuel, some food and camping site costs, and are covered by the holiday budget.

1. Service and check out vehicle.
2. Purchase vehicle consumables, spares and repair parts.
3. Obtain weather forecasts, plan route, obtain maps and verify road conditions.
4. Check out camping equipment and furniture, and if necessary purchase new camping equipment.
5. Reserve camp sites and determine availability of drinking water.
6. Prepare daily activity schedules.
7. Prepare a checklist for personal clothing and equipment for normal, sleeping, sport and leisure activities, as well as for abnormal conditions.
8. Notify neighbours of absence and provide contact details.
9. Arrange garden maintenance during absence.
10. Arrange pet care during absence.
11. Prepare first aid kit.
12. Plan daily menus, and prepare detailed list of food supplies and ingredients.
13. Check cooking utensils and equipment, and purchase new if necessary.
14. Purchase non-perishable food.

Figure 13.6 Camping trip work breakdown structure

13.6.1 Using a product breakdown structure

To develop a WBS, just sit down in front of your PC, sweat some blood and, lo and behold, there it is! Fortunately, if your project has a tangible deliverable, an easier method may be used (see figure 13.7 on the next page). Use a product breakdown structure (PBS) to visualise the project's deliverables and to identify all tasks on the WBS. A PBS may be used in the following situations:

- when the project generates a tangible deliverable, such as in figure 13.7;
- whenever the project's end result may be similarly structured, e.g. for the Olympic Games, the master schedule of sport events may act as the backbone for the WBS, and for a music festival, the music programme could be used (see figure 13.8); and

■ whenever a convenient framework exists around which the WBS can be developed, e.g. the quality assurance tasks for a project may be structured around the requirements from the appropriate ISO 9000 series standard.

Project brief: Design and construct a conference centre for the Polokwane City Council by December 2005. Two conferences, each with 100 participants, need to be accommodated simultaneously. Cost constraint: R150m.

Product breakdown structure
Conference centre
Building
Parking area
Garden
Outdoor entertainment area

Work breakdown structure
1:    Design conference centre
1.1: Design architectural concept and corporate image
1.2: Design garden and outdoor entertainment area
1.3: Design building and parking area
1.4: Complete drawings for all construction work
1.5: Obtain approval from owner
2:    Construct building
2.1: Level terrain, dig foundation, pour concrete and build walls
2.2: Wire electrical cables and install plumbing
2.3: Build roof structure and construct roof
2.4: Complete plumbing, electrical and carpentry
2.5: Install furniture and equipment
3:    Construct outdoor entertainment area
4:    Build parking area
5:    Landscape garden, and plant trees and shrubs
6:    Commission conference centre
6.1: Conduct trial seminar
6.2: Identify and rectify deficiencies

Figure 13.7 A PBS acting as skeleton for the WBS

7: Perform project management*Project brief:* Present a music festival programme in La Scala for the corporate sponsors on 6, 7, 20 and 21 June and 4 and 5 July 2008, in accordance with the approved budget

...

Performance 14: Antonín Dvořák: Concerto for violin and orchestra in A minor, Opus 53, 1879.

Work breakdown structure

Task 14.1: Confirm dates and times of all performances and rehearsals. Decide on which supporting music will be used, including encore. Obtain full musical scores for entire orchestra.

Task 14.2: Define shortlist of soloists. Obtain approval to negotiate.

Task 14.3: Negotiate with potential soloist on the dates for the performance and rehearsals. Negotiate travel and subsistence costs, and performance fees. Develop soloist's contract. Have it signed by all parties. Arrange all appropriate insurances.

Figure 13.8 A music programme acting as a self-evident backbone for a WBS

13.6.2 What a work breakdown structure is not

Do not confuse a work breakdown structure with other planning tools.

- A WBS is not a PBS. The WBS defines all work tasks associated with each element from the PBS. One element from the PBS may have many associated tasks (see figure 13.9). Construct the PBS before developing its WBS. Start each WBS task with a verb to ensure that it describes work.
- A WBS is not a time-phased schedule of work tasks, although each task will be planned on a network. Don't use time logic when designing the WBS. Simply list all tasks.
- A WBS is not an organisation chart of project people, although it is used to assign responsibility for each task. Don't use people logic when designing the WBS. The idea is not to provide work for people, but to accomplish the project.

13.6.3 Work breakdown structure numbering

Each WBS task needs to be uniquely identified by a number or code (see figure 13.9). This identifier must enable data summarisation and reporting – also known as 'roll up' – to any required higher WBS level. The WBS task identifier must be usable for data filtering and sorting. Note that these requirements are valid not only for computerised systems, but also for a manual system.

Project brief: Develop a new passenger vehicle in the R150 000 price class by December 2005.

Product breakdown structure
Vehicle
Engine
Gearbox
...
Work breakdown structure
Task 1: Develop a gearbox in accordance with specification G 238/9.
Task 2: Conduct a critical design review on the gearbox.
Task 3: Manufacture two prototypes of the gearbox.
Task 4: Conduct a reliability growth programme on one gearbox prototype.
Task 5: Conduct test and evaluation of the other prototype in accordance with test procedure TP 238-79.
Task 6: Develop a gearbox technical and maintenance manual in accordance with TMM 238-65.

Figure 13.9 Six WBS tasks are associated with the one PBS element (gearbox)

13.6.4 Project life cycle

Should there be a single WBS for the project? What about the project life cycle? Consider a project in its definition phase. The work in the definition phase WBS has been authorised and is being implemented. However, the implementation phase WBS is being used for planning and has not yet been authorised. The definition phase and the implementation phase are demarcated by a decision-making milestone. If the results of the definition phase are fine, the implementation phase may be started by authorising its WBS. In other words, there may be several project WBSs – the current WBS for execution and the future WBSs for planning.

13.6.5 Using a WBS

Let us further discuss the uses of a WBS.
- **Define work packages:** The WBS divides the project into manageable tasks, or work packages, that can easily be contracted out.
- **Responsibility assignment:** Specific responsibility can be allocated for each work package by means of a linear responsibility chart (see figure 13.10).
- **Project planning:** The WBS forms the framework for the project network and budget. Each work package has an associated schedule and a budget aimed at achieving the task's objective.
- **Performance measurement:** Each WBS task has an objective, a schedule and a budget, and someone responsible for achieving it. Accountability is clearly established and work package performance can readily be measured.
- **Project reporting:** The actual performance of each work package needs to be reported to higher-level managers. Aggregated performance should also be reported by 'rolling up' work package performance data either via the WBS or the organisation chart.

WBS task	Peter Noble	Anne Kennedy	Steering Committee	John Smith
	Music Dept.	Legal Services		Project Manager
141	✓			
142				✓
143		✓		
144			✓	

Figure 13.10 Linear responsibility chart for the music festival

13.6.6 Work packages

Work packages form a management control point where the WBS intersects with the organisation chart. It is the lowest level at which organisational responsibility for individual WBS tasks is allocated. Work packages are the basic building blocks of any project control system. All project management aspects intersect at work packages, including task descriptions, responsibility assignments, budgets, schedules, cost collection, progress assessment, problem identification and corrective action. Work packages are the lowest level at which actual costs are collected and actual performance measurement is done.

13.6.7 The statement of work

The WBS is intended to be a framework containing simple phrases and should preferably be one page in length. However, it is often necessary to communicate to the work package manager or the contractor more details than can be contained in the WBS. The statement of work (often referred to as the scope of work) is a derivative of the WBS that can be used for elaborating and clarifying each task (see figure 13.11).

The statement of work is a primary contracting document. Language usage is very important. 'Shall' or 'shall not' is used for mandatory requirements. 'Should' or 'should not' express non-mandatory statements of desirability or preference. 'May' or 'need not' are used for a non-mandatory suggestion. 'Will' or 'will not' are used for a declaration of purpose or expression of simple futurity.

Work breakdown structure

10: Develop Arrow product specification   11: Construct prototype Arrow
12: Conduct test and evaluation   13: Perform design review

Statement of work

Task 10: Develop Arrow product specification

The contractor shall design and develop the Arrow product specification, document 11-36 issue A, which shall be derived from and traceable to the Arrow development specification, document 10-36 issue 1.

Task 11: Construct prototype Arrow

The contractor shall construct a prototype according to the Arrow product specification document 11-36 issue A.

Task 12: Conduct test and evaluation
The contractor shall test and evaluate the Arrow prototype to verify that all functionality from the development specification has been achieved. The contractor shall compile and deliver a detailed test and evaluation report.

Task 13: Perform design review
The contractor shall perform a design review on the Arrow development specification, document 10-36 issue 1 and the product specification, document 11-36 issue A. The contractor shall prepare a design review data pack, which includes the test and evaluation report, and distribute it to all members of the design review team. The contractor shall also conduct the design review, prepare the design review minutes and action plans, and distribute them to all members of the design review team within ten days. The contractor shall thereafter update and issue the Arrow development and product specifications.

Note 1: A specification is contractually invoked via a statement of work; see task 10. A specification itself is quite meaningless – what work is to be performed in connection with it? Review? Develop? Manufacture?
Note 2: All contractual deliverables are described in the statement of work; see tasks 11 (equipment) and 12 (data).
Note 3: Tasks 10 through 13 are the contractor's implementation of ISO 9001: *Quality systems – Model for quality assurance in design/development, production, installation and servicing*; paragraph 4.4.4: *Design output*, and paragraph 4.4.5: *Design verification. The statement of work is the instrument for customising and tailoring a general standard to a particular contractual situation.*

Figure 13.11 Work breakdown structure and its derived statement of work

13.6.8 The list of deliverables

The list of deliverables is a practical project management instrument. A deliverable is a measurable, tangible and verifiable item, typically hardware or documentation, that the project manager must provide to the customer. Deliverables are often subject to the approval of the customer and a prerequisite for payment.

A list of deliverables defines:
- a specification for the deliverable;
- the quantity to be delivered;
- a delivery date; and
- a delivery address.

The list of deliverables can be used by the project manager and the customer as a checklist for determining the delivery status of the project.

13.6.9 The list of receivables

The customer must provide things to the project, typically known as receivables. Receivables include: equipment, documentation, the decision to proceed with the project, support and access to some or other facility.

There are four aspects to a receivable.

1. A receivable is an indispensable input to the contractor. Without the receivable, the task cannot start.
2. A receivable is beyond the control of the individual or group performing the task.
3. A receivable is within the control of the customer.
4. A receivable does not exist at the start of the project; it is generated during the life of the project.

Example 13.1 Project receivables

Six months into the project a report is to be delivered that will recommend an alternative for further investigation. After a one month study the customer is to formally ratify the chosen alternative and either provide a go-ahead for the rest of the project or terminate the project.

The one month delay may have serious project implications, for instance teams might have to be kept on hold. At project start there is no way of knowing whether or not the customer will actually authorise a go-ahead and, if so, for which alternative.

The best way to handle this uncertainty is to split the project into two separate phases. If that is not possible, then define two receivables as milestones – R1 is the authorisation for project start and R2 is the go-ahead decision. The receivable dates, D1 and D2, are beyond the project manager's control, and will be determined by the date stamp of the customer's registered letter containing the authorisation or go-ahead. The project plan would list the report as deliverable at date D1 + 6 months. The project will continue on date D2. The cost of the delay time has been included in the project budget. For each week shorter than the planned delay time of four weeks, an incentive will accrue to the customer, but for each week longer, a penalty will be levied. If the delay between planned and actual delivery is longer than six weeks, the project will have to be entirely renegotiated.

For an example of receivables, see example 13.1. Project funds are not usually seen as receivables. The list of receivables is the equivalent of the list of deliverables.

Note: The project manager should explicitly define as receivable everything crucial to the success of the project but which is outside his/her control. Never accept responsibility for things that are completely beyond your control.

13.6.10 Contingencies

What about the things that are beyond the control of both the task performer and the customer, e.g. rain on a construction site, fluctuations in currency exchange rates, or changes in tax rates or other statutory costs? The traditional and quite acceptable way to handle this is as contingencies.

13.6.11 Linkages between the components of a project plan

The work breakdown structure, linear responsibility chart, list of deliverables, list of receivables, budget and schedule should all be carefully linked.

- Each task from the work breakdown structure shall have someone responsible to perform it.
- Each deliverable shall be an output from some task from the work breakdown structure.
- Each receivable shall be used as an input by some task from the work breakdown structure.
- Each task from the work breakdown structure shall have adequate cost and adequate time to perform it.
- Each cent and every hour shall be consumed by some task from the work breakdown structure.

The reasons for these linkages are self-explanatory.

13.7 PROJECT TIME MANAGEMENT

Time management attempts to satisfactorily complete a project on time. It consists of five major processes.

- *Activity definition:* Define specific activities to be accomplished.
- *Activity sequencing:* Identify dependencies among activities.
- *Duration estimation:* Estimate the time needed to complete each activity.
- *Schedule development:* Analyse activity sequences and durations, and resource availability to develop a realistic and achievable schedule.
- *Schedule control:* Maintain the project schedule.

An activity is also known as a task. A resource is anything that adds value to a product or service during its creation, production and delivery. Examples of a resource include a group of people, an individual or a machine.

13.7.1 Activity definition

Activities are relatively small and easily manageable WBS elements that form the building blocks for project planning and control. An event, also known as a milestone, is a point of significant accomplishment, a zero-duration activity. All activities have associated events – at least their start and completion. Not all events are necessarily associated with activities. Each activity needs a unique identifier, e.g. A, or 100, or AB201.

13.7.2 Activity sequencing

For activity sequencing, it is necessary to define dependencies, also known as logical relationships or precedence relationships, between two or more activities. An activity-on-the-node project network diagram, also known as a precedence diagram, uses blocks (nodes) for activities and shows the logical relationships between them with arrows.

Figure 13.12 shows the basic relationships between activities and their interpretation.

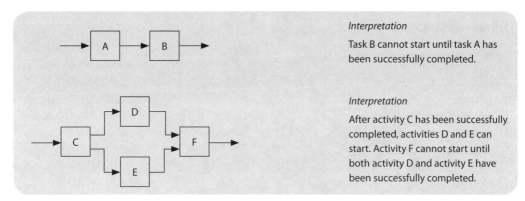

	Interpretation
	Task B cannot start until task A has been successfully completed.
	Interpretation
	After activity C has been successfully completed, activities D and E can start. Activity F cannot start until both activity D and activity E have been successfully completed.

Figure 13.12 Logical relationships in an activity-on-the-node project network diagram

Example 13.2

Construct the activity-on-the-node project network diagrams for the project defined below.

Activity	Preceding ('from') activity	Succeeding ('to') activity
A	–	C, E
B	–	D
C	A	D
D	B, C	F
E	A, G	F
F	D, E	–
G	–	E

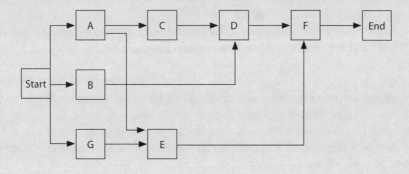

Exercise 13.1

Construct a project network diagram for the following activities:

A   Get out of bed
B   Dress
C   Let in the dog
D   Shave (with a battery shaver)
E   Prepare breakfast
F   Eat breakfast
G   Drive to work

What would Mr Bean's network be?

The rules for network construction are as follows.

- Each activity is represented as a block. The Start and Finish events are also represented as blocks.
- All blocks, with the exception of the Finish block, shall have one or more successors.
- All blocks, with the exception of the Start block, shall have one or more predecessors.
- An arrow specifies only precedence relationships. The arrow length has no duration significance.
- No arrows should be dangling. Every arrow shall have a head and a tail.
- Cycles or closed-loop paths are not permitted. It would imply that an activity is the successor of another activity that depends on it.

There is no single correct project network diagram, although some networks are usually more appropriate than others. The criterion for determining which network is best is the shortest elapsed time. Time is money! The more activities in parallel, the sooner the project will be finished, but the more resources will probably be required. Technological innovations often change constraints and logical relationships. An activity-on-the-node network may have four types of logical relationships.

- *Finish-to-start:* The 'from' activity must be finished before the 'to' activity can start, e.g. Build Walls cannot start before Cast Foundations has been completed, plus a time lag of, say, two days.
- *Finish-to-finish:* The 'from' activity must be finished before the 'to' activity can finish, e.g. Paint Wall cannot be completed until Build Wall has been completed, plus a delay of, say, 10 days to paint the last section.
- *Start-to-start:* The 'from' activity must start before the 'to' activity can start, e.g. Lay Pipe cannot start until Dig Trench has started and the first kilometre has been completed – in other words, a lag of, say, six days.
- *Start-to-finish:* The 'from' activity must start before the 'to' activity can finish. Consider Hire Crane for 30 Days; 30 days after Hire Crane for 30 Days has started, Use Crane must be completed.

Two activities may of course have more than one logical relationship, e.g. Paint Wall cannot start before Build Wall has started (start-to-start with a lag). Start-to-start and finish-to-finish relationships enable fast tracking.

A project network diagram does not know anything about time and does not reflect resource constraints. The network merely shows logical dependencies.

Errors of logic may exist, for instance:
- In a logical loop, the same activity is passed more than once.
- In a logical dangle, an activity either has no 'from' activity or no 'to' activity.

13.7.3 Logical relationships

Mandatory relationships
- It is physically impossible to construct a roof until the foundation has been built.
- It is physically impossible to test a prototype until it has been built.

Discretionary best-practices logic relationships
- Do plumbing and electrical cabling in parallel rather than in sequence.

Discretionary preferential logic relationships
- Use of a unique, one-of-a-kind resource.

External logical relationships
- A new product development project needs to deliver a prototype in time for a tradeshow.
- The project needs to provide financial information to stakeholders.

13.7.4 Duration estimation

Duration estimation predicts the time needed to satisfactorily complete each activity. Duration is expressed in work-days and excludes holidays and other non-working periods. Elapsed time is expressed in calendar-days, and labour content in person-days. Calendar and efficiency estimates are needed to translate work-days to calendar-days.

Example 13.3

Activity P has a duration of 2 work-days, an elapsed time of 4 calendar-days and a labour content of 10 person-days.

We often confuse ourselves by using the term days, without the prefixes 'work', 'calendar' or 'person'. Clear up your thinking by using these prefixes correctly! It is not the world out there that is misleading, but merely our way of thinking about it. (Source unknown).

All duration estimation techniques are based on historical data, and include the following.

■ *Subjective methods (expert opinion):* The more previous experience exists, the more accurate the estimate will be.

■ *Comparative methods (rules of thumb):* Similar activities are likely to have similar durations.

■ *Detailed methods:* Divide the activity into sub-tasks, estimate each sub-task, and aggregate. The smaller the sub-tasks, the more accurate the duration estimate for the activity.

Always include some indication of accuracy for duration estimates, e.g. 10 weeks ±1 week = between 9 and 11 weeks.

13.7.5 Schedule development

Determine the start and completion dates for each activity and thus for the project as a whole. Vital inputs to schedule include:

■ *descriptions of the project resources* – people, equipment and materials. The development of a preliminary schedule may only need broad details, but the master schedule will require very specific resource information;

■ *calendar-defined periods when work is allowed* – project calendars affect the entire project, but resource calendars affect only specific resources; and

■ *constraints that greatly influence the schedule,* e.g. key milestones imposed that are external to the project. Resource or duration assumptions that turn out to be invalid inevitably become problems.

Example 13.4

Consider the following very simple project network diagram (figure 13.13). The duration of each activity in work-days is also shown. Assume that work-days equal calendar-days; in other words, ignore the calendar.

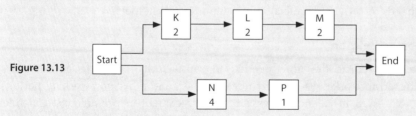

Figure 13.13

When will the project be finished?

After 5 work-days, activity P will be completed, but the project is not yet finished. The logic of the network states that the project will only be finished when both activity M and activity P have been successfully completed, in other words after 6 work-days.

The customer states that the project should be competed earlier, and a young engineer is tasked to speed up the project. After spending R5m, he succeeds in speeding up task N from 4 work-days to 1. This is such a breakthrough that he is awarded the 'Employee of the Year' award and receives a huge bonus! However, does this help the project? Of course not! The duration of the project is determined by the longest path through the network, path KLM, not by the shortest path NP.

A *network path* is any continuous series of connected activities from start to finish. In this case there are only two network paths:

Path KLM: Total duration 6 work-days

Path NP: Total duration 5 work-days.

One of the project's network paths will be the critical path. A project's *critical path* is the network path with the longest duration. The critical path determines the shortest total duration of the project. *Any delay of an activity on the critical path will delay the project completion time.* In this case, if activity N were to be slightly delayed,[1] it would have no impact on the completion date.

All networks have a critical path. There may of course be more than one critical path, e.g. if activity L were to have a duration of one work-day. Is the critical path important? Of course, since it determines the completion date. Furthermore, it allows the project manager to focus. All management revolves around focus. You should focus on the critical path activities, since any delay in a critical activity will delay the end date. The major emphasis of project planning and control should thus be assigned to the critical path.

Since the duration estimates are not perfectly accurate, the project's critical path may change as activities are completed early or late. The critical path may change from time to time as activities are completed ahead of or behind schedule. As the project progresses, the critical path should regularly be re-determined.

What should the schedule of the project be? The Gantt chart of a project displays activities on a timeline and is shown below in figure 13.14.

Activity	Duration (work-days)	Work-day					
		1	2	3	4	5	6
K	2	■	■				
L	2			■	■		
M	2					■	■
N	4	■	■	■	■		
P	1					■	'Spare'

Figure 13.14 Gantt chart

In this schedule, activities N and P have been scheduled as-soon-as-possible. The one work-day 'spare' on path NP has been kept in reserve for the end of the project.

[1] As in calculus, the delay is taken to be infinitesimal.

Activity	Duration (work-days)	Work-day					
		1	2	3	4	5	6
K	2	▓▓	▓▓				
L	2			▓▓	▓▓		
M	2					▓▓	▓▓
N	4	'Spare'	▓▓	▓▓	▓▓		
P	1						▓▓

Figure 13.15 Gantt chart

Activities N and P can also be scheduled as-late-as-possible, but then both paths turn out to be critical; see figure 13.15. The one work-day 'spare' has been consumed before knowing whether it might be needed.

Another alternative schedule is to place the one work-day spare in the middle during work-day 5.

After the critical path has been determined, non-critical path activities still need to be scheduled, either:

▨ as-soon-as-possible; or
▨ as-late-as-possible; or
▨ anywhere in between.

As-soon-as-possible scheduling means that the risk of a non-critical activity becoming critical and delaying the project is low. Selecting the scheduling alternative allows one to:

▨ maximise net present value by scheduling cash-generating activities as-soon-as-possible, and cost-consuming activities as-late-as-possible;
▨ relieve resource constraints when the demand on a resource is larger than its available capacity;
▨ smooth out demand variations on a resource; and
▨ minimise work-in-process inventory by scheduling as-late-as-possible (this is the critical-chain approach).

Question: Do you know how your favourite software project planning application schedules non-critical activities?

13.7.6 Determination of the critical path

The critical path is always determined by using a software project planning application. Nevertheless, it is important to understand a few of the principles involved. The 'spare' day discussed above is technically known as slack.

Determination of the critical path occurs as follows.

▨ The network is traversed from the Start event to the Finish event. This is known as the forward pass. During the forward pass, two values are determined for each activity – the Early

Start and the Early Finish. The Early Start is determined from the logic of the network, and then the Early Finish is calculated:

Early Finish = Early Start + Activity duration

Consider again the example from figure 13.13. Studying the network, it follows that the earliest activity K can start is at time 0. The earliest that activity K can finish is $0 + 2 = 2$. From the network, this also equals the Early Start for activity L. The Early Finish for activity $L = 2 + 2 = 4$. The earliest that activity M can start equals the Early Finish of activity $L = 4$. The Early Finish of $L = 4 + 2 = 6$. This is shown in figure 13.16. Similarly, the forward pass of activities N and P is also shown.

Activity	Duration (work-days)	Early Start	Early Finish = Early Start + Duration	Late Start = Late Finish − Duration	Late Finish	Slack = Late Finish − Early Finish = Late Start − Early Start
K	2	0	2	0	2	0
L	2	2	4	2	4	0
M	2	4	6	4	6	0
N	4	0	4	1	5	1
P	1	4	5	5	6	1

Figure 13.16

- What is the end date of the project? Since from the network both activities M and P need to be completed before the project is complete, it follows that the project end date is the largest of the Early Finishes of activities M and P, or 6 work-days.
- Next the network is traversed from the Finish event to the Start event. This is known as the backward pass. During the backward pass, two values are determined for each activity – the Late Start and the Late Finish. The Late Finish is the latest possible date that the activity can finish without delaying the Finish event. The Late Finish is determined from the logic of the network, and then the Late Start is calculated:

Late Start = Late Finish – Activity duration

Consider again the example from figure 13.13. The Early Finish of activity P is 5. The latest that activity P can finish without delaying the project is 6. Its Late Start = $6 - 1 = 5$. That also equals the Late Finish of N. The Late Start of N = $5 - 4 = 1$. This is also shown in figure 13.16. Similarly, the backward pass for activities K, L and M is also shown.

▪ An activity's *slack* is the time that an activity may be delayed without delaying the project's end date.

Slack = Late Finish – Early Finish = Late Start – Early Start

▪ The critical path consists of all activities having zero slack. As a check, make sure that the two methods of calculating the slack yield equal results. Also verify that the critical path is in fact a network path. Figure 13.16 shows that the critical path is K, L and M.

Exercise 13.3

The example from figure 13.13 is very easy. As an assignment, make sure that you can correctly determine the critical path from figure 13.17, as shown in figure 13.18. As before, assume that all durations are in work-days, and that work-days equal calendar-days.

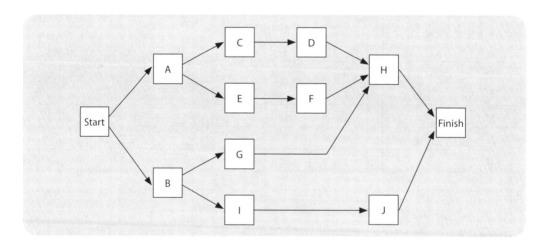

Figure 13.17

Activity	Duration	Early Start	Early Finish = Early Start + Activity Duration	Late Start = Late Finish – Activity Duration	Late Finish (that does not delay end date)	Slack = Late Finish – Early Finish = Late Start – Early Start
A	4	0	4	0	4	0
B	2	0	2	1	3	1
C	2	4	6	14	16	10
D	3	6	9	16	19	10

E	6	4	10	4	10	0
F	9	10	19	10	19	0
G	5	2	7	14	19	12
H	1	19	20	19	20	0
I	7	2	9	3	10	1
J	10	9	19	10	20	1

Figure 13.18

The project will be completed at the end of work-day 20. The critical path thus consists of activities AEFH. There is no slack on the critical path – any delay on a critical path activity will cause a corresponding delay in the project completion date, unless the slippage can be recovered on any downstream activity on the critical path.

Question: If there were a Start-to-Start relationship between activities D and J, how would that influence the critical path?
Question: If there were a lag of 5 days on the Start-to-Start relationship, in other words activity J could start 5 work-days after the start of activity D, what would the critical path be?

Fact: For real-world networks, critical path activities are typically less than 10 per cent of the total number of activities.
Fact: The cost of using network techniques rarely exceeds 2 per cent of the total project costs.

Even relatively simple projects have many network paths. The difficulty of determining the critical path increases much faster than the complexity of the network. Determination of the critical path is thus always done by software!

13.7.7 Project schedules

Gantt charts display activities plotted as bars or lines on a timeline. Open bars are unfinished and shaded bars have been completed. Gantt charts cannot show logical relationships between activities. Gantt charts are often used for summary schedules.

Network scheduling techniques are excellent for planning, but are seldom used for day-to-day project control. However, it is wise to re-determine the critical path on a two-weekly or monthly basis.

Detailed project schedules consist of a hierarchy of schedules, including:
- work package networks;
- integrated contract networks;
- summary networks; and
- interface networks linking the customer, contractor, co-contractors, sub-contractors and vendors.

Self-assessment

13.1 Define the terms project, stakeholders, project brief, project risk.

13.2 Describe how stakeholders may influence the success of a project.

13.3 Briefly describe the four phases of the project life cycle.

13.4 Discuss the challenges and mechanisms of integrative management.

13.5 Discuss the integrative management tools available to project managers.

13.6 Discuss a work breakdown structure (WBS).

13.7 List four requirements that a work breakdown structure should meet.

13.8 Describe the five major processes that comprise project time management.

13.9 Describe how the critical path is determined.

13.10 State the iron law of the project life cycle.

13.11 You have been appointed as the project leader of a project to be undertaken by your company. You have identified the following tasks and made the time estimates for each task as listed in the table below:

Activity	Preceding ('from') activity	Succeeding ('to') activity	Estimated duration (work-days)
A	–	C, E	3
B	–	D	5
C	A	D	3
D	B, C	F	10
E	A, G	F	8
F	D, E	–	14
G	–	E	7

a) Use the above-mentioned table to construct a project network diagram.
b) Determine the critical path for the project and its duration.

13.12 You have been appointed as the project leader of a project to be undertaken by your company. You have identified the following tasks and made the following time estimates for each task as listed in the table below.

Task	Duration (days)	Predecessor
A	6	–
B	10	–
C	5	A
D	14	A
E	20	B
F	4	C
G	2	D

H	21	E
I	17	E
J	11	G
K	30	I
L	40	F, H, J, K

Use the above-mentioned information to do the following.

a) Construct a project network diagram.

b) Determine all the paths through the diagram and their duration.

c) Determine the critical path and its duration.

References

Archibald, R. 1976. *Managing High-Technology Programs and Projects*. New York: John Wiley.

Frame, J.D. 1995. *Managing Projects in Organizations*. San Francisco: Jossey-Bass.

Kerzner, H. 1998. *Project Management: A Systems Approach to Planning, Scheduling and Controlling*. 6th edition. New York: John Wiley.

Project Management Institute (PMI). 2004. *A Guide to the Project Management Body of Knowledge*. 3rd edition. Pennsylvania: PMI.

Simpson, J., Field, L. & Garvin, D. 1988. *The Boeing 767: From Concept to Production*. Harvard Business School Case 688-040. Boston: Harvard Business School Press.

Website

Project Management Institute (PMI): www.pmi.org.

14 Introduction to Accounting, Economics, Financial Management and Budgeting

Wilhelm Nel

Study objectives

After studying this chapter, you should be able to:
- apply the various economic and financial terms described in the chapter
- explain the various factors that influence quantity of products demanded and supplied
- describe the purpose of various financial statements
- develop a budget for a department or area of responsibility
- use budgets to control costs
- motivate budget proposals
- revise budgets

The aim of this chapter is to provide a brief introduction to various economic, accounting and financial management concepts.

14.1 INTRODUCTION

Research has shown that engineers and project managers are involved in the following tasks that require some basic understanding of accounting, economics and financial management (Lock, 1993:18):
- evaluating projects;
- planning and controlling budgets;
- cost estimating; and
- analysing economic risks.

Chapters 14, 15 and 16 attempt to introduce the reader to these topics.

Engineering managers, like most other managers, must develop and implement plans that will maximise the market value of a business. Achieving this aim will ensure business growth and survival, the ability to remunerate employees adequately and good returns for the owners/shareholders and investors. Engineers need to have some knowledge of *engineering economics*, the discipline that is concerned with the cost of alternative engineering solutions to a problem. Engineering economics helps technical people to translate engineering technology into a form that permits evaluation by businesses or investors (Holtzapple & Reece, 2003:654). Engineering economics differs from accounting, which is concerned with the apportionment of costs and the amount of money that was actually spent to achieve an end (Shaw, 2001:358).

How will you be able to determine if a new product, service or project will generate sufficient money or value for your company? A person employed by an engineering company tendering for a contract has to know how soon the project needs to be completed and how much of the various resources to allocate to meet deadlines, the cost of allocating these resources, the risks involved and so on. A person employed by a mining or manufacturing company has to consider many additional factors. What will the price of the mineral or product be for the planned life of the mine, factory or plant? For how long will this mineral or product be in demand? What production rate or scale should be considered? What size capital investment needs to be made? Which technology will result in the most cost-effective operation? The engineer is the person who knows about production rates, capacity required and the machinery or equipment that will deliver the required capacity and therefore the involvement of the engineer is vital in techno-economic analyses.

14.2 MACROECONOMICS AND MICROECONOMICS

Every adult knows that he/she cannot get everything that he/she desires. That is because we do not have the money to buy everything we want – our resources are limited. We have to make choices continuously about what to do with our money, time and energy, because they are limited or scarce. The better we match these limited resources to our needs, the more meaningful our lives can be. Economics is the subject that helps us to match needs with scarce means in the most beneficial way. The idea is to use the least amount of resources to obtain a goal, and to achieve those goals that will have the greatest impact on our personal, organisation's or country's wellbeing.

The broad field of economics is usually divided into macro- and microeconomics. *Macroeconomics* concerns itself with problems at national or regional level such as trade, trade deficits, budget deficits, national debt, unemployment, tariffs, etc. (Shaw, 2001:357). However, the main emphasis of this section is on that part of economics that applies to the individual organisation and individuals, commonly referred to as *microeconomics*. Macroeconomics, although extremely important for policymakers, will only be touched on. *Engineering economics* is a special branch of microeconomics.

14.3 VALUATION AND WEALTH CREATION

The good and bad news to those who may think that wealth creation and destruction work analogously (similarly) to that of energy conservation is that they do not. Energy cannot be either created or destroyed but only changed from one form to another. Fortunately, wealth is not a conserved quantity and therefore much can be done to create more wealth. Some people still think that the economy is a 'zero sum' game in which the gain by one person or group must always result in a loss by others. Have they not heard of the billions that were, and are still created/generated by scientists, engineers and technologists through the development of innovative products and various value-adding processes? Unfortunately, the bad news is that wealth, unlike energy, can be destroyed. It can be destroyed by individuals, by the management of organisations and by governments when bad (investment) decisions are made. If you study this book well you may get to understand something about how to create wealth but also, very importantly, how not to destroy it.

Marketers, product designers, project managers and management often grapple with the question of what it is that consumers, project stakeholders and investors value. This is a difficult question to answer because they may have diverse interests and different value systems. It is, however, possible to put a value on some assets, objects or new projects. The value of an asset is the benefit that a company may derive from its use or application. These benefits are usually expressed in terms of the cash earned by an asset and not only the cost of the asset. The cost of an asset may be a poor indicator of its value. For example, a deep-level, relatively high-capacity mine may cost R5 billion to develop, but it may have relatively very little economic value if its shafts do not provide access to an ore body that can be mined at a profit. Conversely, the R5 billion mine may result in billions of profits for shareholders, billions in taxes paid to government (from which all citizens should benefit) and billions paid in salaries to employees over its life if it provides access to a large, high-grade ore body (not to mention the billions of value that may be added by jewellers and manufacturers to the mineral that is mined). Such a mine should certainly be worth much more than just the R5 billion required to develop it. The following factors need to be considered in evaluating an asset for investment purposes:

- the amount of expected cash flows generated by the asset;
- the timing of the cash flows; and
- the risk involved.

In his book *The Quest for Value,* Bennet Stewart explains that earnings per share (eps) will increase as long as new capital investments earn more than the after-tax cost of borrowing, and should not be used as a performance measure. He says, 'What truly determines stock (share) prices, the evidence proves, is the cash, adjusted for time and risk, that investors can expect to get back over the life of the business.' For now, you should know that value is created by investing in a project with a net present value (NPV) that is greater than zero. Alternatively, value is created when the internal rate of return (IRR) of the project is greater than the hurdle rate of the company. If these requirements are not met, shareholder value is destroyed. These two methods (NPV and IRR) are described in chapter 16.

Economic efficiency is another method of measuring economic viability. Engineers may recognise this definition as being analogous to that of measuring physical efficiencies. Physical efficiencies are defined as the ratios between outputs and inputs, and result in values of less than 1 (or 100 per cent):

Economic efficiency = Worth/Cost

Economic efficiency is defined as economic units of output divided by economic units of input, each expressed in terms of money. Unlike physical efficiencies, economic efficiencies can exceed 100 per cent and must do so for ventures to be successful (Thuesen & Fabrycky, 1993:6). The disadvantage of this method is that the question of how much more than 100 per cent is good is still unanswered and provision is not made for time value of money, a concept that is discussed in chapter 16.

14.4 UTILISING RESOURCES

Resources used by organisations (also called factors of production) can be classified into the following categories:

■ land;
■ labour;
■ capital;
■ enterprise; and
■ knowledge.

Land represents all natural resources, while *capital* includes all the machinery and equipment used to produce something. The people that operate these machines represent *labour*. The resource that organises the other factors of production and takes the risk is called *enterprise* or *entrepreneurship* (see chapter 20). Of all these factors it seems that the most important resource has become *knowledge*. Peter Drucker says: '... knowledge is the primary resource for individuals and for the economy overall. Land, labour, and capital – the economist's traditional factors of production – do not disappear, but they become secondary' (Drucker, n.d.:68).

To remain competitive a company should use its available resources optimally. The *law of diminishing returns* describes the relationship between inputs and outputs of various production systems. This law holds that less and less extra output will be obtained when more of one of the inputs is added while the other inputs are fixed. Table 14.1 shows the diminishing returns obtained as more horsepower (or kilowatts) is used to pump oil through a 30 cm diameter pipe. Twice the amount of oil per day is not produced by the pipeline when 20 000 horsepower (hp) rather than 10 000 hp is applied to pump the oil.

Table 14.1 The relationship between the flow of oil and the power used to pump the oil through a 30 cm diameter pipeline (Samuelson & Nordhaus, 1992:110)

Pumping horsepower	Total product (barrels per day)	Average product (barrels per day per horsepower input)
10 000	43 000	4,30
20 000	57 000	2,85
30 000	67 000	2,23
40 000	75 000	1,88
50 000	82 000	1,64
60 000	88 000	1,47
70 000	93 000	1,33

The law of diminishing returns can also be applied to an engineer that is designing a product. The more time the engineer spends on the design, the better the product becomes. During the early stages of the design, the quality may improve rapidly with each additional hour spent. However, at some point, an additional hour spent does not improve the design at the same rate. The engineer is getting diminishing return on his/her effort (Holtzapple & Reece, 2003:33). Marginal output indicates how much additional output is obtained for each additional resource unit of input that is added. See example 14.1 for another example.

Example 14.1: Diminishing returns and gold recovery

The recovery of gold from ore by means of chemical dissolution depends very much on the size of the ore particles that contain the gold. By grinding such ore finer and finer, more gold can be recovered. Since milling requires energy, there is however an economic turning point where the additional effort required for milling the ore even finer can no longer be justified in terms of the small additional gold output that is obtained (Drunick & Penny, 2005:499). For this reason, the processed ore that is sent to tailings/slimes dams still contains small amounts of gold.

Engineers often have to ask themselves questions about the scale of an operation that they are designing, e.g. 'What advantages or disadvantages will be gained by increasing the *scale of an operation?*' This is because some types of operations have to be of a certain minimum scale for them to be viable (or for economy of scale to be achieved). For example, with current technology, a multi-billion rand shaft system is needed to mine gold ore of a grade less than 12 g/t at a depth of more than 2 000 metres. The production rate should therefore be high enough to get a market-related return on the investment. This implies that such an ore body cannot be mined on a small scale. The three following cases can be distinguished when dealing with returns to scale.

- **Constant returns to scale:** In this case, a change in all inputs leads to an equally large increase in output. If inputs are doubled, so are the outputs. Many handicraft industries show constant returns to scale.
- **Decreasing returns to scale:** In this case, an increase in all inputs leads to a less-than-proportional increase in output. Some productive activities involving natural resources, such as growing grapes or cultivating forests, show decreasing returns to scale (Samuelson & Nordhaus, 1992:111).
- **Increasing returns to scale:** In this case, an increase in inputs leads to a more-than-proportional increase in the level of output. Engineering studies have shown that many manufacturing processes enjoy modestly increasingly returns to scale to a certain point, like the size of many plants in operation today.

We have to make choices continuously because of *scarce resources*. Some choices may affect our lives dramatically, as well as those of our families, colleagues and friends. Mine and factory managers must decide, for example, how much money to spend on safety, health and environment, and in what areas it will have the greatest impact. Government must decide whether a rand spent on basic health care will provide more benefit or not than the same rand spend on heart transplants. Every organisation has to make choices about investments and the type and quantity of goods produced. If more of product A is produced, fewer resources will be at the company's disposal to produce product B, unless the company can raise more capital. The *opportunity cost* of producing more of product A is that of producing less of product B. The following questions must be answered when evaluating projects.

- Which projects will create wealth for the company (i.e. which projects have a net present value greater than zero)?
- Which projects should the company invest in if the necessary capital is not available to invest in all of the ones with a positive net present value?

14.5 WHAT GOODS AND HOW MANY SHOULD BE PRODUCED?

The *law of comparative advantage* states that nations or companies should specialise in producing and selling commodities that they can produce at relatively low cost. Western business strategy says that companies should stay or enter those businesses where there is a good match among competencies, resources and assets acquired and required. In books such as Michael Porter's well-known *The Competitive Advantage of Nations,* the issue of what makes a nation's companies and industries competitive in global markets is discussed. You will find some further information on this topic in chapter 18.

The quantity of a product demanded by consumers and businesses will depend on many factors. Some of these include the price of the product; the (average) level of consumer income or, more correctly, the income of the consumers of that market segment that you are targeting; prices of other products (especially substitutes and complementary products); the tastes of consumers; and how effectively the product is advertised. The quantity demand of a product (let's call it product X) can therefore be expressed as a function of various parameters:

Formula 14.1:

$$Q_{d,x} = Q_{d,x}(P_x, Y, P_1, P_2, ..., P_n, T, A)$$
where:
P_x is the price of product X;
Y is the level of household income;
'$P_1, P_2, ..., P_n$' refers to the prices of related products (substitutes and complements);
T refers to the tastes of consumers; and
A refers to the level of advertising that is done to promote product X.

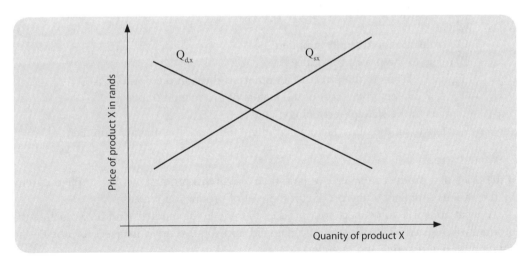

Figure 14.1 The relationship between the quantity demanded, the quantity supplied of a product X and the price of a product when all other variables (e.g. consumer income and taste) stay the same

A small quantity of product X will sell when its price is relatively high. This is because most consumers will rather use substitute products at such high prices or try to manage without the product. For example, when the price of red meat is relatively high then a significant percentage of consumers will adjust their diet and rather eat something that is still relatively cheap at that point in time like chicken, fish, soya or other sources of protein. The higher the average level of household income in a country, the greater the demand for products and services. Households in the 'rich' countries of the world usually spend more money on various types of products and services. The chances are good that the demand for a product will increase when consumers are made more aware of this product by means of advertising. What customers are prepared to pay for a product places an upper limit or ceiling on the price. A larger quantity of product X will be demanded at a relatively low price. This inverse relationship is true for most products. The relationship does not have to be linear, as indicated in figure 14.1. A demand curve (like the one in figure 14.1) is called elastic when a small change in the price of a product would lead to a proportionally large variation in demand.

In the previous paragraph the demand for products and services by consumers and other users is discussed. It is clear that the demand will be high at low prices (see figure 14.1), but will businesses be willing to supply (or produce) such a large quantity at such low prices? Probably not. A business' prime motive for supplying a product is to make a profit, and therefore the quantity supplied by businesses such as mines, factories, plants and retailers will decrease as the price decreases. The cost of manufacturing or mining a product places a lower limit or floor price on such a product at a specific factory, plant or mine. Should the price in the market go lower than this floor price of a specific facility for a lengthy period of time, then the management of such a facility will have to find a way to reduce operating costs or close. The quantity of a product X that businesses will supply can be expressed as a function of various parameters:

Formula 14.2:

$$Q_{s,x} = Q(P_x, P_1, ..., P_n, F_1, F_2, ..., F_m, T, Z)$$
where:
$Q_{s,x}$ refers to the quantity supplied;
P_x refers to the price of product X;
'P_1, ..., P_n' refers to the prices of all other products in the economy;
'F_1, ..., F_m' refers to the cost of the factors of production (inputs);
T refers to the state of technology; and
Z refers to all the other factors that might influence the quantity supplied.

If the price of a product remains constant, then less of this product will be supplied in total by the various suppliers of the product if the cost of production increases.

At what price will you expect your product to trade in the market? And how much of the product will be sold? This is where the demand and supply curves intersect to provide the equilibrium market price and quantity.

14.6 THE DIFFERENCE BETWEEN MANAGEMENT ACCOUNTING AND FINANCIAL ACCOUNTING

Accounting is a means of communicating financial information. The accounting information that may be useful to investors and is contained in financial statements (see section 14.7), such as balance sheets and income statements, falls in the domain of *financial accounting*. Accounting information is needed by project managers and engineers who are trying to control costs (see chapter 15). Given a certain selling price of a specific product, one needs to analyse the various costs of creating such a product to determine whether a company will make a profit. It is therefore important to know something about costs, financial statements, cost-volume-profit analysis and cost estimating.

Management accounting provides financial information about a business for the benefit of internal users. Management accountants provide information on the cost of production, budgets and cash-flow control to management and other internal users. *Financial accounting* provides external users, like the owners and creditors of a business, with information in the form of financial statements. *Financial reporting instruments* like balance sheets, income statements, cash-flow statements and value-added statements tell stakeholders if the company is meeting its financial goals. A reader of company reports must take utmost care when interpreting such reports. Traditional financial reporting instruments are just a model of the real world as it was a day, a quarter, a semester or a year ago. Some of the differences between management accounting and financial accounting are summarised in table 14.2.

Table 14.2 Comparison between management accounting and financial accounting (adapted from Mohr-Swart, 2003)

Factor	Management accounting	Financial accounting
Regulations	Prepared for internal use; no external regulations	Prepared according to GAAP
Range and detail of information	May encompass very detailed and highly aggregated financial, non-financial and qualitative information	Broad based; lacks details; provides an overview of position and performance of an organisation over a period of time
Reporting intervals	Needs of users dictate intervals	Produced annually but may be semi-annually or quarterly
Time period	May include historical and current information	Provide information on the performance and position of the past period, i.e. backward looking
Nature of reports	Specific purpose of reports for specific managers making a particular decision	General purpose reports for shareholders and broad-based stakeholders

You will find information on environmental accounting in chapter 19.

14.7 FINANCIAL STATEMENTS

A complete set of financial statements (depending on the form of enterprise) consists of the following (Von Well & Wingard, 2003:512):

- a balance sheet;
- an income statement;
- a statement of changes in equity;
- a cash-flow statement;
- accounting policies;
- explanatory notes; and
- a directors' report (required by the South African Companies Act).

A *balance sheet* shows the businesses' assets, liabilities and owners' equity at a specific point in time. An *income statement* provides a summary of all income and expenditure over a given period of time. A *cash-flow statement* provides information on the ability of a business to generate cash. *The statement of changes in equity* reflects the increase or decrease in owners' equity and their wealth in the business during the accounting period. Each of these components of a full set of financial statements provides different types of information, and they complement one another. Although most of the components of a complete set of financial statements focus on monetary information and often lack information on customer satisfaction, product and service quality, and quality of management, they should assist investors, creditors and other stakeholders to get a fairly good picture of the company's financial health.

The balance sheet and income statement are briefly discussed below. Note that the balance sheets and income statements of different businesses and types (or forms) of organisations (e.g. close corporations, trusts and public companies) may look different from those listed in figures 14.2 and 14.3 because of factors such as the nature of their business, the nature of various expenses and so on. Public companies that are listed on bourses such as the Johannesburg Securities Exchange (JSE) have additional reporting requirements that should be met.

14.7.1 The balance sheet

A balance sheet shows a company's financial position at a particular point in time. It usually consists of two main sections, namely *assets* and *equity and liabilities*. Assets are further classified as either *current* or *non-current*. Equity and liabilities are reflected by the sub-sections *capital and reserves*, *non-current liabilities* and *current liabilities*. This is illustrated in figure 14.2 on the next page.

All the possessions of the enterprise are reflected under assets. *Current assets* are primarily held for trading purposes and are expected to be realised (converted into cash) within a period of 12 months. *Non-current assets* are not purchased for the primary purpose of resale but rather to assist in the production or sale of goods or services, e.g. factory equipment may have a life of more than five years and is purchased to produce something that will be sold. The factory and factory equipment is therefore indirectly responsible for the generation of cash flows today and in the future. Some assets, e.g. buildings, may be physical or tangible, while others, such as intellectual property (e.g. patents and copyrights), are intangible.

Liabilities are amounts of money that a company owes to banks (e.g. loans, overdraft) and other businesses (creditors). *Current liabilities* are repayable within one year of the balance sheet date. A 20-year loan obtained from a bank for purchasing land and buildings is an example of a long-term liability.

Owners' equity represents funds that the owners contribute to the business and is made up of the initial amount of money the owners invest in the business and undistributed profits that were realised since the inception of the business.

It may be useful to compare the balance sheet amounts of one year with that of the previous year, to see what funds have been injected into the business by shareholders and financial institutions, and to see how these have been used.

ABC Ltd

BALANCE SHEET

	Notes	At 31 Dec. 2005 R'000	At 31 Dec. 2004 R'000
ASSETS			
Non-current assets		000	000
Property, plant and equipment		000	000
Investment property		000	000
Intangible assets		000	000
Financial assets		000	000
Current Assets		000	000
Inventories		000	000
Trade and other receivables		000	000
Cash and cash equivalents		000	000
Total Assets		**000**	**000**
EQUITY AND LIABILITIES			
Capital and reserves		000	000
Issued capital and reserves		000	000
Non-current liabilities		000	000
Financial liabilities		000	000
Retirement benefit obligation		000	000
Deferred tax		000	000
Current liabilities		000	000
Trade and other payables		000	000
Short-term portion of financial liabilities		000	000
Warranty provision		000	000
Taxation		000	000
Total equity and liabilities		**000**	**000**

Figure 14.2 A balance sheet (adapted from Von Well & Wingard, 2003:515)

14.7.2 The income statement

An *income statement* is a summary of all income received and expenses incurred over a given period of time with the object of determining the profit or loss of the business for that period. Typical income and expense items to be found in an income statement are given in figure 14.3. The income comprises both revenue and gains. *Revenue* is generated from the ordinary activities of a business and may include sales, fees, interest, dividends received, royalties and rent, depending on the income generating nature of the business. *Gains*, e.g. on the disposal of non-current assets, may not necessarily arise in the course of the ordinary activities of an enterprise. The *expenditure* of a business is the costs that the business incurs during the creation of income. This may include costs of manufacturing (in the case of a manufacturing organisation) or costs of mining (in the case of a mine) and/or costs of rendering a service, as well as the costs of selling and administering the business.

ABC Ltd

INCOME STATEMENT

	Notes	For the year ended 31 Dec. 2005 R'000	For the year ended 31 Dec. 2004 R'000
Revenue		000	000
Other operating income		000	000
Expenses		(000)	(000)
Raw material and consumables		000	000
Change in inventory of work-in progress		000	000
Change in inventory of finished goods		000	000
Depreciation		000	000
Staff costs		000	000
Advertising		000	000
Transport costs		000	000
Repairs and maintenance		000	000
etc.			
Profit from opertaing activities		000	000
Finance cost		(000)	(000)
Income from associates		000	000
Profit before tax		000	000
Income tax expense		(000)	(000)
Net profit for the year		000	000

Figure 14.3 An income statement (adapted from Von Well & Wingard, 2003:515)

14.7.3 Financial statement analysis

How do we know whether a specific company is doing well or not? Various tools/techniques have been developed to analyse the financial statements of companies. The financial statements of the same company for different financial periods could for example be *compared* and used to identify trends such as the direction of changes, rate of change and the magnitude of change (Nieman & Bennett, 2002:217). Successful companies often point out in the newspapers how profits have increased from the previous year's financial period and so on.

Another tool is that of *ratio analysis*. Various ratios are calculated with information obtained from a company's financial statements and then interpreted and compared with benchmark industry ratios. You will be able to find more information on financial statement analysis from most books on financial management.

14.8 PROFIT AND CASH FLOW

Small businesses often experience cash-flow problems even though they are realising good profits. This is because there is often a difference between the occurrence and recording of certain transactions (e.g. sales and purchases) and cash flow. When a credit sale transaction takes place then it contributes towards profits but the cash inflow may be delayed due to the credit policy of the business – it may only receive payment 30 or 60 days later. A company that pays cash for raw materials needs some money (short-term finance) to 'carry' the business until such raw materials are transformed into products and services that can be sold for cash or until the money for credit sales is received. Table 14.3 lists a number of transactions that affect cash flow in various ways. Sources of (long-term) finance are discussed in chapter 20.

Table 14.3 The impact of various transactions on cash flow

Transaction	Impact on cash flow
Inventory (stock) is increased and paid for in cash.	There is a cash outflow.
More products are sold (to debtors) on credit.	Cash inflow is delayed. Working capital must be increased. More short-term finance is required.
More raw materials are purchased on credit than previously.	Cash outflow is delayed. Creditors effectively provide short-term finance.

A topic related to cash flow, the cash budget, is discussed in the next section.

14.9 BUDGETING

14.9.1 Introduction to budgeting

Budgets describe the responsibilities of managers in monetary terms. They also ensure that the scarce resources of the company are used as efficiently as possible to achieve the objectives of the organisation. Activities inside companies change from time to time as new projects are initiated and outdated products and services are scrapped. A budget should therefore look forward and forecast how much revenue will be generated from future activities and how much money should be spent on various activities in order to meet target profits and other company objectives. Once a budget has been approved it should be used as a means of control against which actual spending is evaluated (see chapter 2 for more information on the control function). When budgets are properly developed and used, the budgeting process should increase profits and reduce unnecessary spending.

An engineer or engineering manager may be responsible for a workshop, maintenance, engineering or manufacturing department, and for its budgeting. An engineer involved in new projects and tender documents will also be involved in cost estimating and budgeting. Cost estimating and budgeting are related activities because all cost/expense items in a budget have to be estimated. Cost estimating is discussed in chapter 15.

Budgets may have different titles and include different expense items and other information due to the form of the organisation (e.g. closed corporation, public company), its nature (e.g. manufacturing, trading), the time frame that is involved and because of personal preferences. The time frame of budgets may vary from short to long term. The annual budget provides a financial plan of action for the financial year ahead, while a project budget is concerned with particular projects that may be of longer duration.

The first step in preparing a budget is to determine its structure – deciding on the subsidiary budgets that must be prepared for each department or functional area and how they should be co-ordinated. Next must be determined the budget items, criteria and norms for each subsidiary budget (Faul *et al.*, 1988: 445). Organisational objectives will strongly impact on these criteria and norms. Historic information and past experience of how costs behave in an organisation or department will also affect these norms (see chapter 15 for more information on how cost estimating is done). In some cases, such as special programmes and projects, no or little data may however be available about the past. In such cases, a *zero-based budget* is prepared without taking any historical data into account. A business plan for a new business that has yet to be established is an example of a zero-based budget (Marx *et al.*, 1998:610).

Variable budgets are used in manufacturing enterprises where the volume of business fluctuates continuously, while *fixed* budgets are used in those organisations such as mines, plants and factories where the facility may be operated at a fixed capacity. Facilities with high fixed costs (overheads) are usually operated at high levels of production close to total (designed) capacity to ensure low unit costs. (see chapter 7 for more information on the design of production capacity).

14.9.2 The impact of organisational objectives on budgets

Budgeting is not an isolated process, but rather the result of a process that starts with strategic and operational planning. An integrated budget can therefore be seen as a company's objectives stated in financial terms. Developing a budget is the process of deciding what amounts should be spent on various activities that will result in the meeting of organisational, departmental or project objectives. The following are examples of company objectives that may have to be achieved by allocating money to appropriate activities.

1. Annual sales must be increased by 20 per cent during the next financial period.
2. Net profit should be more than R1 million.
3. The return on capital employed should be more than 25 per cent before tax.
4. Products X, Y and Z must be developed and should be responsible for at least 20 per cent of sales within three years.

Objectives such as those listed above should be realistic and it therefore requires knowledge of the market and the nature of costs involved in order to develop and manufacture various products and provide various types of services. This knowledge is often vested in different individuals and functional departments inside an organisation. For this reason the main or master budget is usually divided into different budgets for the various functional areas inside an organsiation (e.g. production budget, HR budget) in order to facilitate the various inputs that may be required. These budgets can however not be done in isolation from one another and should form part of an integrated budget system, as illustrated in figure 14.4.

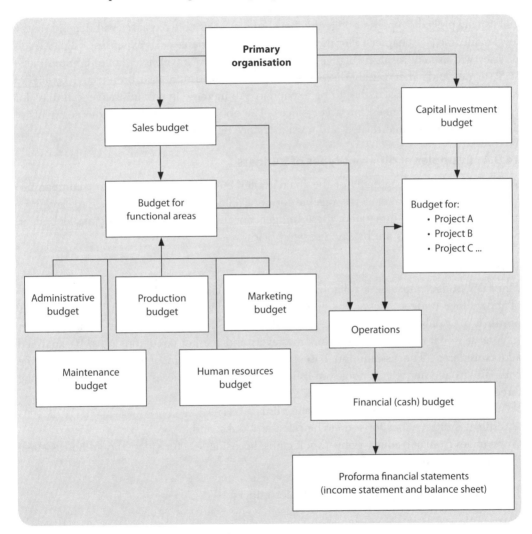

Figure 14.4 The integrated budget system (adapted from Bloom & Boessenkool, 2002:227)

14.9.3 The integrative nature of the budgeting process

When an objective such as objective 1 listed above, 'Annual sales must be increased by 20 per cent during the next financial period', is decided upon, management need to have a good idea of the market's growth potential. Furthermore, several departments will have to co-operate to meet this objective. The marketing department will, for example, have to advertise the products and services of the company to generate a higher level of sales. It will therefore need money to place advertisements, to initiate various promotional projects, to hire staff and for administration. The research and development department will need money for the design of new products that will satisfy market needs and contribute to the increase in future sales. The production or manufacturing department may have to budget for machines that will have to be replaced or will need to be bought to manufacture the new products.

The above example illustrates that those sales and other income forecasts and objectives will usually precede expense budgeting, because many variable expenses will depend on the level of income (see chapter 15 for more information on variable costs/expenses). Higher levels of sales will usually result in higher levels of spending on travel, telephone, administrative staff salaries and other variable expenses. Fixed expenses may also be affected by higher sales when existing capacity has to be expanded. An increase in administrative staff may, for example, result in the need for larger offices. The above illustrates that various segments of the main budget are interrelated and must therefore be well co-ordinated.

14.9.4 Examples of different types of budgets

Various budgets (or segments of the main/master budget) usually exist in a company to ensure that each manager is responsible for controlling spending in his/her department or area of responsibility. Integrated, these budgets are usually called the main/master budget. Some of these budgets are briefly discussed below.

Sales budget

The sales budget must be a reliable forecast of expected sales. It usually forms the basis of most other budgets. Past sales, future market tendencies and the introduction of new products are analysed when volumes and prices are estimated for the sales budget. Factors such as seasonal trends, customer income levels and market conditions must be analysed and considered. The assumption base used for forecasting income will differ from one company to another, e.g. a book publisher could base revenue estimates on the following three factors:

- continuing sales of books on its existing list;
- estimates of revenues for soon-to-be released books; and
- revenues from subsidiary rights (book clubs, licensing fees for video and audio cassettes) (Thomsett, 1988:30).

Table 14.4 is an example of a quarterly sales budget.

Table 14.4 Quarterly sales budget for ABC (Pty) Ltd

PRODUCT	MONTHS							TOTAL
	January		February		March			Rands
	Units	Rands	Units	Rands	Units	Rands		
A @ R50/unit	3 000	150 000	3 500	175 000	4 000	200 000		525 000
B @ R60/unit	2 000	120 000	2 500	150 000	3 000	180 000		450 000
Total		270 000		325 000		380 000		975 000

The production budget

The production budget indicates the number of units that must be manufactured, as dictated by the sales budget. The function of this budget is to ensure that sufficient stock/inventory is always available. It should therefore make provision for seasonal fluctuations. The number of units to be produced can be obtained from the following equation:

Number of units to be produced during period = Expected sales during period + Required opening stock at the beginning of the next period – Closing stock at the end of the previous period

(You will find an example of a production budget on the CD-ROM that accompanies this textbook.)

Other budgets

Budgets other than the sales and production budget exist.
- The *project budget* is concerned with particular projects such as the development of a new product range or the expansion of a plant, mine or factory (Bloom & Boessenkool, 2002:226).
- A *cash budget* is a statement that contains forecasts of cash inflows and outflows for a specific period of time in the future. The cash budget is used to determine whether the organisation has enough funds to continue to operate for a specified period (Bloom & Boessenkool, 2002:226). (You will find an example of a cash budget on the CD-ROM that accompanies this textbook.)
- A *capital budget* consists of the planned investment in land, buildings and equipment (Faul *et al., 1992:253).*

14.9.5 Using budgets to control costs

Budget managers often have to explain to top management when deviations from budgets occur. In some organisations, monthly budget reports are therefore compiled to emphasise significant variances and to allow timeous corrective action where necessary. A *budget report* should comprise the following elements.
- **A comparison of the actual and what was forecast:** The report should point out variances and the degree of the variance, e.g.
  Raw materials used during the quarter:
  Year-to-date R315 000
  Forecast   R240 000
  Variance   R 75 000 (31,3%)
- **An explanation of causes:** Assumptions and budgetary norms used during the budgeting process should be recorded so that they can be revisited when variances between budgets and actual expenses occur in the future. The following assumption/norm could have been made during the budgeting process: 'Raw material forecasts are based on the assumption that products sold will increase by 3 per cent and that the amount of raw materials used will increase by the same percentage in real terms.' The explanation for the 31,3 per cent variance could be that sales increased beyond the expected 3 per cent (quantity increased) and/or that the price of raw materials increased due to a higher-than-expected inflation rate.

- **An analysis of all factors that affect the budget to support the explanation of causes:** Analysis could show that the growth in sales was due to better economic growth, employment and consumer spending in the countries where the company sells its products. A higher inflation rate resulted from the dramatic increase in consumer spending, the high rate of employment and the slow increase in the amount of goods produced by companies.
- **Suggestions for corrective actions and recommendations:** The budget manager may for example recommend that sales volumes should be increased to take advantage of the better economic environment in which the company is operating.

In summary, the budget report should provide top management with good information and possible solutions when actual spending and income varies from what was budgeted. Mistakes should not be repeated when developing budgets in the future. The budget manager should focus on identifying the errors or conditions that caused the variance. Although budgets serve as a standard against which results can be measured, blame should not automatically be assigned for variances. Remember, budgets are the result of forecasts and estimation. It is best to suggest a budget revision if mistakes are found. It is generally more difficult to produce budgets during periods of rapid growth in a company. Companies can improve budget success by learning from past budgets and evaluating their degree of accuracy.

14.9.6 Guidelines for creating successful budgets and budgetary processes

- **Have the right attitude:** Get the budget to work for you, your department and company. Do not see a budget as a time-consuming paper exercise that has been forced on you.
- **Who should develop budgets?** Budgets should be developed by those managers responsible for the various departments, cost centres and profit centres. This will ensure their commitment. The top-down approach whereby senior management dictate the overall budget level is not recommended.
- **Be able to explain each budget item:** Make sure that you can justify your budget to top management. Document all assumptions and norms. Be ready to explain how numbers were calculated. Remember that all budget items (sources of income and various expenses) have logical assumption bases.
- **Don't let the budget procedure frustrate you:** Be proactive. Suggest useful ideas to top management to improve the company's budgeting procedure and reduce frustration with the system.
- **Develop a system:** Develop a system that will help you to create a budget in a short period of time and when you are under pressure. Recognise the absolute relationship between income, volume and certain classifications of general expenses. Resist arbitrary budgeting. Set a standard for logical and intelligent assumptions instead.
- **Keep the budget future-orientated:** Remember that the budgeting process is not an end in itself. The budget will prompt future action when it varies from its predicted course.
- **Decide on an appropriate budget period:** The length of the budget period is usually a financial year. This period should, however, be divided into quarterly or

monthly budgets for control purposes. The shorter periods should allow budget managers to take corrective action when necessary. Budget periods should cover at least one cycle in which seasonal fluctuations occur. A budget that links up with the financial period facilitates the comparison of the actual results with budgeted results.

Self-assessment

14.1 Define the following: engineering economics, microeconomics, macroeconomics, opportunity cost, balance sheet, income statement, assets, liabilities, income, revenue and financial statement analysis.

14.2 Briefly describe the concept of scarcity and explain the consequences of scarcity.

14.3 Your company designs and manufactures interactive television devices.
    i) Explain how the demand for your company's product will be affected if the level of household income is increased.
    ii) Explain whether the supply of such a product will increase or decrease if households have to pay less for it because of a decrease in household income.

14.4 Choose any mineral as an example (e.g. copper) and briefly describe how the following factors may influence the quantity of this mineral that will be demanded:
    i) Price of the mineral
    ii) Average income of consumers
    iii) If a substitute of this mineral (e.g. fibre optics for telecommunications purposes) is significantly reduced in price
    iv) If an important complement of this mineral (e.g. gold when less than 24 carat – copper and gold are used together in some jewellery) is significantly increased in price.

14.5 Differentiate between a balance sheet and an income statement and explain why a company's financial statements have to consist of these and other components.

14.6 You assign an employee the task of developing a budget for the coming year's telephone and fax expenses. Describe how you would instruct that person to calculate the amounts.

14.7 You are co-ordinating the budget for several departments. The production manager has increased all manufacturing and raw material expenses, even though one product has been phased out. Describe how you would suggest a revision.

14.8 Printing, telephone, stationery and salary expenses in your department are running 16 per cent above budget. Identify three possible causes of the problem.

14.9 You want to employ additional employees because your department's workload is greater than expected. List some arguments that you could make to support your request, since the budget does not provide for increased expenditure on salaries.

14.10 A software development company recently introduced a new product. Explain what criteria you would use to forecast income if it is too soon to tell how the new product is selling.

14.11 Should your company increase or decrease equipment and spare parts inventory levels at the store to improve cash flow? Explain your answer.

14.12 Should your company increase or decrease creditors to improve cash flow? Explain your answer.

14.13 Briefly describe how you would set up an annual budget for your department. Explain how you would control performance against this budget.

14.14 A civil engineer has R7 million to spend this year on guardrails for roads. Guardrails cost R300 per metre, so 23,3 km of guardrails can be installed. The guardrails have a service life of about 20 years. The civil engineer is considering spending the money on one of two roads. Road number one is a two-lane road through the mountains that has very steep shoulders. If a car falls off the shoulder, its occupants face certain death. Road number two is a four-lane highway on fairly level ground. If a car drives off the shoulder, there is only a 0,1 probability of death for the occupants. The traffic density on road number one is 20 cars per day and on road number two it is 22 000 cars per day. According to statistics, there are likely to be 0,006 'encroachments' per km per year on road number one and 1,9 'encroachments' per km per year on road number two, given the current traffic densities of the two roads. (Note: An 'encroachment' is when a car drives off the road.) On which road should the civil engineer install the guardrails? Calculate the amount of money spent per life saved. Assume that there is only one occupant per car and that the presence of the guardrail prevents death from a potential 'encroachment' (adapted from Holtzapple & Reece, 2003:49). *See the suggested solution at the back of the book.

References

Annels, A.E. 1991. *Mineral Deposit Evaluation*. London: Chapman & Hall.

Bloom, J.Z. & Boessenkool, A.L. 2002. 'Financial management.' In Nieman, G. & Bennett, A. (eds). *Business Management: A Value Chain Approach*. Pretoria: Van Schaik.

Cronje, G.J., Du Toit, G.S., Mol, A.J., Van Reenen, M.J. & Motlatla, M.D.C. (eds). 1996. *Introduction to Business Management*. Johannesburg: International Thomson.

Drunick, W.I. & Penny, B. 2005. 'Expert mill control at AngloGold Ashanti.' *Journal of the SA Institute of Mining and Metallurgy*, 105, August, pp. 497–506.

Faul, M.A., Pistorius, C.W.I. & Van Vuuren, L.M. 1988. *Accounting: An Introduction*. 3rd edition. Durban: Butterworths.

Faul, M.A., Van Vuuren, S.J. & Du Plessis, P.C. 1992. *Fundamentals of Cost and Management Accounting*. 2nd edition. Pietermaritzburg: Butterworths.

Garrison, R. & Noreen, E. 1994. *Managerial Accounting: Concepts for Planning, Control, Decision Making*. 7th edition. Burr Ridge: Irwin.

Glahe, F.R. & Lee, D.R. 1989. *Microeconomics Theory and Applications*. 2nd edition. Fort Worth: Harcourt Brace Jovanovich.

Holtzapple, M.T. & Reece, W.D. 2003. *Foundations of Engineering*. 2nd edition. Boston: McGraw-Hill.

Lock, D. (ed.). 1993. *Handbook of Engineering Management*. 2nd edition. Boston: Butterworth-Heinemann.

Marx, S., Van Rooyen, D.C., Bosch, J.K. & Reynders, H.J.J. (eds). 1998. *Business Management*. 2nd edition. Pretoria: Van Schaik.

Mohr-Swart, M. 2003. 'Environmental management accounting in the South African mining industry: Setting the scene.' Proceedings of the Mining and Sustainable Development Conference, Sandton, South Africa.

Naismith, A. n.d. *Financial Valuation of Mining Projects*. Course material, University of the Witwatersrand.

Nel, W.P. 1999. Deuteronics Computerised Tutor 2000 Software.

Nieman, G. & Bennett, A. (eds). 2002. *Business Management*. Pretoria: Van Schaik.

Prince, L.J. 1963. *The Use of Present Values in Mining Economics Problems*. Johannesburg: SAIMM.

Riggs, J., Bedworth, D. & Randhawa, S. 1996. *Engineering Economics*. 4th edition. New York: McGraw-Hill.

Robbins, Lord. 1932. *Essay on the Nature and Significance of Economic Science*. Recorded in *The Concise Oxford Dictionary of Quotations*. 3rd edition. Oxford: Oxford University Press.

Samuelson, P.A. & Nordhaus, W.D. 1992. *Economics*. 14th edition. New York: McGraw-Hill.

Shaw, M.C. 2001. *Engineering Problem Solving: A Classical Perspective*. Norwich: William Andrew Publishing and Noyes Publications.

Sloan, D.A. 1983. *Mine Management*. New York: Chapman & Hall.

Smith, L.D. 1995. 'Discount rates and risk assessment in mineral project evaluations.' *CIM Bulletin*, 88 (989), April, pp. 34–43.

Stewart, G.B. 1991. *The Quest for Value: The EVA Management Guide*. USA: Harper Business.

Thomsett, M.C. 1988. *The Little Black Book of Budgets and Forecasts*. New York: American Management Association.

Thuesen, G.J. & Fabrycky, W.J. 1993. *Engineering Economics*. 8th edition. Englewood Cliffs: Prentice-Hall.

Ullmann, J.E. 1976. *Quantitative Methods in Management*. New York: McGraw-Hill.

Von Well, R. & Wingard, C. 2003. *GAAP Handbook 2004*. Durban: LexisNexis Butterworths.

Wanless, R.M. 1983. *Finance for Mine Management*. New York: Chapman & Hall.

15 Cost Estimating, Cost Engineering and Cost Management

Wilhelm Nel

Study objectives

After studying this chapter you should be able to:
- differentiate among different types of costs
- apply the cost-volume-profit method in decision making
- prepare a capital cost estimate

The aim of this chapter is to introduce the reader to cost engineering, cost estimating and cost management.

15.1 INTRODUCTION

When designing new products, processes or facilities such as hospitals, mines, factories and plants, engineers have to come up with new ideas, satisfy the laws of nature as well as the needs of humankind and do it in such a way that it is affordable or economically viable. The latter is often one of the biggest challenges for technical people to meet. Henry Ford is the well-known example of somebody who achieved that. A number of innovators designed and built motor cars before Henry Ford did, but his vision was to do it in such a way that it was affordable by a much bigger group of people than was previously the case. Today there are many products that engineers know how to design and manufacture and that many people would love to have but engineers have not yet solved the problem of how to produce them in such a way that a bigger group of people can afford them. Space tourism, space mining, hydrogen as transportation fuel, and fuel cell vehicles are examples of emerging industries and products of which the successful evolution and diffusion (or spreading) will depend much on further cost reduction (see chapter 11 for more information on the diffusion of technology). Most new technologies struggle to compete with dominant, incumbent (or established) technologies for a number of reasons. One such reason is that the products/services in which such technologies are imbedded (or form part of) are not yet manufactured in large volumes. Once products are manufactured in large volumes they are continuously improved and costs are usually reduced due to economies of scale and learning effects.

Cost plays an important role in established and mature products and industries. Mass-produced, unspecialised products (commodities) have to be produced at as low a cost as possible and sold at competitive prices. The producers of commodity products are often price takers. Due to the nature of their products (which are characterised by the inability of consumers to differentiate them from those of competing products) and the market (consisting of many producers), many mining companies and producers of commodity products cannot determine the prices of their products. The only way for such companies to make a decent profit is to ensure that the difference between the price and cost of production remains sufficient. Price pressure may be slightly less in companies that operate like semi-monopolies because they manufacture differentiated products that may be protected by patents.

Cost management, control and estimating is of importance to both the project and operational stages of a facility. One of the methods used by organisations to allocate resources and control costs is the operational budget. This is discussed in chapter 14. A budget is forward-looking and tries to ensure that resources will be available to meet the future objectives of the organisation. Because of its forward-looking nature a degree of forecasting and estimating of future income and expenses are required. Various capital cost estimating techniques are used by firms to estimate capital expenditure. Such capital expenditure or investment in a project to establish a facility like a plant mine or factory can then be compared with forecasted future cash flows that may be generated by such a facility to evaluate whether the project is economically viable or not. Cash flows generated by a facility depends largely on prices received for goods manufactured/mined and its operational cost structure. Project evaluation is discussed in chapter 16. In this chapter the focus is on estimating capital and operating expenditure, which will feed into financial valuation.

Another reason for doing cost estimating is that many products and services offered by engineering workshops vary in size and complexity and a cost estimate and quote has to be prepared for clients. The quoted amount will usually include the estimated cost, contingency costs and profit. All companies must determine as accurately as possible what it will cost them to offer various products and services so that they can price them correctly. Some understanding of the nature of costs is required for both the costing of products and services and the estimating of capital expenditure. An understanding of the nature of costs will also assist with cost analysis and control.

15.2 THE NATURE OF COSTS

15.2.1 Variable costs

Variable costs vary directly with volume. For example, if the volume of gold produced by a mine increases by 5 per cent, the variable cost will increase proportionally. For a cost to be variable, it must vary according to something. That 'something' is its activity base or cost driver. Machine-hours, units produced (e.g. tons of ore mined, kilograms of gold produced), and units (e.g. kilograms of platinum, tons of aggregate, tons of iron ore) sold are examples of common activity bases that drive cost. To plan and control variable costs, a manager must be acquainted with the various activity bases of the company.

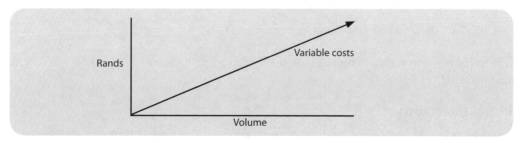

Figure 15.1 Representation of variable costs

On a mine the following costs are normally variable to the number of tons mined: explosives used, mine support installed, stope labour, power consumed, lubricants used, and so on.

15.2.2 Fixed costs

Fixed costs do not vary with volume, but could increase due to the passage of time rather than the level of activity, e.g. the amount of a supervisor's salary for three months is three times that for one month regardless of the volume or output. It is important to note that costs can only be classified as fixed within the boundaries of a fixed production capacity (relevant range). This means that the fixed costs of a mine will probably increase when its capacity is significantly expanded, say from 120 000 tons to 220 000 tons per month for an underground gold mine. In many companies there is a trend towards greater fixed costs. Increased automation at companies often results in increased investment in machinery and equipment. This results in higher depreciation or lease charges, which contribute towards higher fixed costs in the cost structure of such companies.

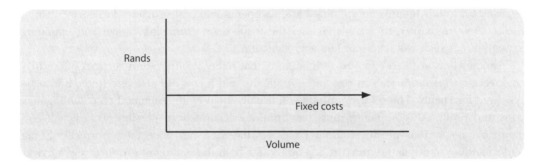

Figure 15.2 Representation of fixed costs

15.2.3 Semi-variable costs and unit costs

Semi-variable costs vary with volume, but not directly so. Examples of semi-variable costs are: indirect labour, maintenance, clerical costs and telephone costs. Where possible, semi-variable costs are split into their fixed and variable components, and incorporated into these cost categories. Telephone costs often consist of a fixed component, line rental, and a variable cost component that depends on how much the phone was actually used.

Costs expressed in unit terms – e.g. per ton of ore hoisted – are very different from those expressed in total terms – e.g. a monthly or quarterly cost. As a rule:
- variable cost per unit of volume is constant; and
- fixed cost per unit decreases as volume increases.

Unit cost (UC) = Total cost/Number of units = TC/N

Unit variable cost or unit marginal cost is the change in total cost if an additional unit is manufactured.

15.2.4 Direct costs

Direct costs are those that can be directly associated with the work accomplished. Examples of direct costs are direct labour, material and equipment costs. Direct labour costs are the total amounts paid to field personnel (e.g. carpenters, labourers, masons, printers) who perform the actual work. Labour costs include basic wages, fringe benefits and items such as taxes and insurance that employers must pay by law. Direct material costs include all material that is essential to constructing, manufacturing or operating a product or facility. The engineering drawing bill of materials lists the quantity and type of direct materials used in manufacturing processes and the cost of this material can be obtained from suppliers. The direct costs related to the manufacturing or production of a specific product should be allocated to that product only and not to the cost of other products, in order to calculate the cost of products accurately. It may be relatively easy to identify the direct costs that are involved in the production or manufacturing of a specific product. A company that produces a number of products for different clients may however find it difficult to decide how to allocate indirect costs (that are not related to any single product) to a specific product so that the cost of each product can be calculated accurately and in a fair manner. Indirect costs are discussed in the next section.

15.2.5 Indirect costs

Indirect costs are indirectly related to the work done during a job or project. Indirect costs include the salaries of supervisors and managers, selling and distribution expenses, research and development costs, indirect materials such as repair materials, and maintenance. Tax is an example of indirect cost, since the amount of tax paid may depend on the tax status of the owner or the country in which the company operates. Overhead costs are also examples of indirect costs. Home-office or head-office rent is a fixed expense incurred in the course of doing business, regardless of the amount of work completed or contracts received. Escalation is another example of an indirect cost (this is discussed in chapter 16).

15.2.6 Cost allocation

The cost of a specific product that is manufactured or produced by a company will consist of direct costs and indirect costs. This is illustrated in figure 15.3.

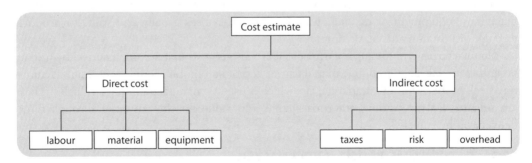

Figure 15.3 Components of a cost estimate

Example 15.1 Cost allocation in a furniture factory

A small furniture factory that produces furniture for clients on order may find it relatively easy to determine the direct cost of a specific piece of furniture. It is relatively easy to determine the quantity of wood that is typically used to produce a specific piece of furniture and the average labour time involved. The total direct cost of a specific piece of furniture can then be calculated by multiplying the quantity of wood used by the unit price of that type of wood (e.g. rands per cubic metre) and then add it to the product of the labour rate and average time required to produce the piece of furniture.

To calculate the indirect cost that should be correctly proportioned to this specific piece of furniture may however not be that easy. How should a cost such as factory rental or the salary of a supervisor be divided amongst the various items that may be manufactured during a typical month?

One means of dividing overhead cost in general and in example 15.1 is to use direct labour costs as a basis. For example, to keep it simple, let's say the small furniture factory in example 15.1 manufactured only two pieces of furniture during the month. The one piece took 200 man-hours to complete and the other took 120 man-hours. 200/320 of the indirect costs will then be allocated to the piece of furniture that required 200 man-hours and 120/320 of the indirect costs will be allocated to the other.

The above approach of apportioning indirect costs may work well for relatively labour-intensive businesses and industries. There are however examples of businesses, e.g. in the area of advanced manufacturing, where labour costs are less than 10 per cent of total manufacturing costs. Activity-based costing techniques have been developed to ensure better allocation of overhead costs to different products. *Activity-based costing* is an accounting technique that allocates overhead costs in actual proportion to the overheads consumed by the production activity. Cost drivers such as machine hours, beds occupied, computer time and flight hours are used as a means for allocating overheads (Aquilano *et al.*, 2003:185).

15.2.7 The cost structures of manufacturing and trading concerns

A trading enterprise buys and sells manufactured goods. These enterprises buy finished goods and therefore do not incur any manufacturing costs such as raw materials, labour, machinery, equipment or maintenance. Trading enterprises usually add value to products through distribution and marketing.

Manufacturing and mining enterprises incur various costs that are not incurred by trading organisations, e.g. an automobile manufacturer and/or automobile components manufacturers buy steel, rubber, aluminium, plastic, etc. which is used to manufacture motor vehicles. These are sold to dealers (trading enterprises) that add value for the customer through direct delivery of the vehicle.

15.3 COST-VOLUME-PROFIT (CVP) ANALYSIS

15.3.1 Application of CVP

Production activities can be modelled in terms of cost- and revenue-generating characteristics. Cost-volume-profit (CVP) analysis, also called break-even analysis, is widely used in financial studies because it is simple and extracts useful insights from a modest amount of data. Break-even techniques could be used to get answers to the following questions.

- What volumes are required to break even, i.e. at what volume of production are costs and income equal?

- What volumes are required to achieve the planned profit?
- What profit will a given sales volume yield?
- How much should I ask or get per item or ton of product to break even if I have a certain production capacity?
- How will expansion affect sales, prices and volume?
- How will a change in costs affect profits? How will modernising affect sales, prices and volumes if fixed costs and variable costs are substituted?

15.3.2 Assumptions

Several assumptions are inherent in CVP-type models.

- *Costs can be divided into variable and fixed components.* In practice, it may not always be that easy to do so.
- *Cost-to-volume relationships are linear and can be represented by straight lines.* In practice, this may not be the case.

In practice, prices and revenues change continuously and therefore changes must be made to present current scenarios.

15.3.3 CVP relationships

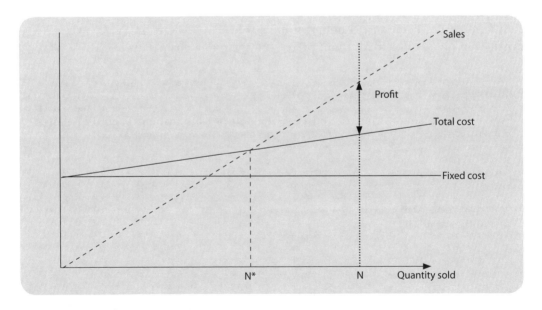

Figure 15.4 Ilustration of the CVP model

Notes:

N   Number of units produced and sold
N*   Break-even number of units

Abbreviations used in CVP calculations:

CM   Contribution margin (also called marginal income)
FC   Fixed cost
P   Profit generated
R   Unit price (price received for one unit)
S, I   Sales, total income or total revenue generated
TC   Total cost
UFC   Unit fixed cost
UTC   Unit total cost
UVC   Unit variable cost
VC   Variable cost (total not per unit)
DOL   Degree of operating leverage

CVP formulas:

Income or sales = Unit price x Number of units sold
Profit = Income – Total cost
Contribution = Income – Variable cost
Contribution margin = Unit price – UVC
Total cost = Fixed cost + Variable cost
Variable cost = UVC x Number of units sold
Unit total cost = Total cost/Number of units sold   UTC = TC/N
Break-even units = Fixed cost/(unit price – unit variable cost; N* = FC/(R – UVC)

Example 15.2

Pete's hot-dog factory has a fixed cost of R450 per month and each unit (hot-dog) has a variable production cost of R1,70. Each unit sells for R2,50. Calculate:

1. Break-even volume per month.

2. The total profit (or loss) made per month when the following number of units are sold:

   a) 440 units

   b) 940 units.

3. The increase in profits if the monthly sales of 940 units is increased by 10 per cent.

Answers:

1. FC = R450,00; UVC = R1,70; R = R2,50; therefore,

   N* = FC/(R – UVC) = 450/(2,50 – 1,70) = 562,5; say 563 units

Alternative method:

At break-even point, the total cost will equal the income or sales because no profit is made at break-even. Therefore, TC = Sales

Let N* = break-even number of units, then

Total cost, TC = FC + UVC x N* = 450 + 1,7 x N*

Income = Rx N* = 2,5N*

Therefore, 450 + 1,7N* = 2,5N*

Make N* the subject, then N* = 450/0,8 = 562,5; say 563 units

2   a)   TC at 440 units = FC + UVC x N = R450 + 1,70 x 440 = Rl 198

I = R x N = R2,50 x 440 = Rl 100

Income < Total cost, therefore a loss is made

Loss = Rl 198 – Rl 100 = R98,00

2   b)   TC at 940 units = FC + UVC x N = R450 + 1,7 x 940 = R2 048

I = R x N = R2,50 x 940 = R2 350

Profit = I – TC = R2 350 – R2 048 = R302,00

Note: A loss is incurred as soon as less units than the break-even number of units are sold. At break-even no profit or loss is made. A profit is realised as soon as more than the break-even number of units are sold.

3.   940 x 1,10 = 1 034 units

TC (1 034 units) = FC + UVC x N = 450 + 1,7 x 1 034 = R2 207,80

I = N x R = 1 034 x 2,50 = R2 585,00

Profit, P = I - TC = 2 585 - 2 207,80 = R377,20

Increase in profit = (Profit at 1034 units – Profit at 940 units)/Profit at 940 units

Increase in profit (as percentage) = [(377,2 – 302) x 100]/302 = 24,9%

Alternative method:

Degree of operating leverage at 940 units = Contribution/P

Degree of operating leverage at 940 units = (940 x 0,8)/302 = 2,49

Increase in profit = Increase in units sold x DOL = 10 x 2,49 = 24,9%

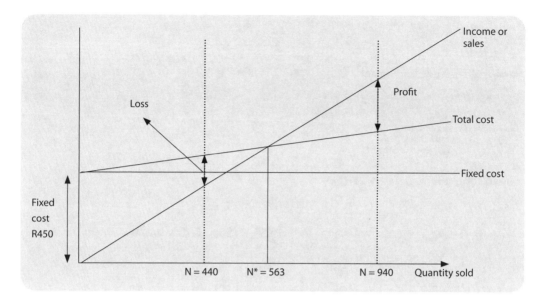

Figure 15.5 Illustration of the CVP model for Pete's hot-dog factory – see example 15.2 (not to scale)

The above example illustrates the concept of operating leverage, i.e. increasing profit by spreading fixed costs over a higher volume. Unit fixed costs become smaller as the number of units is increased. This is one of the reasons why one should try to run a mine, factory or any production process as close as possible to its designed capacity, assuming that there is a market to sell everything produced. In this case, the volume increased by 10 per cent, but total profit increased by 24,9 per cent.

15.4 COST ESTIMATING AND COST EVALUATION

Engineers have to estimate costs when preparing bids, when trying to determine the price of a new product or job, and when they evaluate projects or new business opportunities. They also often have to evaluate the estimated costs on proposals and bids received from contractors (Ertas & Jones, 1996:187).

It can be a challenging task to commit an organisation financially and contractually at the tender or quotation stage when the information used to estimate the cost of the project may be limited. A number of estimating methods can be used depending on the amount, type and detail of information that are available.

Some cost estimating methods rely mostly on historic cost data derived from previous projects. Estimating the costs of a new project will be more accurate if direct and indirect costs are recorded in separate databases. This is because indirect costs are more influenced by the unique conditions of a new project. Using historic, direct costs of past projects from the database to estimate the direct costs of new projects may offer a greater degree of success. Cost estimates of capital investment can be based on the cost of equipment required. Many of the serious errors in capital investment estimation are usually due to omissions of equipment, services or auxiliary facilities rather than errors in costing.

A list of all the quantities of work involved in the manufacturing of a product or when executing a project is needed to determine direct costs associated with the project or product. In the case of a project, the scope documents will provide some information. Table 15.1

provides a checklist that can be used to ensure that all possible indirect costs are included in the cost estimate.

Table 15.1 Construction job indirect cost checklist (adapted from AACE, 1992)

Salaries: Supervisors	Project director, construction manager, general superintendent, excavation superintendent, concrete superintendent, carpenter superintendent, rigging superintendent, welding superintendent, electrician superintendent, chief warehouseperson, etc.
Salaries: Engineering	Chief engineer, office engineer, cost engineer, schedule engineer, materials engineer, draftspersons, field engineers, survey party chiefs, instrument person, etc.
Salaries/ wages: Other	quality assurance (qa) engineers, safety engineers, mechanics, plant operators, first aid workers, Secretaries and clerks, Computer operators, Warehousepersons, Guards, Janitors, etc.
Transport	Cars, pickups, ambulances, tractor-trailers, special-purpose vehicles, fuel/lube trucks, etc.
Buildings and major equipment	Project office, warehouses, change house, carpenter shop, pipe fabrication shop, welder test facility, electrical shop, rigging loft, first aid station, tool sheds, powder house, cap house, resteel fabrication, machine shop, maintenance shop, concrete batch plant, quarry, hoisting equipment, training building, etc.
Temporary horizontal construction	Access roads, construction bridges, drainage structures, rail spurs, laydown areas, fencing and gates, parking areas, environmental protection, etc.
Support systems	Water supply, compressed air, electrical site, site communication, inert gas, oxygen, etc.
General expenses	Office furniture, engineering supplies, engineering equipment, printing/ reproduction, computer terminal, CPM scheduling, phone/telegraph/radio, utilities (electricity and water), portable toilets, signs, safety equipment, permits and licences, advertising, contributions, job travel expenses, testing and laboratory, legal fees, audit fees, medical supplies, progress photos, sanitary facilities, building rental, building and grounds, maintenance, exterior lighting, drinking water and ice, taxes, insurance, payroll burden costs, back-charges from others, consultants, weather protection, general office, etc.

15.5 CAPITAL COST ESTIMATING METHODS

A variety of capital cost estimates exists. A method can be chosen or developed depending on the information that is available. These methods vary from methods where only the magnitude of a proposed project may be known to a detailed estimate at the other end of the spectrum where estimates are prepared from completed drawings and specifications. Some of these methods are briefly discussed. It is important to remember that estimating is not a science (i.e. it is not exact) but rather an art and that the accuracy of the estimate will depend on the quality of the information and the method used.

15.5.1 Order-of-magnitude (conceptual or ball-park) estimates

Order-of-magnitude estimates are also known as ball-park estimates, conceptual estimates or pre-design estimates. Such estimates are made without detailed engineering data. Cost capacity curves, scale-up and scale-down factors, and ratio estimates are examples of this type of estimate, which could normally be expected to be accurate to within +50 per cent or –30 per cent. These estimates can be used to determine the feasibility of a project quickly or to screen several types of design. These methods are therefore important to ensure that money is not wasted on studying projects any further when there is not enough potential. They are also used when time constraints do not allow for a detailed estimate to be prepared. These estimates are generally prepared with basic criteria such as desired output, total square metres or number of units. There are several order-of-magnitude estimating techniques:

- end-product units method;
- scale-of-operations method;
- ratio or factor method; and
- physical dimensions method.

The first three are briefly discussed below:

> The end-product units method is used when enough historic data is available for a particular type of project, to relate end-product units to construction or manufacturing costs. Assume, for example, that Eskom recently constructed an electricity-from-coal generating plant. The plant generates 1 800 MW of electricity and was constructed at a cost of R20bn. If you have to provide an estimate of how much it will cost to construct a similar plant with double the capacity (3 600 MW) then you will just double the cost of the already constructed plant (R40bn). Table 15.2 illustrates some relationships between cost and end-product units that could be used for estimating. The results obtained from this method should be used cautiously because escalation and economy of scale are not considered.

Table 15.2 Some relationships between cost and product units that can be used for estimating

Type of project	Endproduct units	Example
Construction of a luxury hotel	Number of rooms	A luxury hotel was constructed six months ago with 200 rooms at a cost of R13,5m. The cost of building a new luxury hotel with 300 rooms is estimated at R20,25m. (13,5 x 300/200).
Construction of a hospital	Number of beds	A hospital was constructed six months ago with 340 beds at a cost of R13,6m. The cost of building a new hospital with 280 beds is estimated at R11,2m. (13,6 x 280/340).
Apartment building	Number of apartments	A building was constructed six months ago with 360 apartments at a cost of R11,8m. The cost of a new building with 480 apartments is estimated at R15,7m. (11,8x480/360).
Parking garage	Number of parking spaces	A parking garage was constructed three months ago with 1 200 parking spaces at a cost of R5,6m. The cost of building a new garage with 800 parking spaces is estimated at R3,7m (5,6 x 800/1 200).

The *scale-of-operations estimating method* (one of the various order-of-magnitude estimating methods) incorporates economy of scale, unlike the end-product units estimating method. The *cube-square law* is used here as an example to explain why scale matters when (material and manufacturing) costs are estimated. The cube-square law says that as an object gets smaller, its volume decreases much faster than its surface area (Holtzapple & Reece, 2003:31). In other words, as an object gets bigger, its volume increases much faster than its surface area. The surface-area-to-volume ratio of a sphere is $3/r$ [surface area of sphere = $4\pi r^2$; volume of sphere = $4/3\pi r^3$]. Say for example you have a metal container that can store 100 000 litres of petrol, then the amount of metal used to built a container that can store twice as much petrol will not double. This is because the 200 000 litre container has less surface area per volume than the 100 000 litre container – see last column of table 15.3. The cost of manufacturing the 200 000 litre container should therefore not be double that of manufacturing the 100 000 litre container, since the material cost did not double. This is illustrated in table 15.3 for a spherical container.

Table 15.3 Relation between volume and surface area for a spherical container

Radius, r (m)	Volume, V (m³)	Surface area, A (m²)	V/A	A/V (3/r)
2,879	V_1 = 100 (or 100 000 litres)	104,2	0,96	1,04
3,628	V_2 = 200 (or 200 000 litres)	165,4	1,21	0,83
Comparison:	V_2/V_1 = 2	A_2/A_1 = 1,59 (< 2)		

The cost of manufacturing the 200 000 litre spherical container will probably be somewhere between 1,59 and 2 times that of the 100 000 litre spherical container. The 200 000 litre container wall may be slightly thicker, however, to withstand greater pressure and therefore the amount of material used will probably be more than 1,59 times that of the smaller container. The labour and handling cost associated with the 200 000 litre container may like the material costs also be less than double of that of the 100 000 litre container. The scale-of-operations method uses historically derived empirical equations to obtain an estimate for different sized industry facilities of the same type.

The following is an example of such an equation:

$$C_2 = C_1 \times (Q_2/Q_1)^Y \qquad \textbf{Equation 15.1}$$

Where: C_2 = cost of desired plant or piece of equipment;
C_1 = known cost of plant or piece of equipment;
Q_2 = capacity of desired plant or item;
Q_1 = capacity of known plant or item; and
Y = constant (sometimes called the cost capacity factor), usually between 0,6 and 0,8.

Equation 15.1 is known as the six-tenths factor rule when a Y-value of 0,6 is used (Peters *et al.*, 2003:242). The equation can be used to estimate the cost of a specific piece of equipment when the price of that particular size or capacity is not known.

Example 15.3

A small container manufacturing company knows from recent experience that it cost R15 000 to manufacture a 100 000 litre spherical container. Use equation 15.1 to estimate the cost of a 200 000 litre spherical container if historic data supports an $Y = 0,7$.

Solution:

$C_2 = C_1 \times (Q_2/Q_1)^Y \qquad = R15\ 000 \times (200\ 000/100\ 000)^{0,7}$

$\qquad = R15\ 000 \times 1,6245 = R24\ 367,57$

Example 15.4

Should a large capacity engine cost twice of that of an engine of half the capacity to manufacture? From at least a materials cost perspective, a 3 litre engine should not cost twice of that of a 1 500cc engine – twice the material is definitely not used because simplistically speaking the 3 litre engine has bigger holes in the engine block.

Example 15.5

The known cost of an 88 kW cooling plant is R5 500. Estimate the cost of a 151 kW plant if empirical data shows the following relationship between costs and capacity: $C_2 = C_1 \times (Q_2/Q_1)^{0,7}$.

Solution:

$C_2 = 5\ 500 \times (151/88)^{0,7} = R8\ 026$

A project or product may consist of several individual parts and it may not always be physically or economically possible to estimate the costs of all of these items (e.g. small parts such as nuts and bolts). In such a case, emphasis should be placed on high-value items that contribute most of the overall value. In some cases, the cost of specialised equipment may make up a major portion of the total project cost. *Pareto's principle* states that 20 per cent of the items are likely to account for 80 per cent of the total value. In such cases the estimator may use *ratio or factor methods*. Approximate project costs may be estimated by multiplying the cost of major items of equipment by a ratio obtained from either historical data or other reliable resources (see example 15.6).

Example 15.6

The total cost of all specialised equipment at a chemical manufacturing plant is R72m. Estimate the total plant cost if the ratio between total plant cost and specialised equipment cost is found to be about 4,8 from historic data for this type of plant.

Solution:

Total plant cost = R72m x 4,8 = R345,6m

326

15.5.2 Detailed estimate

Historical data is used in all types of estimates, but specifically in detailed estimates. Detailed estimates are particularly appropriate for items of high value or when high volumes of a product are produced. These estimates are prepared from very defined engineering data and are appropriate when detailed drawings are available. In the building industry, these methods will be used when complete plans and elevations, piping and instalment diagrams, single-line electrical diagrams, equipment data sheets and quotations, structural sketches, soil data and sketches of major foundations, buildings, and a complete set of specifications are available.

The following steps may be followed when producing a detailed estimate.

- List the materials to be used. Calculate quantities and allow for wastage.
- Determine the design time required.
- Establish the sequence of operations for each component and allocate labour time.
- List sub-contract work required.
- List equipment required.
- List special tooling and test equipment requirements.

These steps clearly illustrate why people with technical knowledge must be involved in cost estimating.

15.5.3 Other factors to consider when producing cost estimates

In a number of types of cost estimates historic data is often used as a basis. Due to *inflation* (the phenomenon that prices may increase over time) or *deflation* (the phenomenon that prices may decrease over time) this information may not be that relevant anymore today. (You will find some information on inflation in chapter 16). If you know what the cost was of constructing a hotel 10 years ago and you want to construct an exact replica of it today then you may produce a fairly good cost estimate for the new hotel by simply factoring in the increases or decreases in price of materials, labour and so on. Various industry-specific cost indexes can be used instead of the more general producer's price index (PPI) or consumer price index (CPI) that is often referred to as the inflation rate. Some indexes can be used for estimating equipment costs while others apply to labour, construction, materials or other specialised fields. No cost index can take all factors into account such as technological advancements (construction methods are improved over time) or local conditions.

The accuracy of cost estimates can be improved if not one person but representatives from departments such as purchasing, engineering, manufacturing and accounting produce a joint estimate. The advantages of this method are speed and the pooling of information at minimum cost.

Cost estimating of long projects may result in a less accurate estimate and a greater degree of contingency will have to be built into the estimate.

Example 15.7 Example of a detailed cost estimate record for part of a manufacturing project (Lock, 1993:261)

COST ESTIMATE

Part name: Motor components (prototype); Job number: A1300

PART NAME	OPERATION	STANDARD HOURS	MATERIAL	LABOUR	TOOLS AND EQUIPMENT	OVER-HEADS	TOTAL
	Design	25		125		100	225
	Materials:		80				80
	– housing		40				40
	– stator		65				65
	– rotor						
	Machining:						
	– stator	6		30		60	90
	– rotor	15		75		150	225
	Outside heat treatment		30				30
			215	230		310	755

15.6 LEARNING CURVES AND (LABOUR) COST ESTIMATING

Whenever some repetitive work is involved then a cost estimate for doing that work has to consider the learning curve phenomenon/effect. As people learn, the time required for them to do a given task decreases. If a task is repeated a number of times then a person will do it faster. This phenomenon, called the learning-curve phenomenon, was first observed in airframe manufacturing. It was observed that the second plane required 80 per cent as much direct labour as the first; the fourth 80 per cent as much as the second and so on. In this case the 80 per cent is known as the *learning rate*. The gain experienced by the learning rate is achieved every time the number of repetitions doubles. Learning curves apply to individuals, teams and organisations (Finch, 2003:406). The outcome of learning effects is that production rates may rise over time with no change in capacity level. A learning curve follows the following general formula:

$$T = T_1 n^r$$

Equation 15.2

Where: T = Time required to produce the nth unit
n = Unit number
T_1 = Time required to produce/manufacture the first unit
r = (Logarithm of learning rate)/(Logarithm of 2)
The learning curve pattern for a learning rate of 92% is tabled in table 15.4 and illustrated in figure 15.6.

Table 15.4 Learning curve pattern for a learning rate of 92% (see figure 15.6)

Task number	Time to complete (hours)	Task number	Time to complete (hours)
1	22,00	32	14,5
2	20,2	64	13,3
4	18,6	128	12,3
8	17,1	256	11,3
16	15,8	516	10,4

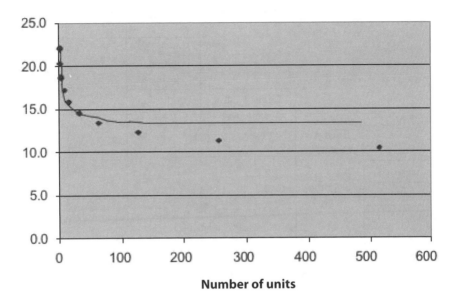

Figure 15.6 Learning curve pattern for a learning rate of 92% (see table 15.4)

The learning rate is usually predicted, since nobody knows in advance how fast a person or team will learn.

15.7 PRICING

The pricing of a product may be influenced by a number of factors. A number of methods can be used to determine the price of a product (Bennett, 2002:186). Note that the following methods refer to the pricing of a manufactured product and cannot be used to determine the monetary value of a business that can generate money now and in the future. One or a combination of the following methods can be used to determine the price of a product.

■ *Mark-up pricing:* A fixed percentage is added to the cost of a product to determine the basic price. See figure 15.7.
■ *Target-return pricing:* A price is based on the desired return on investment.
■ *Perceived value pricing:* Why ask R100 for a product that cost R80 to produce if a significant percentage of customers are willing to pay much more? This method uses consumers' perception of value as the main input in the calculation of the basic price of the product.
■ *Value pricing:* A relative low price is asked for a high-quality product.
■ *Going-rate pricing:* The prices of similar competing products are considered when the basic price is determined. Can you justify asking a similar or higher price than that of your competitors? Does your product offer similar or more value than theirs?

					Mark-up rate	Selling price
				Selling and distribution expenses	Total cost	
			Administrative expenses	Production cost		
	Factory expenses	Factory cost				
Direct materials	Direct cost or Prime cost					
Direct labour						
Direct engineering						
Direct expenses						

Figure 15.7 Determining the minimum price of a product (AACE, 1992:4-3)

The (cost) items are accumulated from left to right to get to the selling price at the right-hand top of the figure. In other words:

Direct cost = Direct materials + Direct labour + Direct engineering + Direct expenses
Factory cost = Direct cost + Factory expenses
Production cost = Factory cost + Administrative expenses
Total cost = Production cost + Selling and distribution expenses
Price = Total cost (per unit) + Mark-up rate.

No estimating method provides a 100 per cent accurate estimate all the time. Risk is the chance of not obtaining a certain return or profit, and is compensated for by adding an amount (contingency) to the price of a product or project tender.

Self-assessment

15.1 Describe the difference between fixed and variable costs.

15.2 Explain the difference between costing and pricing.

15.3 List the various estimating methods and compare them in terms of application and accuracy.

15.4 Differentiate between direct and indirect costs, and give examples of each.

15.5 You are a member of a close corporation that manufactures and sells solar cells and related components. State whether the following expenses are examples of direct or indirect cost;

i) the cost of material that is used to manufacture the solar cells

ii) the salary of the supervisor who oversees the activities in the small solar cell manufacturing plant.

15.6 You have just returned from a meeting with the local government officials of a large city. The city is contemplating constructing a 500-unit public housing project. The officials need a rough estimate for a meeting with the mayor the following morning. You know that the city has just completed a 350-unit project for a total cost of R22,75m. The constant factor for large public housing projects is 0,8. The preliminary information you received from the local government officials indicates that the design of the proposed project will be similar to the one recently completed. What figure will you give to the officials?

15.7 Calculate and tabulate the area/volume ratios of 100 000, 200 000, 300 000 and 400 000 litre spherical containers. (Hint: See table 15.2.) Comment on the probable cost of manufacturing a 400 000 litre container compared to that of the 100 000 container.

15.8 Your company has determined in great detail that it cost them R13 300 to refurbish a 37 kW electric motor for customers. Use the relationship $C_2 = C_1 \times (Q_2 / Q_1)^Y$ to estimate the likely cost of refurbishing a 54 kW electric motor for customers. Assume a cost capacity factor of 0,8.

15.9 Blue Sky Airways provides a daily service between Johannesburg and Durban. The aircraft has a capacity of 220 passengers and each trip costs the company R70 000, regardless of the number of passengers. Additional costs are R60 per passenger (baggage, cabin services and booking costs).

    i)   If Blue Sky Airways charge R450 per passenger, how many passengers do they need to break even with each flight?

    ii) Research showed that planes are actually carrying an average of 80 per cent of its capacity of passengers. What profit (or loss) is realised at 80 per cent capacity and R450 per ticket?

15.10 The graph illustrates the fixed cost per ton of aggregate produced by a quarry. The total fixed cost of the quarry per month is R640 000 and it has the capacity to produce 100 000 tons of aggregate per month. Assume that the average market price for a ton of aggregate is R32,00. What deductions can be made from this graph?

Fixed cost per unit

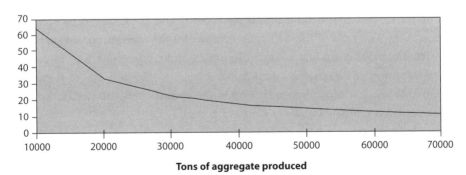

Tons of aggregate produced

15.11 Company XYZ specialises in reworking old gold mine dumps. Since it is using better metallurgical processes (compared to those used years ago) it extracts on average 0,28 grams of gold per ton of waste dump material processed. Its plant has the capacity to handle up to 180 000 tons per month. At this capacity their fixed cost per month is about R540 000, while variable costs are about R8 per ton of material processed. Calculate the break-even tonnage per month. Assume a gold price of R51 700 per kg of gold.

15.12 Develop a fixed-capital cost estimate for a process plant if the purchased equipment cost is R10 m. The plant will handle both solids and fluids with a high degree of control automation. It will be an essentially outdoor operation. Use the information in the table below (adapted from Peters *et al.*, 2003:240).

Table: Typical percentages of fixed-capital investment values for direct and indirect cost segments for multi-purpose plants

Component	Range of fixed-capital investment (%)
Direct costs	
▇ Purchased equipment	15 – 40
▇ Purchased equipment installation	6 – 14
▇ Instrumentation and controls (installed)	2 – 12
▇ Piping (installed)	4 – 17
▇ Electrical systems (installed)	2 – 10
▇ Buildings (including services)	2 – 18
▇ Yard improvements	2 – 5
▇ Service facilities (installed)	8 – 30
▇ Land	1 – 2
Indirect costs	
▇ Engineering and supervision	4 – 20
▇ Construction expenses	4 – 17
▇ Legal expenses	1 – 3
▇ Contractor's fee	2 – 6
▇ Contingency	5 – 15

15.12* Develop a ball-park cost estimate for a new ore treatment plant with a proposed production capacity of 160 000 tons per month. This plant should be completed by June 2008. The ratio of fixed to variable costs for the construction of such a plant is 40:60. The ABC plant that is similar in design and construction with a capacity of 140 000 tons per month was commissioned in June 2005 at a cost of R180 million. The SEIFSA index for price escalation in the construction and steel industry for this period is estimated at 7 per cent per annum.

* See suggested solution at the back of the book.

References

AACE (American Association for Cost Engineering). 1992. *Skills and Knowledge of Cost Engineering*. 3rd edition. Morgantown: AACE.

Aquilano, N.J., Chase, R.B. & Davis, M.M. 2003. *Fundamentals of Operations Management*. Boston: Irwin/McGraw-Hill.

Bennett, J.A. 2002. *Business Management: A Value Chain Approach*. Pretoria: Van Schaik.

Ertas, A. & Jones, J.C. 1996. *The Engineering Design Process*. 2nd edition. New York: John Wiley.

Faul, M.A., Pistorius, C.W.I. & Van Vuuren, L.M. 1988. *Accounting: An Introduction*. 3rd edition. Durban: Butterworths.

Faul, M.A., Van Vuuren, S.J. & Du Plessis, P.C. 1992. *Fundamentals of Cost and Management Accounting*. 2nd edition. Pietermaritzburg: Butterworths.

Finch, B.J. 2003. *OperationsNow.com: Process, Value, and Profitability*. Boston: McGraw-Hill.

Garrison, R. & Noreen, E. 1994. *Managerial Accounting: Concepts for Planning, Control, Decision Making*. 7th edition. Burr Ridge: Irwin.

Holtzapple, M.T. & Reece, W.D. 2003. *Foundations of Engineering*. 2nd edition. Boston: McGraw-Hill.

Lock, D. (ed.). 1993. *Handbook of Engineering Management*. 2nd edition. Boston: Butterworth-Heinemann.

Nel, W.P. 2005. Engineering Management Tutor (Software).

Ostwald, P.F. 1992. *Engineering Cost Estimation*. 3rd edition. Englewood Cliffs: Prentice-Hall.

Pansini, A.J. 1996. *Engineering Economic Analysis Guidebook*. Lilburn: Fairmont Press.

Peters, M.S., Timmerhaus, K.D. & West, R.E. 2003. *Plant Design and Economics for Chemical Engineers*. 5th edition. Boston: McGraw-Hill.

Riggs, J., Bedworth, D. & Randhawa, S. 1996. *Engineering Economics*. 4th edition. New York: McGraw-Hill.

Secrett, M. 1993. *Mastering Spreadsheet Budgets and Forecasts*. New York: Pitman.

Thomsett, M.C. 1988. *The Little Black Book of Budgets and Forecasts*. New York: American Management Association.

Thuesen, G.J. & Fabrycky, W.J. 1993. *Engineering Economy*. 8th edition. Englewood Cliffs: Prentice-Hall.

Welsch, G.A. 1988. *Budgeting: Profit Planning and Control*. 5th edition. London: Prentice-Hall.

16 Introduction to Time Value of Money and Project Selection

Wilhelm Nel

Study objectives

After studying this chapter, you should be able to:
- define the terms interest, time value of money, inflation and escalation
- apply time value of money concepts and formulas
- apply net present value (NPV) and internal rate of return (IRR) methods to evaluate projects

The aim of this chapter is to introduce readers to project evaluation and underlying concepts such as time value of money.

16.1 INTRODUCTION

This chapter introduces you to the subject of time value of money, as well as various areas where time value of money principles are used. Engineering economics, the discipline that is concerned with the cost of alternative engineering solutions to a problem, is one such area where time value of money is often considered. A number of qualitative and quantitative criteria can be used when selecting projects. These are briefly discussed in this chapter. The emphasis is, however, on long-term projects where methods based on time value of money (or discounted cash flow) such as net present value (NPV) and internal rate of return (IRR) are often used.

A typical big company may be looking at a number of projects at any one time. This is because the assumption can be made that only a certain percentage of them may eventually be implemented. The implementation or execution of big projects such as the sinking of a new shaft system or the development of a new product such as a new medicine/drug could cost millions of rands (see example 16.1). Successful projects will ensure the long-term survival and success of the company. The challenge therefore is to select the right projects to pursue. Whether successful or not, projects tie up company resources and the selection of one project may preclude your company from pursuing another, perhaps more successful project. The concept of opportunity cost, as explained in chapter 14, is the key to understanding this. Research has shown that companies that do not have a formal process for choosing among various product development projects may pursue more projects than they may have resources for and this may result in poor productivity, the slipping of deadlines and rising product development costs (Wheelwright & Clark, 1992:71).

Example 16.1 The cost of big projects

One cost estimate for the development and building of the Pebble Bed nuclear demonstration reactor is R14,5bn. The cost of developing the South Deep mine, including the development of the world's deepest single-lift shaft system, will be in excess of R5bn before the mine runs at full production capacity. The initial estimate for the Gautrain project was R7bn. A more recent estimate is R20bn. The cost of developing a new prescription drug/medicine is about US$800m and may take 10 to 15 years (Schilling, 2005:131).

16.2 INTEREST AND THE TIME VALUE OF MONEY CONCEPT

Charging rental for the use of money is an old practice. The amount of money borrowed from a bank or another type of financial institution is called the *principal* (P) and the *interest* (I) is the 'rent' paid for use of the principal. When investors invest money in projects, they expect to be paid back from the earnings generated by the project. Money that is paid by companies to shareholders is called *dividends*. The rate of gain received from an investment is known as the *interest rate*. The rental rate for a sum of money is usually expressed as a percentage to be paid for its use during a one-year period and is also called the *interest rate* (i). Different approaches exist for quoting interest rates and therefore for determining the amount of interest earned. Interest becomes due at the end of the interest period. The interest period is therefore the length of time after which interest becomes due. The term 'interest' is often used to mean either the cost of borrowed money or the rate of return on an investment to an investor. Interest and repayments of the principal amount are usually based upon a legal contract between the lender and the borrower. It is therefore important to study the terms and conditions specified in such a contract.

R1 000 received today is expected to be worth more than R1 000 received one year from now because it can be invested to earn interest. If R1 000 received today is invested in a savings account it will earn interest of R70 if the interest rate is 7%. This example illustrates the earning power of money. One would therefore expect the future value of money to be always greater than its present value. In this example, the future value (FV) of R1 000 is R1 070. The term *time value of money'* refers to the opportunity of earning interest on an investment. The purchasing power of money may change with time and contribute to the time value of money effect. If the rate of *inflation* is 6%, for example, it means that the average household is spending 6% more on the same amount of groceries, housing, travel, etc. than one year ago. Inflation therefore causes the purchasing power of money to decline and that you would need 6% more income after tax than a year ago to maintain your standard of living. In an environment where interest rates are zero and the expected inflation rate is 8%, R100 will be paid today to purchase an item. In a year's time you may need R108 to purchase the same item.

16.2.1 Simple interest

Under conditions where simple interest rates are applied, interest owed upon repayment of a loan is proportional to the length of time the principal sum has been borrowed. *Simple interest* refers to an interest payment that is based only on the original principal amount that was borrowed – no interest is paid on interest, as is the case for *compound interest* (when more than one interest period is involved). The formula for determining simple interest earned is:

I = Pni   **Equation 16.1**

Where:

I = interest earned   P = principal amount
i = (simple) interest rate   n = interest period

Example 16.2 Simple interest rate calculation

ABC Ltd borrows R10 000 from a bank at a simple interest rate of 14% per annum (year). The contract between ABC and the bank specifies that the interest and principal payment have to be made after a period of three years. What amount must be paid by ABC Ltd to the bank at the end of three years?

Answer:

$I = Pni = R10\ 000 \times 3 \times \frac{14}{100} = R4\ 200$ (Remember, % means 'out of 100')

Total sum to be paid to the bank = principal + interest

= 10 000 + 4 200 = R14 200

This is illustrated in figure 16.1.

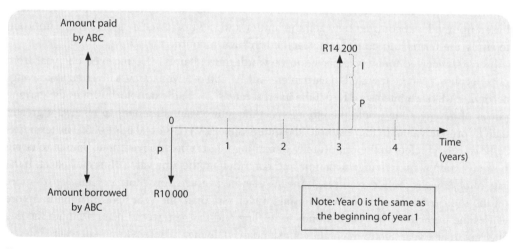

Figure 16.1 Money borrowed and paid by ABC Ltd

16.2.2 Compound interest

In the case of simple interest, the interest is usually not paid until the loan is terminated. This leaves the lender deprived of the interest money that he/she could have loaned out in the interim. There are two ways to overcome this: (1) specify in the contract that the interest payments are to be made at the end of each interest period; or (2) the unpaid interest is treated as unpaid principal and collects interest in the same way as the principal (Holtzapple & Reece, 2003:658). Scenario 2 is known as compound interest. Table 16.2 illustrates the first scenario when interest is calculated and payable at the end of each interest period, when a loan is made for several interest periods. The payments on a three-year loan of R10 000 at 14% interest per annum is shown in table 16.1 and figure 16.2.

Table 16.1 Interest scenario (1) when interest is paid annually

Year	Amount owed at beginning of year	Interest to be paid at end of year	Amount owed at end of year	Amount to be paid by borrower at end of year
1	R10 000	R1 400	R11 400	R1 400
2	R10 000	R1 400	R11 400	R1 400
3	R10 000	R1 400	R11 400	R11 400

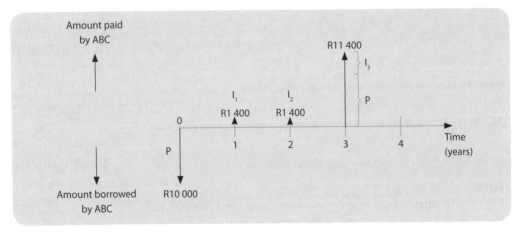

Figure 16.2 Graphical illustration of interest scenario when paid annually

If the borrower does not pay the interest earned at the end of each period, and is charged interest on the total amount owed (principal + interest), the interest is said to be compounded. This scenario (2) is illustrated in table 16.2 and figure 16.3.

Table 16.2 Interest scenario (2) when interest is accumulated and interest is paid on interest

Year	Amount owed at beginning of year	Interest to be paid at end of year	Amount owed at end of year	Amount to be paid by borrower at end of year
1	R10 000	$10\ 000 \times 0,14$ = R1 400	$10\ 000 \times 1,14$ = R11 400	R0,00
2	R11 400	$11\ 400 \times 0,14$ = R1 596	$11\ 400 + 1\ 596$ = R12 996 or $10\ 000 \times 1,14^2$	R0,00
3	R12 996	$12\ 996 \times 0,14$ = R1 819,44	$12\ 996 + 1\ 819,44 =$ R14 815,44 or $10\ 000 \times 1,14^3$	R14 815,44

Table 16.2 illustrates why some people refer to compound interest as interest paid on interest. In year 2, for example, interest is paid on the R1 400 interest that 'stood over' from year 1. If you compare the simple interest case where R14 400 is owed to the bank (example 16.2) with the compound one where R14 815 is owed to the bank, then you will notice that the sum to be repaid is higher when the interest is compounded.

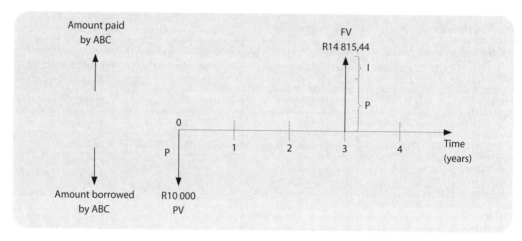

Figure 16.3 Graphical illustration when interest is accumulated

The general formula for the calculation of a future value (FV) is:

$$FV = PV(1 + i)^n \text{ or } FV = PV \times FVIF(i\%, n) \qquad \textbf{Equation 16.2}$$

Where:

FV	=	a single sum of money at some designated future date
PV	=	a present single amount of money
i	=	the interest rate or compound rate per period (e.g. per year or per month)
n	=	the duration of the investment, measured in number of periods (e.g. years or months)
FVIF	=	the future value interest factor $= (1 + i)^n$. This value can also be obtained from financial tables for certain values of i and n.

Example 16.3 Compound interest example

Calculate the future value of R1 000, invested at an interest rate of 7% per annum (compounded) for three years.

Answer:

$$FV = PV(1 + i)^n = 1000(1 + 0{,}07)^3 = R1\ 225 \text{ (Use equation 16.2.)}$$

Note that the value of i is 7% or 0,07. Do not use a value of 7 in the equation but 0,07. This is a mistake that many learners make.

Alternatively: FV = PV $\times$ FVIF(7%, 3) = 1000 $\times$ 1,225 = R1 225

Note that FVIF(7%, 3) can be obtained from financial tables or the tutor programme that is included on the compact disk (CD-ROM) at the back of this book.

Equation 16.2 acts as a 'time-machine'. It tells you what the present value of R1 000 will be worth in three years' time if a compound interest rate of 7% is applied. It is, however, possible to determine what a future value will be worth in the present, given the interest or discount rate.

16.2.3 Present value

The *present value* of a future amount is the amount that, if invested at a given interest rate, would grow to the same future amount at the end of n periods. The general formula for the determination of the present value of a single amount to be received n years from now is:

$$PV = FV/(1 + i)^n \text{ or } PV = FV \times PVIF(i\%, n) \qquad \textbf{Equation 16.3}$$

Equation 16.3 can be obtained from equation 16.2 by means of simple mathematical manipulation.

Example 16.4 Present value of a single amount

Three years from now you expect to receive R1 225. Calculate the present value of R1 225 if the discount rate is 7%.

Answer:

$$PV = \frac{1\,225}{(1+0,07)^3} = R1\ 000; \text{ or } PV = FV \times PVIF(7\%, 3) = 1\ 225 \times 0,816$$

The PVIF can be looked up in financial tables.

The 'time-machine' equations (16.2 and 16.3) for taking money to the future or bringing it from the future to the present have an interesting consequence, the *principle of equivalence*. The present value of R100 and its future value three years later, R122,50, are equal when there are investment opportunities offering a return of 7%. *This use of interest rates or discount rates to relate present money to future or past money is the cornerstone of engineering economics.* The comparison of the financial merits of two or more projects or other investment possibilities, with different cash flow streams, is meaningful only if the individual cash flow amounts for each investment alternative are adjusted to present values, at a common point in time, using an appropriate discount rate. However, the specific point in time selected as the 'zero point in time' for the comparison of alternative investments is not critical, providing all the alternatives are compared at the same point in time.

16.2.4 Annuities

It is sometimes necessary to find the single future value that would accumulate from a series of equal payments occurring at the end or beginning of succeeding interest periods. This is called an annuity and more specifically the *future value of an annuity*. More generally, an *annuity* is a series of equal payments occurring at equal time periods that last for a finite length of time. A *present value of an annuity* will be used when an amount was borrowed and repaid in a series of equal amounts. The mathematical formulas for calculating the future and present values of ordinary annuities follow.

$$FVA = A \times FVIFA(i, n); \text{ where } FVIFA(i, n) = \left[\frac{(1+i)^n - 1}{i}\right]$$

$$\text{or } FVIFA(i, n) = \left[\frac{(1+i)^n - 1}{i}\right]$$

Equation 16.4

$$PVA = A \times PVIFA(i, n); \text{ where } PVIFA(i, n) = \left[\frac{(1+i)^n - 1}{i(1+i)^n}\right]$$

$$\text{or } PVIFA(i, n) = \left[\frac{(1+i)^n - 1}{i(1+i)^n}\right]$$

Equation 16.5

In equations 16.4 and 16.5, the following abbreviations are used:

FVA = future value of annuity
PVA = present value of annuity
A = the amount that is invested or paid out (depending on the type of annuity) during each interest period
FVIFA = future value interest factor of an annuity
PVIFA = present value interest factor of an annuity

Payments are made or received at the end of each period for normal or *ordinary annuities*. Annuities where payments are made at the beginning of each interest period are called *annuities due* and their mathematical formulas are not provided here. Equations 16.4 and 16.5 can be used to solve a variety of annuity problems. Examples are listed in table 16.3.

Table 16.3 Examples of problems that can be solved with annuity formulas

Problems that can be solved with equation 16.4 (FVA)	Problems that can be solved with equation 16.5 (PVA)
Sinking funds that are used to build capital (save money) for the planned replacement of equipment, for planned maintenance, for the rehabilitation of a mine that has reached the end of its life, etc.	**Installment loans** (e.g. mortgage and payment of vehicles, machinery, etc.)
Retirement plans (e.g. retirement fund)	

Example 16.5 Future value of an annuity (FVA)

Calculate the future value three years from now of a uniform series of Rl 000 investments made at the end of each year for the next three years if interest is 10% per year compounded annually.

Answer:

This exercise can be solved by using either equation 16.2 or 16.4. It will however be much easier to use equation 16.4 when more than just three payments are involved.

$$\text{FVA} = A \times \text{FVIFA}(i, n) = A \times \left[\frac{(1+i)^n - 1}{i}\right] = 1\ 000 \times \left[\frac{(1+0{,}1)^3 - 1}{0{,}1}\right]$$

$$= 1\ 000 \times 3{,}31 = \text{R3 310,00}$$

Note that:

■   A is the annual investment amount of R1 000.

■   10% = 0,1.

■   n = 3, because three payments were made to the financial institution where the money was invested.

■   Remember to start your calculation with the innermost brackets $(1 + 0{,}1)^3$ in this case.

■   Round your answer to the nearest cent. When working with money, it does not make sense to provide an answer with three or more digits after the decimal comma.

■   It is possible to obtain the value for FVIFA(i, n) from financial tables. It is however better to learn how to use the formula, as financial tables do not contain all the possible variations of interest rates and periods, especially when the interest rate is not a natural number. The value of 3,310 can be obtained from the table where the interest = 10% column and n = 3 row intersect.

Alternatively, each R1 000 can be taken to the future (end of year 3) by means of equation 16.2. This is illustrated in table 16.4.

Table 16.4 Solving an annuity problem with a single sum equation

End of year	Investments	Years left until end of year 3	FV of investments
1	1 000	2	$1\ 000 \times (1{,}1)^2 = 1\ 000 \times 1{,}12 \quad = 1\ 210$
2	1 000	1	$1\ 000 \times (1{,}1)^1 \qquad\qquad\qquad = 1\ 100$
3	1 000	0	$1\ 000 \times (1{,}1)^0 = 1\ 000 \times 1 \quad\; = 1\ 000$
			Total: R3 310

The money flows are illustrated in figure 16.4. Note that the last R1 000 does not earn any interest because the financial institution pays the investor the full annuity amount on that day. In the case of an annuity due, the investments will be made at the beginning of each period and therefore interest will be earned on the last payment.

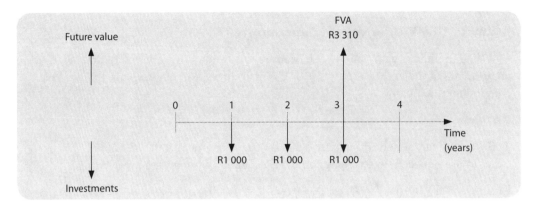

Figure 16.4 Money flows for an annuity

Example 16.6 Present value of an annuity (PVA)

Calculate the present value of a series of Rl 000 payments to be made at the end of each year for three years if interest is 12% per year compounded annually.

Answer:

This exercise can be solved by using either equation 16.3 or 16.5. It will, however, be much easier to use equation 16.5 when much more than just 3 payments are involved.

$$PVA = A \times PVIFA(i, n) = A \times PVIFA(i, n) = A \times \left[\frac{(1+i)^n - 1}{i(1+i)^n}\right]$$

$$PVA = 1\ 000 \times \left[\frac{(1+0,12)^3 - 1}{0,12(1+0,12)^3}\right] = 1\ 000 \times 2,40183 = R2\ 401,83$$

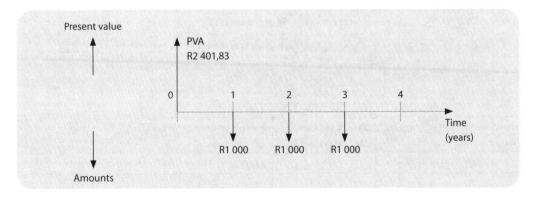

Figure 16.5 Present value of an annuity example

The above example can also be interpreted as follows: if R2 401,83 is paid to a financial institution today, then this institution may be able to pay three amounts of R1 000 each back to the investor periodically over a period of three years, as illustrated in figure 16.5.

16.2.5 The installment loan

The installment loan is one example (see table 16.3) where the present value annuity equation (equation 16.5) can be applied. An *installment loan* comes into existence when someone borrows a lump (or single) sum of money and agrees to make equal periodic payments until the loan is paid off. Everyday examples of installment loans are mortgages and vehicle loans.

Example 16.7 Installment loan annuity

A company plans to purchase a truck for R330 000. This amount is borrowed from a financial institution and will be paid off by means of 36 equal monthly instalments. What will the monthly payment be if the interest rate is 14% per year?

Answer:

$PVA = 330\ 000 = PVIFA \times$ monthly payment

where $PVIFA = \left[\dfrac{(1+i)^n - 1}{i(1+k)^n}\right]$

Monthly interest rate $= \dfrac{14}{12} = 1{,}1667\%$ or $0{,}011667$ (Note: payments are made monthly, therefore the annual interest rate must be changed to a monthly interest rate.)

Number of payments to be made to the financial institution $= 3 \times 12 = 36$

$PVIFA = \left[\dfrac{(1+i)^n - 1}{i(1+i)^n}\right] = \dfrac{(1+0{,}011667)^{36} - 1}{0{,}011667(1 + 0{,}011667)^{36}}$

$PVIFA = 29{,}2589$

Monthly payment $= \dfrac{330\ 000}{29{,}2589} = $ R11 278,62

Note that the amount paid back to the financial institution ($36 \times 11\ 278{,}62 = $ R406 030,24) is more than the amount that you borrowed (R330 000) from it. You have compensated the financial institution for taking the risk of loaning you money. It will also realise a profit in the process by borrowing money at a lower rate at which it loaned money to you.

16.2.6 Sinking fund

Amortisation is the name given to the process of returning to the investor the amount that has been invested in a wasting asset. This amount can be accumulated by creating a sinking fund or redemption fund. Installments are periodically paid into such a fund and earn interest for a certain number of years, with the aim of redeeming or replacing the sum of money borrowed or used.

Equations 16.3 or 16.4 can be used to calculate the annual installment to be paid into the fund, depending on whether the present or future amount is known.

Example 16.8 Sinking fund

It is estimated that the owners of a quarry will have to spend R650 000 in 10 years' time on rehabilitation at closure. Calculate the uniform series of equal payments made at the end of each year for 10 years that are equivalent to a R650 000 payment 10 years from now, if interest is 16% per year compounded annually.

Answer:

$$A = 650\,000\left[\frac{0,16}{(1+0,16)^{10}-1}\right] = \text{R30 485, 70}$$

Alternatively,

$$A = \frac{FV}{FVIFA(i\%,n)}(i\%,\,n) = \frac{FV}{FVIFA(16\%,10)}$$

$$A = \frac{650\,000}{21\,321} = \text{R30 486,37}$$

16.2.7 Mortgage bonds

Equation 16.5 can be used to calculate the monthly repayment amount of a mortgage bond. Such loans are usually paid monthly, so remember to change the annual interest rate to a monthly one, if applicable. See example 16.9.

Example 16.9 Mortgage bond

Question: Pete recently bought a house at a price of R650 000. He obtained a (100%) house bond for the whole amount from XYZ Bank. The interest rate that XYZ charged him was recently reduced from an annual rate of 15% to 14%. How much less will Pete have to pay each month? Pete has 20 years to repay the bank.

Answer:

Step 1: Decide whether this is a single sum or an annuity. In this case it is an annuity, since 240 (n = 20 × 12 = 240 periods) equal monthly payments have to be made by Pete.

Step 2: Is it a PVA or FVA? In this case it is a PVA, since the current purchasing price of the house is known. Therefore equation 16.5 has to be used.

Step 3: How often does compounding takes place? In this case, it is monthly. Convert annual interest rate to a monthly one.

$$i_{old} = \frac{0,15}{12} = 0,0125, \text{ whereas } i_{new} = \frac{0,14}{12} = 0,011667$$

$$PVIFA(0,0125;240) = \frac{(1+0,0125)^{240}-1}{0,0125(1+0,0125)^{240}} = 75,9422$$

Similarly, PVIFA(i = 0,011667; n = 240) = 80,4168

$$\text{Old monthly payment} = \frac{650\,000}{\text{PVIFA}}\,(i = 0{,}0125;\ n = 240) = \text{R8 559,13}$$

$$\text{New monthly payment} = \frac{650\,000}{80{,}4168} = \text{R8 082,89}$$

$$\text{Saving} = \text{R476,24 per month}$$

16.3 PROJECT SELECTION

Numeric project selection models include scoring and financial models. Financial numeric project selection models quantify projects in terms of payback time and return on investment (ROI), whether their net present values (NPVs) are positive or whether their internal rates of return (IRRs) are bigger than required hurdle rates. Non-numeric project selection models may consider issues such as market share, client retention, skills requirements and availability, compliance with standards and environmental issues, just to name a few (Burke, 2003:56). The simplistic weighted scoring model in example 16.9 utilises a combination of financial and other considerations.

Example 16.10

Question: Use a weighted scoring model to choose between three mining projects (A, B, C). A score of 1 represents unfavourable, 2 satisfactory and 3 favourable. The ore bodies are in three different countries.

Category/Criteria	Weight	Project		
		A	B	C
Country risk and socio-economic environment	0,2	3	2	1
Labour productivity	0,2	3	2	2
Labour supply	0,1	2	2	3
Transportation infrastructure	0,15	3	2	1
Financials (ROI, IRR and NPV)	0,35	2	3	2

Answer:

Category/Criteria	Weight	Projects			Weighted average		
		A	B	C	A	B	C
Country risk and socio-economic environment	0,2	3	2	1	0,6	0,4	0,2
Labour productivity	0,2	3	2	2	0,6	0,4	0,4
Labour supply	0,1	2	2	3	0,2	0,2	0,3
Transportation infrastructure	0,15	3	2	1	0,45	0,3	0,15
Financials (ROI, IRR and NPV)	0,35	2	3	2	0,7	1,05	0,7
Total					2,55	2,35	1,75

Conclusion: Choose project A.

16.3.1 Cash flows during a project's life cycle

You may find it useful to first read the section on the life cycle of a facility such as a mine, factory or plant and have a look at figure 16.6. It is important to note that money is usually spent and not earned during the project phase of such a facility. During the project stage of a deep-level mine's life, millions of rands are spent to sink shafts and develop tunnels that will provide access to the ore body. Only then can production start (there may be exceptions such as mid-shaft loading and early shaft pillar extraction from which some income may be generated). During the operational (production) phase of the mine, minerals will be extracted, processed and sold. So it is usually only during the operational phase of such a facility that income will be generated. At the end of a mine's life, money has again to be spent (on rehabilitation). This again results in an outflow of money. This is illustrated in figure 16.6. Note that the cash outflows and inflows will only balance at some point after production has started. This is called the break-even period. Smaller projects such as the expansion of a facility and the maintenance of a facility that take place during the operational stage may be financed by means of profits generated.

Note that a project is evaluated before it is implemented/constructed, therefore the cash outflows and inflows during the life of a facility must all be estimated/forecast and considered when such a project is evaluated. Estimating the cost of a project and future cash inflows is the difficult part of financial project selection. Applying the financial project selection methods is the easy part, as spreadsheets and other software packages can be used to do that. These cash flows may take place throughout the months and years during which the facility is built/constructed and in operation. To simplify matters, the assumption is often made that cash flows occur only at the end of the period (month or year) that is considered.

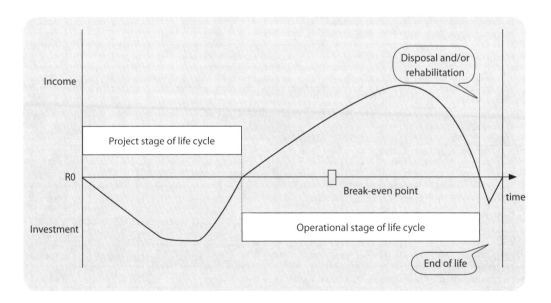

Figure 16.6 Cash flow during the economic life of a facility (adapted from Burke, 2003:72)

16.3.2 Payback period

The payback period is the time taken (measured in months or years) to gain a financial return equal to the original investment (Burke, 2003:60). This means that the non-discounted payback period for 'project' A in table 16.5 will be four years, as the initial (estimated) investment of R88m will be recovered by the end of year 4 after four (estimated) cash inflows of R22m each occurred. The payback period for 'project' B is three years and that of 'project' C is four years. The payback period method or technique states that project B should be selected because it will recover its capital outlay one year sooner than that of projects A and C. Note that the amounts in brackets at the beginning of year 1 (or year 0) indicate (estimated) cash outflows during the project stages of the facilities. The project stages of 'projects' A, B and C take place in less than a year and the investments made at the beginning of year 1 then result in the realisation of cash flows for seven years afterwards when the end of their operational lives have been reached. 'Projects' A, B or C could for example be machines that will be purchased to manufacture plastic bottles . The bottles will be sold over a period of seven years, after which the machines would be at the end of their lives and will then be sold for scrap. The estimated price at which the machines will be sold should be included in the last cash flow that will be realised during year 7.

Table 16.5 Projected cash flows for three facilities, products or projects

(end of) Year	Project A ('000')	Project B ('000')	Project C ('000')
0	(88 000)	(45 000)	(160 000)
1	22 000	5 000	40 000
2	22 000	20 000	45 000
3	22 000	20 000	40 000
4	22 000	5 000	35 000
5	22 000	5 000	32 000
6	22 000	12 000	28 000
7	18 000	10 000	29 000

The payback period technique has some advantages, such as that it is easy to use and that it reduces uncertainty about future cash flows. It also has disadvantages, e.g. it does not look at the total project. Projects or facilities that may generate good cash flows much later (e.g. after year 3 in this example) may therefore be penalised in favour of projects that generate reasonable to good cash flows much earlier. The undiscounted payback period technique does not consider time value of money, which may be a serious draw-back for especially long-term projects. In spite of its disadvantages, the payback period technique can be successfully used as an initial filter to determine whether a project shows some promise or not.

16.3.3 Return on investment (ROI)

ROI is a popular investment and project appraisal technique that – unlike the payback period – considers the whole project. ROI is calculated by means of the following two-step process.

Step 1: Calculate the average annual profit

This is done by deducting the total project outlay from the total gains and dividing the result by the number of years over which the venture will run.

$$\text{Average annual profit} = \frac{[(\text{Total gains}) - (\text{Total outlay})]}{\text{Number of years}} \qquad \textbf{Equation 16.6}$$

Step 2: Divide the average annual profit by the initial investment and convert that to a percentage by multiplying by 100.

$$\text{ROI} = \frac{\text{Average annual profit}}{\text{Original investment}} \times 100 \qquad \textbf{Equation 16.7}$$

Example 16.11 Calculate the ROI for project A (in table 16.5)

$$\text{Average annual profit for project A} = \frac{(22m \times 6 + 18m - 88m)}{7} = 8{,}857m$$

$$\text{ROI} = \frac{8{,}857}{88} \times 100 = 10{,}06\%$$

The ROI technique does not consider time value of money. Discounted cash-flow techniques such as net present value and internal rate of return are discussed next.

16.3.4 Net present value (NPV)

In the NPV technique, each forecast future cash flow is discounted to get its value at the beginning of the project. Cash inflows and (negative) outflows are added to see if the net effect will be positive or not. If positive, then the project should be considered, but if negative, the project should be abandoned.

This technique involves the following steps.
- Estimating future cash flows that may be generated by the project.
- Determining the risk profile of the project and decide on an appropriate discount rate to reflect the project's risk.
- Discounting the cash inflows by the discount rate (use equation 16.3) and added. Cash outflows such as the initial investment are also discounted if necessary and subtracted from the discounted cash inflows to obtain the NPV. If the result is positive, the project should result in an increase of shareholder's wealth and therefore be initiated by the company if the necessary finance can be obtained.

NPV of a project = present value of cash inflows − present value of cash outflows

The NPV of a project with a single cash outflow at the beginning of the project can be calculated by means of the following equation:

$$NPV= \sum_{t=1}^{n} \frac{CF_t}{(1+k)^t} - I$$

Equation 16.8

Where:

CF_t = cash flow at time t

I = the initial outlay or investment or cost of the project at the beginning of the project

k = discount rate

Example 16.12 Determining the NPV of project A using a discount rate of 14%

$$NPV_A = -88 + \frac{22}{(1,14)} + \frac{22}{(1,14)^2} + \frac{22}{(1,14)^3} + \frac{22}{(1,14)^4} + \frac{22}{(1,14)^5} + \frac{22}{(1,14)^6} + \frac{18}{(1,14)^7} =$$

R4,744m or as illustrated below.

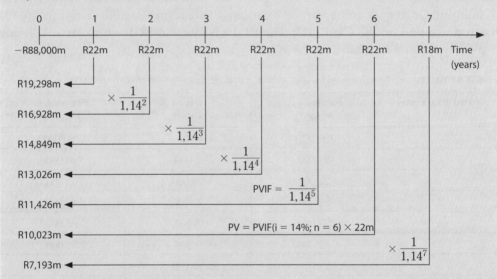

R4,744m net present value, discount rate = 14%

This can also be done by means of a spreadsheet – see the CD-ROM at the back of the book.

In practice, it may be difficult to estimate future cash flows. Just think how the prices of oil, platinum, coal and paper change over time and how that may impact on the viability of wells, mines and factories that produce such goods over relatively long periods of time (e.g. decades). A mining project may have a negative value at some point in time. This may, however, change if mineral prices increase, enabling the realisation of better future cash flows. Costs may also change significantly with time. Investors usually prefer stable inflation rates and exchange rates (especially if a significant percentage of their products are exported), since this makes decision making easier. In practice, several options will be investigated to improve the feasibility of a project.

16.3.5 Internal rate of return (IRR)

The IRR is the discount rate that causes the present value of all future cash flows of a project to equal the present value of the cost of the investment. It can therefore be said that the IRR is the rate for return for which the NPV of a project will be zero. If r is the internal rate of return and all initial cash outflows take place within one period at the beginning of the project, then:

$$\sum_{t=1}^{n}[CF_t(1+r)^t] - I = 0 \qquad \textbf{Equation 16.9}$$

Once the IRR of a project has been calculated, it must be compared with a *hurdle rate*, which is similar to the discount rate used in NPV calculations. A project will add value for shareholders if the IRR is greater than the hurdle rate. The IRR can be found through trial and error, or by using a spreadsheet as illustrated in the file that you will find on the CD-ROM. The ROIs, NPVs and IRRs of projects A, B and C are summarised in table 16.6.

Table 16.6 Projected cash flows for three facilities, products or projects and selection of the projects

(end of) Year	Project A ('000')	Project B ('000')	Project C ('000')
0	(88 000)	(45 000)	(160 000)
1	22 000	5 000	40 000
2	22 000	20 000	45 000
3	22 000	20 000	40 000
4	22 000	5 000	35 000
5	22 000	5 000	32 000
6	22 000	12 000	28 000
7	18 000	10 000	29 000
Payback period:	4 years	3 years	4 years
ROI:	10,06%	10,16%	7,95%
NPV:	R4,744m	R3,295m	– R1,599m
IRR:	16%	16%	14%

Example 16.13 Determining IRR through trial and error

Question:

A project is expected to generate the following cash flow stream:

Year	Cash flow (R'000)
0 (Investment)	– R1 500
1	R900
2	R800
3	R500

Determine the internal rate of return for this project. Is the project acceptable if the hurdle rate is 17%?

Answer:

By definition, *the internal rate of return is the discount rate for which the net present value of the cash-flow stream is equal to zero.* The IRR can be determined by trial and error according to the following method.

Determine the net present value at a discount rate of, say, 15% in order to determine whether the cash flow stream is negative or positive. The cash flow, at a discount rate of 15%, is still positive (see following table), which means that the internal rate of return is in excess of 15%. The discount rate is now increased to, say, 25%. The cash flow stream at a discount rate of 25% is negative, which means that the internal rate of return is somewhere between 15% and 25%.

Year	Discount rate = 0% (R'000)	Discount rate = 15% (R'000)	Discount rate = 25% (R'000)	Discount rate = 24% (R'000)
0	(1 500)	(1 500)	(1 500)	(1 500)
1	900	782,61	720,00	725,81
2	800	604,91	512,00	520,29
3	500	328,76	256,00	262,24
NPV	700	216,28	– 12,00	8,34

Because the NPV is close to 0 when a discount rate of 25% is used, a slightly smaller rate of 24% is used next. Since this rate produces a negative NPV, the IRR is between 24% and 25%. The IRR can be determined to a further degree of accuracy if the above method is further applied. The project will be accepted, because the IRR is in excess of the hurdle rate of 17%. (The IRR can also be solved by means of a spreadsheet – see the CD-ROM at the back of the book.)

The difference between NPV and IRR can be described as follows (Schilling, 2005:122): NPV asks, 'Given a particular level of expenditure, particular level(s) and rate of cash inflows, and a discount rate, what is the project worth today?' IRR asks instead, 'Given a particular level of expenditure and particular level(s) and rate of cash inflows, what rate or return does this project yield?' Because project expenditures have to be estimated and future cash inflows have to be forecast, it makes sense to use both best-case, worst-case and other 'what-if' scenarios (e.g. for a gold mining project you will ask: 'What is the best price that gold will probably reach over the next 30 years and what is the lowest price?').

16.3.6 Deciding on a discount rate and hurdle rate

In example 16.11 on page 348, a discount rate was provided. The question is, however, how to decide on a discount rate for a specific project. The risk associated with each project is somewhat unique, since most projects are unique. One thing is sure, however: a discount rate should be bigger than the interest rate provided by a bank on money invested. If that is not the case, why would investors invest their money in risky projects if the expected return is not higher than investing it at much lower risk with a bank? When a firm is faced with two investment opportunities with equal returns, it will select the project with the lower risk profile, because there is a greater chance that the riskier project may not deliver the expected returns. Companies (and individuals) require greater (expected or potential) returns for investment opportunities of greater risk.

The discount rate is the opportunity cost of capital for the project in question and the expected rate of return that could be earned from an investment of similar risk (Afuah, 2003:230). One method of determining the discount rate for a project is to use the capital asset pricing model (CAPM). The weighted cost of capital (WACC) of a company can also be used to estimate the discount rate. A company's cost of capital can be used as a hurdle rate against which the IRR of a project can be compared. This can only be done, however, if the project risk is comparable to the risk of the company as a whole. A further premium has to be added to the company's cost of capital if the project risk is greater than that of the company. Neither the CAPM nor the WACC methods will be discussed here in further detail. Those readers who would like to learn more about them should be able to find more information in most books on financial management.

16.3.7 The impact of taxation and depreciation on cash inflows

In the previous NPV and IRR examples, cash flows were provided. But how are they obtained? A project cash flow for a specific period is basically the income for that period minus the expenditure that was incurred during that period. Some of the other factors that impact on the cash flow are discussed below.

When an organisation buys equipment, that equipment will lose some of its value over its lifetime through wear and tear, until it is worth very little or nothing. At the end of its useful operational life, such equipment may only be worth the value of its scrap metal or salvageable parts. As equipment depreciates in value over its lifetime, such annual depreciation losses may be used to offset taxes on profits. Depreciation schedules can be obtained from income tax legislation (e.g. Income Tax Act 58 of 1962), tax interpretation notes and the South African Revenue Services (SARS). This specifies the amounts of depreciation that may be written off against company profits before the amount of tax to be paid to SARS is calculated. Taxation is a major expense for most companies. It is therefore important to determine the extent to which a project's cash flows will be affected by tax charges. The cost of assets is usually written off over a number of years in the form of a wear-and-tear allowance that is usually based on a straight-line method. Taxation issues could become fairly involved, especially for gold mining companies. Although example 16.14 does not reflect all the issues that need to be considered when taxation is brought into account, it does illustrate some of the additional issues that must be considered.

Example 16.14

Develop the cash-flow schedule for a small high-capacity mine based on the following information.

- Development period   2 years
- Production period   10 years
- Capital expenditure   R6m in first year
  (CAPEX)   R8m in second year
- Annual revenue   R5m in the first year of production
                                         R9m thereafter
- Annual operating   R2m in the first year of production
  expenditure (OPEX)   R3m thereafter
- Tax rate   40%

Answer:

End of year	CAPEX	Revenue	OPEX	Unredeemed CAPEX	Taxable profit	Tax	Cashflow
0	–	–	–	0	–	0	–
1	6	0	0	6	0	0	(6)
2	8	0	0	14	0	0	(8)
3	0	5	2	11	0	0	3
4	0	9	3	5	0	0	6
5	0	9	3	0	1	0,4	5,6
6	0	9	3	0	6	2,4	3,6
7	0	9	3	0	6	2,4	3,6
8	0	9	3	0	6	2,4	3,6
9	0	9	3	0	6	2,4	3,6
10	0	9	3	0	6	2,4	3,6
11	0	9	3	0	6	2,4	3,6
12	0	9	3	0	6	2,4	3,6

16.3.8 Considering inflation and escalation when selecting projects

Possibly no other single consideration is as important to the magnitude and relevance of economic evaluation results as the handling of inflation and escalation effects on project costs and revenues, especially in a high-inflation environment. The term 'inflation' refers to the average annual percentage increase in the prices of all goods and services that make up a Consumer Price Index-type basket of goods. Inflation can be accounted for in two ways: firstly by including it in the project discount rate (use a nominal discount rate) or by excluding it from the project discount rate (use a real discount rate). If the first approach is followed, all costs and expenses should be inflated too. This approach is often used, because taxation and its effects are more accurately determined. The price of an item, such as raw materials, should be increased every year due to inflation. The price of some items may increase at a rate higher than the inflation rate. Since the inflation rate is, by definition, the average price increase of goods produced (Production Price Index – PPI) or goods consumed (Consumer Price Index – CPI), it follows that the price of some goods may be above or below this rate. Escalation refers to changes in prices of specific goods and services, usually above the inflation rate, if a nominal discount rate is used. In low-inflation environments, where real discount rates are used, escalation refers to the increase in the cost of goods and services over time.

Interest rates offered by financial institutions should be higher than the inflation rate for investors to make an actual return. The real rate of return can be calculated by means of equation 16.10.

$$i_{real} = \frac{(1 + i_{nominal})}{(1 + i_{inflation})} - 1 \qquad \textbf{Equation 16.10}$$

Example 16.15 Real interest rate

A financial institution (e.g. a bank) tells you that it pays 8% interest on money invested with it. You know that the current inflation rate is 3,8%. Calculate the real interest rate that you will receive on your investment with the institution.

Answer:

From equation 16.10, $i_{real} = \frac{(1 + i_{nominal})}{(1 + i_{inflation})} - 1 - 1 = \frac{(1 + 0,08)}{(1 + 0,038)} - 1 = 0,04$ or 4,0%

16.4 THE SELECTION OF NEW PRODUCT DEVELOPMENT PROJECTS

Discounted cash-flow techniques (e.g. NPV and IRR), as discussed in the previous section, are not always the best way of selecting among new product development projects. One reason is that a radical innovation (see chapter 18 for a definition) is very different from other projects and it may be difficult to decide on an appropriate discount rate to use. DCF techniques may also discriminate against projects that are long and risky, while such projects may often be required for strategic reasons. Such projects may also result in new organisational capabilities and learning, which may result in future returns and options that may be difficult to predict now. To better incorporate strategic implications in the new product development investment decision, some people use option evaluation methods (Afuah, 2003:231; Schilling, 2005:125).

Self-assessment

The following questions can be used to test your knowledge. You will find suggested solutions for those questions marked by an asterisk (*) at the back of the book. You will also find some software that generates questions and model answers as well as some spreadsheet examples on the CD-ROM.

16.1   Explain the meaning of 'time value of money'.

16.2   Explain why money has earning power.

16.3* You have received a 13th cheque and want to invest a single amount of money with a financial institution. The aim is to grow this amount to R20 000 in five years' time. Calculate how much money must be invested now if an annual interest rate of 14% (compounded) applies.

16.4   Calculate how much must be invested now at 14% compounded annually so that R2 000 can be received five years hence.

16.5   A survey contractor is considering the acquisition of a piece of equipment that may have the following impact on his company's cash flow:

Year	Disbursements (cash outflows)	Receipts
0	50 000 (CAPEX)	0
1	10 200 (OPEX)	18 000
2	10 200 (OPEX)	28 000
3	5 500 (OPEX)	32 000
4	2 800 (OPEX)	24 000

If money is worth 10%, should the investment be made? The effective tax rate of the company is 38%. Capital expenditure may be written off over three years.

16.6   Determine the replacement cost of a piece of equipment in five years' time. The cost of this equipment now is R92 000 and the escalation rate is projected to be 7,6%. What uniform sum must be set aside each year if you want to establish a sinking fund that will retire this obligation in five years' time? Your company receives an interest rate of 14% from the bank on money invested.

16.7   A mining company plans to purchase a 35m³ truck for R270 000, which will be paid for in five equal installments at 16% interest. What will the payment amount be?

16.8   The projected cash flows for a deep underground goldmine from the first month of production until the end of its life are as follows:

Months after shaft sinking is completed	Cash flow per month (after tax)
1 to 12	R2m
13 to 24	R4m
25 to 36	R7m
37 to 48	R13m
49 to 300	R23m

Two shaft sinking and equipment companies have tendered for the sinking project. How much more would you be willing to pay company A if it plans to finish the sinking project nine months earlier than company B, which is planning to take 72 months? Assume that cash inflows and outflows are incurred at the end of each month. Use a discount rate of 12% per annum, compounded monthly.

16.9 The management of ABC Engineering works is considering overhauling a certain type of production machine that would reduce projected maintenance costs by an estimated R30 500 per machine per year over the next five years. If money costs 10%, what is the maximum amount that can be spent on an overhaul per machine?

16.10 A mining company plans to purchase a small LHD for R240 000, which will be paid for in five equal installments at 18% interest. What will the payment amount be? Determine the replacement cost of a piece of equipment in five years' time. The cost of this equipment now is R140 000 and the escalation rate is projected to be 7,6%. What uniform sum must be set aside each year if you want to establish a sinking fund that will retire this obligation in five years' time? Your company receives an interest rate of 14% from the bank on money invested.

16.11 Joe Smart is offered the alternative of receiving R4 000 today or R6 700 at the end of 10 years. If he accepts the R4 000 today, he will deposit it in a savings account. If the bank pays 5% interest, which alternative will Joe prefer?

References

Afuah, A. 2003. *Innovation Management: Strategies, Implementation and Profits*. New York: Oxford University Press.

Annels, A.E. 1991. *Mineral Deposit Evaluation*. London: Chapman & Hall.

Burke, R. 2003. *Project Management: Planning and Control Techniques*. Available at http://www.burkepublishing.com

Correia, C., Flynn, D. Uliana, E. & Wormald, M. 1993. *Financial Management*. 3rd edition. Cape Town: Juta.

Cronje, G.J., Du Toit, G.S., Mol, A.J., Van Reenen, M.J. & Motlatia, M.D.C. (eds). 1996. *Introduction to Business Management*. Johannesburg: International Thomson.

Holtzapple, M.T. & Reece, W.D. 2003. *Foundations of Engineering*. 2nd edition. Boston: McGraw-Hill.

Naismith, A. n.d. *Financial Valuation of Mining Projects*. Course material. University of the Witwatersrand.

Nel, W.P. 1999. Deuteronics Computerised Tutor 2000 Software.

Prince, L.J. 1963. *The Use of Present Values in Mining Economics Problems*. Johannesburg: SAIMM.

Riggs, J., Bedworth, D. & Randhawa, S. 1996. *Engineering Economics*. 4th edition. New York: McGraw-Hill.

Schilling, M.A. 2005. *Strategic Management of Technological Innovation*. Boston: McGraw-Hill Irwin.

Smith, L.D. 1995. 'Discount rates and risk assessment in mineral project evaluations.' *CIM Bulletin*, 88 (989), April, pp. 34–43.

Thuesen, G.J. & Fabrycky, W.J. 1993. *Engineering Economy*. 8th edition. Englewood Cliffs: Prentice-Hall.

Wanless, R.M., 1983. *Finance for Mine Management*. New York: Chapman & Hall.

Wheelwright, S.C. & Clark, K.B. 1992. 'Creating project plans to focus product development.' *Harvard Business Review*, March/April, pp. 70–82.

 Business and Technology Strategy

Jannie Lourens

Study objectives

After studying this chapter, you should be able to:

- define and differentiate among the concepts of strategy, strategic management, strategic planning, strategic thinking, vision, mission, SWOT analysis and competitive analysis
- identify external environmental factors that will impact on business or personal endeavours
- assess the impact of external environmental factors on the short-, medium- and long-term planning of an organisation
- apply the knowledge gained from this chapter to your work setting or organisation
- do basic strategic planning
- analyse the industry or product market in which your organisation is active or intends to become active
- visualise where and how you could contribute to the strategic direction of a company, institution, department, unit, team or own business

The aim of this chapter is to introduce the reader to strategic management issues and the factors that impact on strategic decisions.

17.1 THE ROLE OF THE SCIENTIST, TECHNOLOGIST OR ENGINEER IN STRATEGIC MANAGEMENT

Scientists, engineers and technologists may believe that strategy and strategic management are not for them. They may argue that they are scientific professionals and that any notions of strategy are for business people, therefore their contributions are limited. This is far removed from the truth, especially in today's world where technology is changing at an unrelenting pace and is an essential driver of many strategies. Ignoring technological change may soon render a business extinct.

Scientific professionals are playing essential roles in many business areas. The production and services functions in most businesses are based on some core technology. This could range from brick making or bread baking to microchip manufacturing or Internet service provision. The point is that all of the operations require a technology to operate and to do business. These technologies need to be updated, maintained, renewed or replaced with new, innovative methods in order to supply the same or new services or products of a higher quality and at better prices. It is the scientific professional's role to ensure technologies are developed, maintained or replaced in the most effective manner. In strategic planning, it is of paramount importance that the inputs of the technological professional are incorporated into the process.

In another important business function – research and development (R&D) – the essential role of the scientific professional is the development of new, innovative products and processes. In the absence of these, the company would be unable to replace obsolete products and processes or grow markets. Lack of innovation or rapid product development may cause a company to lose out on business, particularly if its competitors offer new products and services. Again, it is essential that the scientific professional's knowledge of new products, markets and process developments is incorporated into environmental scanning exercises.

Many technological developments have had dramatic influences on companies and have created new corporate structures and cultures. A case in point is the way information technology, intranets and the Internet have led to flatter and leaner organisations world-wide.

In activities such as engineering, construction and production, many scientific professionals play an important role in establishing cost-effective plant and operations. Cost effectiveness is usually an essential strategic driver in any business and an area where the scientific professional can play a significant role.

Many scientific professionals are employed in marketing and technical support services. Their knowledge of products, markets and associated developments in the industry is important in strategy formation and the vision-setting process. Close relationships are essential between the technical marketing professional and business counterparts in the strategic planning process. The scientific professional may wish to start a business, based on knowledge about markets, technology or innovative methods. Knowledge and the application of strategic planning, vision setting and environmental scanning will determine the difference between a successful, less successful or failing new enterprise.

17.2   WHAT IS STRATEGIC MANAGEMENT?

Why do some companies or organisations prosper in challenging, competitive times, while others, with products and services that are as good or even better, fail? The answer lies in the ability to out-think, out-plan and out-manoeuvre competition with a strategically successful positioning of the organisation.

There is no perfect model of strategic management for all organisations. There is a vast list of models and methods to choose from, but ideas must suit their time and situation. Any organisation's survival and future prosperity depend on its ability to formulate, implement and update effective strategies. Change places a premium on management that is responsive to threats and opportunities. Just as important, intensified international competition, which has become the norm, increases the pressure to have the 'best' strategy possible.

Strategic management of an organisation can be likened to a car journey along a highway to the corporate vision of where an organisation wants to be in five or ten years (Schultz, 1991). It proceeds with clear, concise, operational definitions of the vision, purpose, principles and values, and mission statements.

The vision statement will identify in a concise, clear and inspiring manner where the organisation will be in the planning horizon. The principles and value statements give direction for action. They are the 'kerbs' on the road, beyond which movement is not permitted. The mission statement identifies what must be done and why the organisation exists – its reason for being, to achieve the vision.

An organisation's vision is achieved through the implementation of certain activities. These activities are driven by critical objectives that must be achieved during the year. The critical objectives are derived from strategies, tactics and critical success factors that give direction on how to reach the vision. Critical objectives may, for example, be the reduction of waste and rework, the reduction of costs or improving processes to achieve greater customer satisfaction. Critical objectives should be based on thorough analysis of current conditions.

The path to the vision is not necessarily a straight one. The environment changes; 'potholes' or 'ruts' may appear in the road that need to be steered around. However, the vision must always be kept in sight. Once priorities have been established, the next step is to carefully plan the methods or means to achieve the goals. One must answer the questions why? when? where? who? and how? If these questions are answered sufficiently, an effective implementation programme has been created.

17.3 STRATEGY

Strategy is a blueprint of all entrepreneurial, competitive and functional area actions to be taken in pursuing organisational objectives and positioning an organisation for sustained success (Thompson & Strickland, 1987).

A strategy is a set of decision-making rules to guide organisational behaviour (Ansoff & McDonnell, 1990). There are four distinct types of rule:

- yardsticks against which the present and future performance of the company are measured. The quality of these yardsticks is usually called objectives and the desired quantity is called goals;
- rules for developing the company's relationship with its external environment: what products the firm will develop, where and to whom the products will be sold, and how the firm will gain advantage over competitors. This set of rules is called the product-market or business strategy;
- rules for establishing the internal relations and processes within the organisation. This is frequently called the organisational concept; and
- the rules by which the company conducts its day-to-day business, called operating policies.

A strategy has several distinguishing characteristics.

- The process of strategy formulation results in no immediate action. Rather, it sets the general directions in which the company will grow and develop.
- Therefore, strategy must be used to generate strategic projects through a search process.
- Strategy becomes unnecessary whenever the historical dynamics of an organisation will take it where it wants to go.
- Strategy formulation is based on highly aggregated, incomplete and uncertain information about different alternatives.
- The success of strategy requires strategic feedback.
- A strategy that is valid for one set of objectives may lose its validity when the objectives of the company are changed.

- Strategy and objectives are interchangeable at different points in time and at different levels of an organisation. Elements of strategy at a higher managerial level become objectives at a lower one.

Strategy is an elusive and somewhat abstract concept. Its formulation typically produces no immediate productive action. Above all, it is an expensive process in terms of both finance and managerial time.

17.3.1 Primary determinants of strategy

Many factors must be considered when formulating strategy. Six broad categories usually dominate strategic design:
- market opportunity, industry attractiveness and competitive forces;
- what a company's skills, capabilities and resources allow it to do best;
- emerging threats to the company's wellbeing and performance;
- the personal values, aspirations and vision of managers, especially those of the most senior executive;
- social, political, regulatory, ethical and economic aspects of the external environment in which the enterprise operates; and
- the organisation's culture, core beliefs and business philosophy (Thompson & Strickland, 1987).

The issues strategy must address are:
- how to respond to changing conditions, e.g. what to do about shifting customer needs and emerging industry trends; which new opportunities to pursue; and how to defend against competitive pressures and other externally imposed threats;
- how to allocate resources across the organisation's business units, divisions and functional departments (making decisions about the allocation of capital investment and human resources to the strategic plan is always critical);
- how to compete in the industry in which the organisation participates, e.g. decisions about how to develop customer appeal, positioning the firm against rivals, emphasising some products and de-emphasising others, and meeting specific competitive threats; and
- what actions and approaches to take in the major functional areas and operating departments in order to create a unified and more powerful strategic effort throughout the business unit.

Strategy formulation is largely an exercise in entrepreneurship (Thompson & Strickland, 1987). The content of a strategic plan reflects entrepreneurial judgements about the long-term direction of the organisation. It determines needs for major new initiatives and actions aimed at keeping the organisation in a position of sustained success. Specific entrepreneurial aspects of the strategy formation process include:
- searching actively for innovative ways in which the organisation can improve on what it is already doing;
- finding new opportunities for the organisation to pursue;

- developing ways to increase the company's competitive strength and place it in a stronger position to cope with competitive forces;
- devising ways to build and maintain competitive advantage;
- deciding how to meet threatening external developments;
- encouraging individuals throughout the organisation to offer innovative proposals and championing those that have promise;
- directing resources away from areas of poor results, towards areas of high or increasing results;
- deciding when and how to diversify; and
- choosing which businesses (or products) to abandon, which to emphasise and which to add.

17.4 STRATEGIC MANAGEMENT

Strategic management is the process whereby managers establish an organisation's long-term direction; set performance objectives; develop strategies to achieve these objectives, considering all the relevant internal and external circumstances; and implement the chosen action plans.

Strategic management involves more than strategic planning. It is based on the fundamental ideas of strategy and management and includes a range of management processes. In addition to planning, it includes implementation and supervision (Hahn, 1991). After thorough analysis and projection into the future, the main steps required are: identification of critical issues, generation, evaluation, selection, implementation and the control of strategic alternatives. Strategic management usually includes the definition of goals and objectives. These corporate goals need to be based on the shared values of top management, who largely determine the corporate culture. So, strategic management can be defined as involving the development and communication of an organisation's corporate goals, strategic plans, corporate philosophy and corporate culture.

An integrated strategic management process incorporates three components: the strategic plan, the operational plan and results management. It is tied together through the concepts of integration and communication, with clear recognition that planning is an ongoing people process (Below *et al.*, 1987).

The three components serve distinctly different purposes. The strategic plan focuses on the basic nature (mission) and direction (strategy) of the organisation. The operational plan concentrates on how to implement the strategic plan and produce short-term results. The results management component is concerned with comparing performance with plan (both strategic and operational) and ensuring the achievement of results. Although each component serves a different purpose, they are all highly integrated (see figure 17.1).

Strategic management activity is concerned with establishing objectives and goals for an organisation. It is also concerned with maintaining a set of relationships between the organisation and the environment. These relationships will enable the organisation to pursue its objectives, are consistent with the organisational capabilities and continue to be responsive to environmental demands (Ansoff & McDonnell, 1990).

One product of strategic management is the potential fulfilment of the organisation's objectives. These objectives consist of the input to the organisation, such as finance availability, personnel, information and raw materials. At the output end, developed products and/or services are tested for their potential profitability. Finally, a set of social behaviour rules exists that permit the organisation to meet its objectives.

Another product of strategic management is an internal structure and dynamics capable of continued responsiveness to changes in the external environment. In a business, this requires a managerial capability of interpreting environmental change coupled to a capability of conceiving and guiding strategic response, and an operational capability of conceiving, developing, testing and introducing new products and services.

While strategic management activity is concerned with creating a strategic position that assures the future environmental viability of the organisation, operations management is concerned with exploiting the current strategic position to achieve organisational objectives. In an organisation, strategic management is concerned with continued profitability potential; while operations management is concerned with converting the potential into actual profit. In operations management, the major activity is to establish levels of organisational output that will best contribute to the objectives. The product of operations activity is delivery of products and/or services to the environment in exchange for rewards.

17.4.1   Tasks and responsibilities of strategic management

The critical tasks and responsibilities of strategic management can be defined as those of:
- formulating the company's mission, purpose, philosophy and goals;
- developing an internal company profile;
- assessing the company's external environment;
- analysing the company's options by matching its resources with the external environment;
- identifying the most desirable options by evaluating options according to the company's mission;
- selecting a set of long-term objectives and grand strategies;
- implementing strategic choices by means of budgeted resource allocations; and
- evaluating and controlling the success of the strategic process as input for future decision making.

As these tasks indicate, strategic management involves the planning, directing, organising and controlling of a company's strategy-related decisions and actions. These are the basic managerial functions that were covered in chapter 2. Here, however, the emphasis is on strategic rather than general management.

17.4.2   Benefits of strategic management

Using the strategic management approach, managers at all levels of an organisation interact in planning and implementing. Therefore, the benefits of strategic management are similar to those of participative decision making.
- Subordinates should be aware of the needs of strategic planning and should aid management in planning, and in monitoring and forecasting responsibilities.
- The strategic management process results in better decisions, because group interaction generates a variety of strategies and forecasts based on the range of perspectives.

- Employee involvement in strategy formulation improves their understanding of productivity–reward relationships in every strategic plan and therefore heightens motivation.
- Resistance to change is reduced. Greater awareness of the parameters that limit options makes employees more likely to accept strategic decisions.

17.4.3 Strategic management process

Businesses vary in the processes they use to formulate and direct strategic management activities. Sophisticated planners have developed more detailed processes than less-formal planners of similar size. Small businesses that rely on the strategy formulation skills and limited time of an entrepreneur typically exhibit more basic planning models than those of larger businesses in their industries. Understandably, businesses with multiple products, markets or technologies tend to use complex strategic management systems. Despite differences in detail and the degree of formalisation, the basic components of the models used to analyse strategic management operations are very similar. A generic model for the strategic management process is shown in figure 17.1.

Figure 17.1 A generic model for the strategic management process

17.4.4 Strategic management as a process

A process is the flow of information through interrelated stages of analysis towards the achievement of an objective. The strategic management model in figure 17.1 depicts such a process. The flow of information involves historical, current and forecast data on the organisation's operations and environment.

Viewing strategic management as a process has important implications (Pearce & Robinson, 1991). Firstly, a change in any component will affect several or all of the other components, e.g. forces in the external environment may influence the nature of a company's mission. In turn, the company may affect the external environment and heighten competition in its realm of operation.

A second implication of viewing strategic management as a process is that strategy formulation and implementation are sequential. The process begins with the development or re-evaluation of the company's mission. This step is associated with, but essentially followed by, the development of a company profile and an assessment of the external environment. Then follow, in order, strategic choice, definition of long-term objectives, the design of the grand strategy, definition of short-term objectives, the design of operating strategies, implementation of the strategy, and review and evaluation.

A third consequence of viewing strategic management as a process is that feedback should be obtained from the implementation of the process. Such feedback must be used to evaluate the early stages of the process. Feedback can be defined as the collection of post-implementation results to enhance future decision making. Therefore, as indicated in figure 17.1, strategic managers should assess the impact of implemented strategies on external environments.

A fourth implication of viewing strategic management as a process is that it has to be regarded as a dynamic system. The term 'dynamic' characterises the constantly changing conditions that affect interrelated and interdependent strategic activities. Since change is continuous, the dynamic strategic planning process must be monitored constantly for significant shifts in any of its components as a precaution against implementing an obsolete strategy. The strategic management process must undergo continual assessment and updating. Although the elements of the basic strategic management model rarely change, the relative emphasis that each element receives varies with the decision makers that use the model and with the company's environment.

17.4.5 Components of strategic management models

The main components of most strategic management process models are listed below and discussed in detail later in this chapter:
- company vision;
- company mission;
- external and internal environmental assessment or scan;
- strategic analysis and choice;
- long-term objectives;
- grand strategy or generic strategy;
- annual or short-term objectives;
- functional strategies;
- policies;
- implementation of the strategy/ies; and
- control and evaluation of strategic plan/s.

17.5 STRATEGIC PLANNING AND STRATEGIC THINKING

Strategic planning consists of the following primary building blocks:
- defining the organisation's business and strategic mission;
- assessing the company's current position;

- identifying the company's desired position;
- evaluating the strategic gap between the two and the critical issues that need to be resolved to close the gap;
- establishing strategic objectives and performance targets; and
- formulating strategies and action steps to achieve targeted objectives and resolve critical issues.

It is clear that strategic planning is an important part of the strategic management process. It encompasses the three top phases of the strategic management model shown in figure 17.1. Strategic planning should not be confused with strategic management, but should be recognised as an essential part of the process.

17.5.1 Differences between strategic planning and strategic thinking

According to Mintzberg (1994), strategic planning is not strategic thinking. Indeed, according to him, strategic planning often spoils strategic thinking, causing managers to confuse real vision with the manipulation of numbers. This confusion lies at the heart of the issue: the most successful strategies are visions, not plans. Strategic planning, as it has been practised, has really been strategic programming – the articulation and elaboration of strategies or visions that already exist. In contrast, strategic thinking is about synthesis, and involves intuition and creativity. Loewen (1997) makes a distinction between strategic planning and thinking, as shown in table 17.1

Table 17.1 Differences between strategic planning and strategic thinking

STRATEGIC PLANNING	STRATEGIC THINKING
- Senior executives only - Head-office planners do the thinking and distribute 'The Plan' - Generic strategy process applied to any culture and business situation - Structured planning period, like budget period - Structured sessions with agenda - Established format - Correct answers and compliance with senior management - Plans and specific steps - Control with normative measurements; only financial - Formal	- Draws on all levels of the company - Strategy developed by people involved at client level - Strategy process to suit business needs - Continuous process, no timed output - Loose process, theme-based - Great variety, each focusing on a different angle of business - No immediate answer; combination of intuition and facts - Vision, creative, intuitive, project-based - Innovative measurement, linked to business plan, customer satisfaction and human resources - Chaotic, informal

Harvard Business School professor Michael Porter (1987) comments:

> The need for strategic thinking has never been greater .... What has been under attack is not these questions [about the direction of competition, needs of customers, importance of gaining competitive advantage, and so on], but the techniques and organisational processes which companies used to

answer them. There are no substitutes for strategic thinking. Improving quality is meaningless without knowing what kind of quality is relevant in competitive terms.

Nurturing corporate culture is useless unless the culture is aligned with a company's approach to competing. Entrepreneurship unguided by a strategic perspective is much more likely to fail than succeed. And, contrary to popular opinion, even Japanese companies plan.

17.6 STRATEGIC PLANNING MODEL

In figure 17.2 on the next page, a generic strategic planning process model is presented (Holman & Farkas, 1991). Each component – i.e. strategic vision, mission, internal and external environment assessment, strategic gaps and strategic issues – is then discussed in detail. It represents the sequential nature of the strategic planning process to reach the goal of strategic and implementation plans. However, as discussed previously, the process is, in reality, inherently iterative, with many feedback and feed-forward process loops.

17.6.1 Vision

Vision is the strategic intent which informs the whole organisation, not just individual components. It decides on the future state to which you want to move. This can include a description of how the organisation will look, or what it will achieve.

The function of leadership – i.e. the primary responsibility of a leader – is to form a clear and shared vision for the company and to secure commitment to a vigorous pursuit of that vision. This is a universal requirement of leadership, and no matter what his/her style, a leader must perform this function (Collins & Lazier, 1993). Vision is not necessary to make money; one can certainly create a profitable business without it. However, if you want to do more than just make a lot of money – if you want to build an enduring, great company – then you need a vision.

The leader spends time articulating, communicating and explaining the vision to everyone in the organisation. The vision becomes the magnet of everyone's activities and efforts (Robert, 1993). The vision gives answers to the following basic questions.
▓ Where is the organisation going?
▓ What will it 'look' like in the future?

Organisational direction setting should always begin with a clear concept and vision of what business the organisation is in and what path its development should take. The question, 'What is our business and what will it be?' can be answered in at least eight different ways (Thompson & Strickland, 1987), in terms of:
▓ the products or services provided;
▓ the principal ingredient in a line of products or services;
▓ the technology that spawns the product;
▓ the customer groups served;
▓ the customer needs and wants being met;
▓ the scope of activities within an industry;
▓ creating a diversified enterprise that engages in a group of related businesses; and
▓ creating a multi-industry portfolio of unrelated businesses.

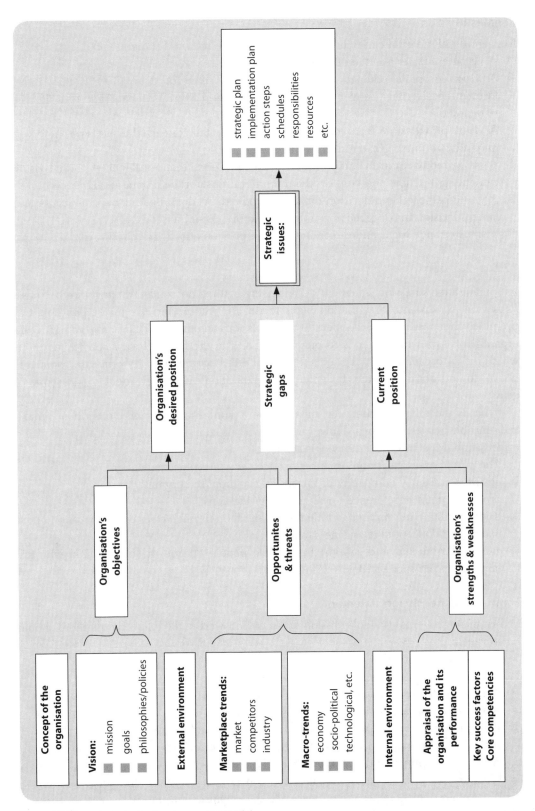

Figure 17.2 Generic strategic planning process model

Benefits of vision

Managing with a vision can benefit an organisation in five ways (Lipton, 1996).

- A vision enhances a range of performance measures.
- A vision promotes change. It serves as a road map for companies as they move through accelerated change, and eases the transition from a bureaucratic to a flexible organisation.
- A vision provides the basis for a strategic plan. A plan conceived as a vision, however, may prove a much greater incentive to action.
- A vision motivates individuals and facilitates the recruitment of talent. A shared vision can energise people and connect them to the purpose of the organisation.
- A vision helps to keep decision making in context. As organisations become leaner and flatter, decision making becomes inherently more decentralised.
- Visions provide focus and direction, and can provide effective control.

17.6.2 Mission

An organisation's mission statement describes the nature and concept of its future business. It establishes what the organisation plans to do, for whom, and the major philosophical premises under which it will operate. The mission statement specifies the activities the organisation as a whole intends to pursue now and in the future. It says something about what kind of organisation it is now and what it will become, and, by omission, what it is not to do and not to become. It depicts an organisation's character, identity and scope of activities.

What is the difference between mission and vision? Many organisations do not make a distinction between the two. However, vision is the grand goal or future purpose of the organisation. Mission is how the goal is achieved.

The primary reasons for an organisation to have a mission statement are to:

- ensure consistency and clarity of purpose throughout the organisation;
- provide a point of reference for all planning decisions;
- gain commitment from those within the organisation through clear communication of the nature and concept of its business; and
- gain understanding and support from those people outside the organisation who are important to its success (Below *et al.*, 1987).

Formulating the mission statement

In formulating an organisation's mission statement, answers to the following questions should be considered.

- What business should we be in?
- Why do we exist – what is our basic purpose?
- What is unique or distinctive about our organisation?
- Who are our principal customers, clients or users?
- What are our principal products or services, present and future?
- What are our principal market segments, present and future?
- What are our principal distribution channels?
- What are our principal economic/financial concerns and how will they be measured?
- What philosophical issues are important to our organisation's future?

- What is different about our business?
- What considerations do we have regarding our stakeholders?

17.6.3 Business environment assessments

The assessment of an organisation's current position answers two fundamental questions.

- What business is the organisation currently in?
- How well is the organisation doing in this business?

Making this assessment involves rigorous appraisal of the firm's operating performance or evaluation of the internal business environment. Secondly, it needs an analysis of the industry in which the organisation is competing, and a comprehensive analysis of the macro-economic, social, political and technological environments (or the external environment) in which the company operates.

This appraisal allows strategic planners to identify the important internal factors that contribute to the company's success and to look at these factors in terms of the company's primary strengths and weaknesses. Evaluation of the external environment produces a clear picture of current marketplace opportunities and threats. The mapping of company strengths and weaknesses against current opportunities, threats, risks and success determinants provides a concise assessment of the company's position.

A host of external factors influence an organisation's choice of direction. These factors, which constitute the external environment, can be divided into three interrelated sub-categories: the remote environment, the industry environment and the operating environment. Figure 17.3 on the next page suggests the interrelationship between the organisation and its remote, industry and operating environments. Combined, these factors form the basis of the opportunities and threats a firm faces in its competitive environment (Pearce & Robinson, 1991).

External environment

The external environment comprises factors originating beyond, and usually irrespective of, any single organisation's operating situation (economic, social, political, technological and ecological factors). The environment presents threats, opportunities and constraints, but rarely does a company exert any meaningful reciprocal influence on that environment.

Economic factors

Economic factors concern the nature and direction of the economy in which a firm operates. Because consumption patterns are affected by the relative affluence of various market segments, each firm must consider economic trends in the segments that affect its industry in its strategic planning. On both a national and an international level, it must consider the general availability of credit, the level of disposable income and the propensity of people to spend. Prime interest rates, inflation rates and trends in gross national product growth are other economic factors to consider.

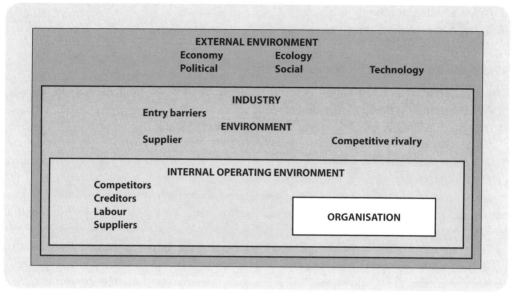

Figure 17.3 External, industry and internal environments of an organisation (Pearce & Robinson, 1991)

Social factors

The social factors affecting a company involve the beliefs, attitudes, opinions and values lifestyles of people in the external environment, developed from cultural, ecological, demographic, religious, educational and ethnic conditioning. As social attitudes change, so too does the demand for various types of clothing, books, leisure activities and so on. Like other forces in the remote external environment, social forces are dynamic, with constant change resulting from the efforts of individuals to satisfy their desires and needs by controlling and adapting to environmental factors.

Political factors

The direction and stability of political factors is a major consideration for managers in formulating company strategy. Political factors define the legal and regulatory parameters within which companies must operate. Political constraints include laws, tax programmes, wage legislation, pollution and pricing policies, and many other actions aimed at protecting employees, consumers, the general public and the environment. Since such laws and regulations are most commonly restrictive, they tend to reduce potential profits. However, some political actions are designed to benefit and protect businesses. Such actions include patent laws, government subsidies and product research grants. Political factors may either limit or benefit the companies they influence.

Technological factors

To avoid obsolescence and promote innovation, a business must be aware of technological changes that might influence its industry. Creative technological adaptations can suggest possibilities for new products, improvements in existing products, or new manufacturing and marketing techniques. A technological breakthrough can have a sudden and dramatic effect on the business environment. It may spawn sophisticated new markets and products or significantly shorten the anticipated life of a manufacturing facility. All firms, particularly those in turbulent growth industries, must strive for an understanding of existing technological advances and probable future advances that may affect their products and services.

Ecological factors

The most prominent factor in the remote environment is often the reciprocal relationship between business and ecology. The term 'ecology' refers to the relationships among humans and other living things, and the air, soil and water that support them.

Industry environment

Porter (1980) propelled the concept of industry environment into the foreground of strategic thought and business planning. The nature and degree of competition in an industry hinge on five forces: the threat of new entrants, the bargaining power of customers, the bargaining power of suppliers, the threat of substitute products or services (where applicable), and the jockeying among current contestants. To establish a strategic agenda for dealing with these contending currents and to grow in spite of them, a company must understand how they work in its industry and affect its particular situation.

Porter's Five Forces Model

Industry environment analysis investigates the industry's competitive process: the sources of competitive pressures, how strong these pressures are, what competitors are doing, and what future competitive conditions will be like (Thompson & Strickland, 1987).

Four areas of enquiry are essential.
1. What competitive forces exist and how strong are they?
2. What are the relative cost positions of rival firms in the industry?
3. What are the competitive positions and relative strengths of key rivals (what are their strategies, how well are they working, why are some rivals doing better than others, and who has what kind of competitive edge)?
4. What moves can key rivals be expected to make?

This phase of business strategy analysis is particularly important, because competitive forces shape strategy and because the strategies of rival firms shape competitive forces. The Five Forces Model explains the nature and intensity of competition in an industry.

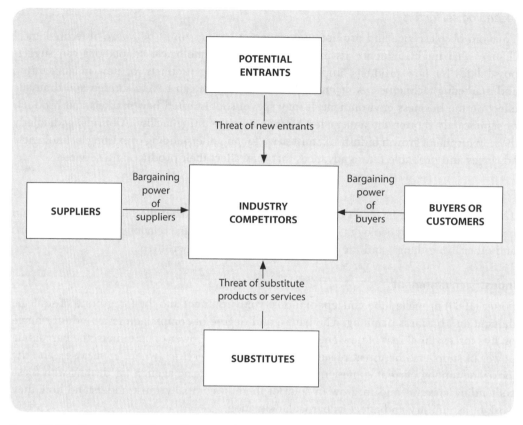

Figure 17.4 The five competitive forces (Porter, 1980)

Generally, competition in the marketplace is a function of five competitive forces:
▪ the jockeying for position among rival companies that flows from their strategic moves and countermoves to gain competitive advantage;
▪ the competitive intrusions and threats from the substitute products of companies in other industries;
▪ the potential entry of new competitors;
▪ the economic power and bargaining leverage of suppliers; and
▪ the economic power and bargaining leverage of customers.

a) Rivalry among companies

The competitive arena is dominated by the competitive manoeuvring of rival companies. An organisation's competitive strategy is the portion of business strategy dealing with its action plans for achieving market success and gaining a competitive edge. Many factors influence rivalry, of which the following are the most common.
▪ Rivalry tends to intensify as the number of competitors increases and as they become equal in size and capability.

- Rivalry is usually stronger when demand for the product is growing slowly.
- Rivalry is more intense when competitors are tempted by industry conditions to use price cuts or other competitive methods to boost unit volume.
- Rivalry is stronger when the products and services of competitors are so weakly differentiated that customers incur low costs in switching from one brand to another.
- Rivalry tends to be more vigorous when it costs more to get out of a business than to stay in and compete. The higher the exit barriers (and the more it costs to abandon a market), the stronger the incentive for firms to remain and compete.
- Rivalry becomes more volatile and unpredictable the more diverse competitors are in terms of their strategies, personalities, corporate priorities, resources and countries of origin.
- Rivalry is increased when strong companies outside the industry acquire weak firms in the industry and launch aggressive, well-funded moves to transform the newly acquired competitor into a major market contender.

b) Potential new entry

New entrants to a market bring new production capacity, the desire to establish a secure place in the market and, often, large resources with which to compete. Just how serious the competitive threat of entry is to a particular market depends on two factors: barriers to entry and the expected reaction of companies to the new entry. A barrier to entry exists whenever it is hard for a newcomer to break into the market. If the economics of the business places a potential entrant at a price/cost disadvantage compared to its competitors, this too is a barrier. There are several entry barriers.

- **Economies of scale:** Important scale economies deter entry, because the potential entrant is forced to enter the market on a large-scale basis or accept a cost disadvantage (and lower profitability). Scale-related barriers may be found in advertising, marketing and distribution, financing, after-sales customer service, raw materials purchasing, research and development, and production.
- **Learning- and experience-curve effects:** When achieving lower unit costs is a function of experience in producing the product and other learning curve benefits, a new entrant is at a disadvantage when competing with established firms with more accumulated know-how.
- **Brand preferences and customer loyalty:** When the products of rival sellers are well differentiated, buyers usually have some degree of attachment to existing brands.
- **Capital requirements:** The greater the investment needed to enter the market successfully, the more limited the pool of potential entrants.
- **Cost disadvantages independent of size:** Existing firms may enjoy cost advantages that are unavailable to potential entrants, regardless of the entrant's size. Sources of cost advantage include access to the best and cheapest raw materials, possession of patents and proprietary technology, and the benefits of any learning- and experience-curve effects.
- **Access to distribution channels:** When a product is distributed through established market channels, a potential entrant may face the barrier of distribution access. Some distributors may be reluctant to take on a product that lacks buyer recognition.

■ **Government actions and policies:** Government agencies can limit or even bar entry by instituting controls over licences and permits. Regulated industries all feature government-controlled entry. Entry can also be restricted, and certainly made more expensive, by stringent government-mandated safety regulations and environmental standards.

c) The power of substitute products

Businesses in one industry are often in close competition with those of another industry, because their respective products are good substitutes, e.g. consider the manufacturers of packaging methods and materials. The producers of plastic containers are in competition with the makers of glass bottles and jars, the manufacturers of paperboard cartons, and the producers of tin and aluminium cans.

The competitive force of closely related substitute products affects sellers in several ways. Firstly, the price and availability of acceptable substitutes places a ceiling on the price producers of the product can charge. At the same time, the ceiling price places a limit on profit potential.

Secondly, unless the sellers of the product can upgrade quality, reduce prices via cost reduction, or otherwise differentiate their product from substitutes, they risk a low growth rate in sales and profits because of the inroads substitutes may make. The more sensitive the sales of the product are to changes in the prices of substitutes, the stronger the competitive influence of the substitutes.

Thirdly, the competition from substitutes is affected by the ease with which buyers can change over to a substitute. A key consideration is usually the buyer's switching costs – the one-time cost the buyer faces in switching from a product to a substitute.

d) The power of suppliers

The competitive impact suppliers can have on an industry is a function of how significant the input they supply is to the buyer. The extent to which this potential impact is realised depends on a number of factors. In general, a group of supplier firms has more bargaining power when:
■ the input is important to the buyer;
■ suppliers' products are differentiated to the extent that it is difficult or costly for buyers to switch from one supplier to another;
■ suppliers of input do not have to compete with the substitute inputs of suppliers in other industries;
■ one or more suppliers pose a credible threat of forward integration into the business of the buyer industry; and
■ the buying firms show no attempt toward backward integration into the suppliers' business.

The power of suppliers can be an important economic factor in the marketplace because of the impact they can have on their customers' profits. Powerful suppliers can squeeze the profits of a customer industry through price increases, which the latter is unable to pass on fully to the buyers of its products.

e) The power of customers

Just as powerful suppliers can exert a competitive influence over an industry, so too can customers. The leverage and bargaining power of customers tends to be greater when:

- customers are few and when they purchase large quantities;
- customers' purchases represent a sizeable percentage of the selling industry's total sales;
- the supplying industry comprises many relatively small sellers;
- the item being purchased is standardised so that customers can not only find alternative sellers, but can also switch suppliers easily;
- customers pose a credible threat of backward integration; and
- sellers pose little threat of forward integration into their customers' product market.

A company can enhance its profitability and market standing by seeking out customers who are in a comparatively weak position to exercise power. Rarely are all buyer groups in a position to exercise equal bargaining power, and some may be less sensitive to price, quality or service than others.

The analytical contribution of the Five Forces Model is the systematic way it exposes the make-up of competitive forces. Analysis of the competitive environment requires that the strength of each of the five competitive forces be assessed. The collective impact of these forces determines competition in a given market and, ultimately, the profits industry participants will be able to earn.

It makes sense for a business to search for a market position and competitive approach that will insulate it as much as possible from the forces of competition. It can also attempt to influence the industry's competitive rules in its favour, to give it a strong position from which to compete.

Internal operating environment

The internal operating environment – sometimes called the competitive or task environment – comprises factors in the competitive situation that affect a company's success in acquiring resources or in profitably marketing its goods and services. Among the most important of these factors are the firm's competitive position, its customer composition, its reputation among suppliers and creditors, and its ability to attract capable employees. The operating environment is typically much more subject to the firm's influence or control than the remote environment.

SWOT analysis

Strategic managers gauge the strategic significance of a company's internal competencies based on the opportunities and threats present in its competitive industry environment. The essence of a well-formulated strategy is that it achieves an appropriate match between a company's opportunities and threats, and its strengths and weaknesses.

SWOT is an acronym for the internal strengths and weaknesses of a firm, and the environmental opportunities and threats facing that firm (Pearce & Robinson, 1991). SWOT analysis is a systematic identification of these factors and the strategy that represents the best match between them. It is based on the assumption that an effective strategy maximises strengths and opportunities, and minimises weaknesses and threats. Accurately applied, this simple assumption has powerful implications for the design of a successful strategy.

- **Strengths:** A strength is a resource, skill, or other advantage relative to competitors and the needs of the markets a company serves or expects to serve. A distinct competence gives the firm a comparative advantage in the marketplace. Strengths may include financial resources, image, market leadership and buyer/supplier relations.
- **Weaknesses:** A weakness is a limitation or deficiency in resources, skills or capabilities that seriously impedes a company's effective performance. Facilities, financial resources, management capabilities, marketing skills and brand image can be sources of weaknesses.
- **Opportunities:** An opportunity is a favourable situation in a company's environment. Key trends are one source of opportunities. Identifying a previously overlooked market segment, changes in competitive or regulatory circumstances, technological changes, and improved buyer or supplier relationships could represent opportunities for the company.
- **Threats:** A threat is an unfavourable situation in a company's environment. Threats are key impediments to the company's current or desired position. New competitors, slow market growth, increased bargaining power of key buyers or suppliers, technological changes, and new or revised regulations could represent threats to a company's success.

SWOT analysis can be used in many ways to aid strategy analysis. The most common way is to use it as a logical framework guiding systematic discussion of a company's situation and the basic alternatives that can be considered. What one manager sees as an opportunity, another may see as a potential threat. Similarly, a strength for one manager may be a weakness for another.

In this way, different assessments may reflect differing factual perspectives. The key point is that systematic SWOT analysis ranges across all aspects of a company's situation. Consequently, it provides a dynamic and useful framework for strategic analysis.

A second way in which SWOT analysis can be used is where opportunities and threats are systematically compared with internal strengths and weaknesses in a structured approach. The objective is to identify matches between a firm's internal and external situations.

Two other essential areas of internal environment assessment are the concepts *of core competencies* and *critical success factors* (CSFs). Identifying these, or the lack of them, will lead to the identification of the relevant strategic gaps, issues and/or niches an organisation should either develop or exploit in formulating and implementing its strategies.

Assessment of core competencies

The term 'core competence' refers to a skill or activity a company does especially well compared to its rivals. Superior capability in some competitively important aspect of creating, producing or marketing the company's product or service can be the vehicle for establishing a competitive advantage and then leveraging this advantage into improved business performance. The test for core competencies is an ability to meet the following criteria (Tampoe, 1994). They must be:

- essential to corporate survival in the short and long term;
- invisible to competitors;

- difficult to imitate;
- unique to the corporation;
- a capability the organisation can sustain over time; and
- essential to the development of core products.

The CSF concept

Critical success factors are those things that must go well to ensure success for a manager or an organisation. They represent managerial or enterprise areas that must be given special and continual attention to bring about high performance. CSFs include issues vital to an organisation's current operating activities and to its future success (Boynton, 1984).

Several questions are important here (Priest & Ritsema, 1993).
- What distinguishes a company's products or services from its competition?
- Which functional activities should the company be able to perform to perfection, to maintain its competitive edge?
- Which trends and developments in the company's environment can undermine its competitive advantage?

17.6.4 An organisation's desired position

An organisation's desired position is developed by mapping its strategic vision of itself against the opportunities, threats and risks that shape its business environment over the planning horizon (Holman & Farkas, 1991). This is then translated into a set of longer-term objectives that, if realised, will successfully exploit the opportunities and eliminate or manage the threats and opportunities. The organisation should describe its desired position in terms of objectives in the following areas:
- industry and marketplace position;
- revenue size and growth rate;
- products and services produced;
- financial performance;
- technological objectives;
- organisational structure and culture, etc.

The objectives are stated in quantifiable, measurable terms, against specific deadlines, and allocated to responsible individual managers. Progress can be evaluated and management can clearly determine when the objectives have been met.

17.6.5 The strategic gap and critical issues

Figure 17.2 on page 367 shows that the strategic gap is the difference between a company's current position and its desired strategic position (Harrison & Pelletier, 1998). Operationally, the gap is a measure of the difference between the stated target values and current values. Within the gap are the critical issues that must be resolved to move the company to its desired position. A critical strategic issue is a forthcoming development, either inside or outside of the organisation, that is likely to have an important impact (positive or negative) on the ability of the enterprise to meet its objectives (Ansoff & McDonnell, 1990). Critical issues must be ordered according to their overall importance and time frame for resolution.

17.6.6 Formulating strategies and action steps

The final step in the strategic planning process requires an organisation to develop strategies to manage successfully the identified critical issues. Strategies are necessary for each critical function of the organisation, such as financial, technology, marketing, operations or manufacturing, and personnel strategies. Each of the functional strategies must align with the generic business strategy selected by the organisation. It requires the organisation to formulate specific actions that will be carried out to implement each strategy. Technology strategy and how it relates to the business strategy are discussed in section 17.6.7, below.

There are two basic generic business strategy options to deal with the business environment and to compete with competitive rivals in the market. They are called the *cost leadership* and *differentiation* strategies. A cost leadership strategy is followed where competitiveness is achieved through low-cost manufacturing and supply of generally commodity-type products that can easily be compared with those of another company in the same industry, e.g. cement, motor cars, furniture, detergents, paints, etc. However, by following a differentiation strategy, an organisation differentiates its products through, for example, adding other features or extra functionality that differentiates its products from its competitors. Consumers who value the additional features or functionality of the differentiated product may therefore be willing to pay a bit more for it as long as they perceive it to be 'good value for money' (or higher quality).

A cost leadership strategy always require the continuous pursuit of cost reductions from previous experience, improved operations, improved technology development and application, tight cost control, and cost containment in all functions such as marketing, personnel, R&D, advertising, services and sales expenses. A lower cost of manufacturing the product or delivering a service relative to competitors' products or services becomes the theme of the strategy, although quality, support services and other functional areas to assist the effective supply of the product or service cannot be ignored. It may require the design of products for easy manufacturing, maintaining a wide line of related products to spread overhead costs, and large up-front capital investments to invest in the latest equipment and technology, and aggressive marketing and pricing to build the market share necessary to compete.

A differentiation strategy requires the continuous pursuit of product or service diversification to stay in front of competitors who may want to follow a similar strategy or who have copied your product idea. It means the creation of a product or service that is perceived as unique or different by the industry and the customer. Achieving differentiation will require a trade-off with cost if the activities required in creating it are inherently costly, such as extensive R&D, application of the latest technology, innovative product designs, high-quality materials, or intensive customer services and support. Differentiated products are commonly more expensive compared to the commodity equivalents due to their enhanced features or additional services and are therefore normally sold at a premium (e.g. compare what you pay and what you get when buying a small basic Ford 1600 as opposed to a large BMW 7 series motor vehicle). Not all customers are able and willing to pay the required higher prices, but what the organisation loses in terms of volume it often gains in extra product margins.

Often a third generic strategy called a *focus strategy* is followed. It is a combination of the cost leadership and differentiation strategies. By following this strategy, the organisation attempts to deliver differentiated products or services and achieve lower costs as well in serving a target market. It may not achieve this position within the industry as a whole, but may achieve it in its selected target market. In achieving this position, the organisation may earn above-average financial returns for its industry. Clearly, to be successful in this type of strategy, the organisation will have to be superior in most aspects of its product delivery.

Having selected a particular generic strategy the strategic action steps are detailed in terms of:

- the specific content;
- time frame;
- the individual's responsible to complete the actions;
- resources required, and;
- measurement standards.

Utilising this format, the organisation can monitor the action steps and further refine them as it implements its strategic business plan and subordinate functional strategies over time.

17.6.7 Relationships between generic business strategy and technology strategy

In section 17.6, the generic strategic planning model is explained. This model can be applied at varies levels in the organisation. Its primary function is to assist with the planning of the overall business strategy, but it can also be applied effectively in planning other functional strategies in the organisation such as the technology strategy. Figure 17.5 is an adaptation of figure 17.2 to illustrate how the generic strategy planning model is applied for formulating technology strategy.

The technology strategy model illustrates that the business strategy and the technology strategy are mutually dependent upon each other. Certain business strategies may require that the technology strategy has to be adjusted to accommodate the business strategy, while the available technology base often determines the extent and speed at which a predetermined business strategy can be executed. For example, a low-cost strategy usually requires process innovation, as cheaper and more efficient processes are required to manufacture the same or similar products. A generic differentiation strategy, on the other hand, usually requires product (or product group) innovation to provide additional features/functionality or totally new products that address the needs of consumers (or market segments) better than existing products.

The technology strategy of the organisation involves decision making regarding the following:

- the organisational R&D, licensing and technology acquisition policies and practices;
- the selection of technology portfolio components, based upon a thorough technology audit that will determine the organisation's technology strengths, weaknesses, threats and opportunities. The gaps so analysed through an audit will assist in determining what technology to develop and how to exploit the organisation's technology portfolio;
- the sources from which technologies are acquired – whether it is internal R&D or acquisition from external technology suppliers through licensing, joint development or straight acquisition;
- the level of R&D investment needed to support the business and other functional strategies;

- the desired level of competence in each technology. Depending on which technologies will be acquired and which technologies will be internally developed, the company will determine the level of skills and competencies necessary to either acquire or develop for it to effectively exploit its technology portfolio; and
- the competitive timing of market entry of a new products or services. This is dependent upon the marketing strategy of the business and the readiness of the technology.

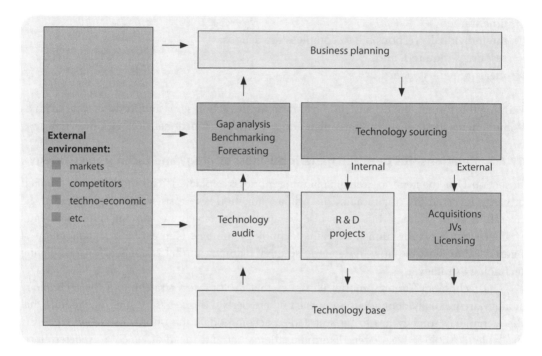

Figure 17.5 Technology Strategy Model

This section describes the link between business and technology strategy. Technology strategy itself is discussed in chapter 18.

17.6.8 Evaluation and control

Once a strategy is implemented, one cannot simply assume that its outcome will attain the managerial objectives. A system of follow-up, evaluation and control is essential to ensure that the results agree with those expected at the time the strategy was formulated. This system involves three steps:

- establishing standards derived from the managerial objectives;
- measuring actual performance against the established standards; and
- management initiating timely corrective action to return actual performance to an acceptable level of congruence with the standards, if actual performance reflects an unacceptable variance from the standards.

In this way, the implemented strategy is more likely to produce an outcome that attains the objectives and reflects effective strategic management.

Self-assessment

17.1 Describe the role of engineers in strategic management.

17.2 Use Michael Porter's Five Forces Model to analyse the current state of any industry of your choice. Examples that may be used are the gold mining industry, motor industry and energy industry.

17.3 Define business strategy.

17.4 List the factors a company has to consider when formulating strategy.

17.5 List the tasks and responsibilities of strategic management.

17.6 Differentiate between strategic planning and strategic thinking.

17.7 What is the vision and mission of your company? Does it meet the criteria described in this chapter?

17.8 List the five forces upon which the nature and intensity of competition in industry hinges.

17.9 Define the core competence of an organisation.

17.10 Define critical success factors.

17.11 Explain the link between business strategy and technology strategy.

17.12 Briefly describe the generic business strategies of cost leadership and differentiation. Give examples of companies that have implemented these strategies.

References

Ansoff, H.I. & McDonnell, E.J. 1990. *Implanting Strategic Management.* 2nd edition. New York: Prentice-Hall.

Below, P.J., Morrisey, G.L. & Acomb, B.L. 1987. *The Executive Guide to Strategic Planning.* San Francisco: Jossey-Bass.

Boynton, A.C. 1984. 'An assessment of critical success factors.' *Sloan Management Review*, Summer, pp. 17–27.

Collins, J.C. & Lazier, W.C. 1993. 'Vision.' *IEEE Engineering Management*, 21 (1), pp. 61–75.

Hahn, D. 1991. 'Strategic management: Tasks and challenges in the 1990s.' *Long Range Planning*, 24, pp. 26–39.

Harrison, E.F. & Pelletier, M.A. 1998. 'Foundations of strategic decision effectiveness.' *Management Decision*, 36 (3), pp. 147–59.

Holman, W.R. & Farkas, A.L. 1991. 'A step-by-step model for strategic planning.' In Glass, H.E. (ed.). *Handbook of Business Strategy.* 2nd edition. Boston: Warren, Gorham & Lamont.

Lipton, M. 1996. 'Demystifying the development of an organisational vision.' *Sloan Management Review*, Summer.

Loewen, J. 1997. *The Power of Strategy: A Practical Guide for South African Managers.* Johannesburg: Creda Press.

Mintzberg, H. 1994. 'The fall and rise of strategic planning.' *Harvard Business Review*, Jan.–Feb., pp. 107–14.

Pearce, J.A. & Robinson, R.B. 1991. *Formulation, Implementation and Control of Competitive Strategy.* 4th edition. Boston: Irwin.

Priest, B. & Ritsema, H. 1993. 'Corporate strategy: Implementation and control.' *European Management Journal,* 11 (1), pp. 122–30.

Porter, M.E. 1980. *Competitive Strategy: Techniques for Analyzing Industries and Competitors.* New York: Free Press.

Porter, M.E. 1987. 'The state of strategic thinking.' *The Economist,* May.

Robert, M. 1993. *Strategy Pure and Simple.* New York: McGraw-Hill.

Schultz, L.E. 1991. 'The rings of management: The new management theory.' *Human Systems Management,* 10, pp. 11–17.

Tampoe, M. 1994. 'Exploiting the core competencies of your organisation.' *Long Range Planning,* 27 (4), pp. 66–77.

Thompson, A.A. & Strickland, A.J. 1987. *Strategic Management: Concepts and Cases.* 4th edition. Piano: Business Publications.

18 | Managing Technology and Innovation

Darryl Aberdein

Study objectives

After studying this chapter, you should be able to:

- define and describe the discipline of technology and innovation management
- classify and understand the differences between the various types of innovation
- understand the various models of technological and industry evolution and describe the patterns that occur
- understand what organisational factors support the creation and maintenance of an innovative organisational environment
- describe the elements of a technology strategy and understand how these are related to a business strategy
- understand the challenges that face marketers of new technological products and how these challenges can be overcome
- understand the factors that contribute to successful new-product development and be able to describe the main process used when developing new products
- describe the ways in which the benefits of innovation can be appropriated
- understand the mechanisms and processes involved in the various statutory protection mechanisms

The aim of this chapter is to introduce readers to the body of knowledge that is concerned with the commercialisation and development of new technologies.

> One should recognise and manage innovation as it really is – a tumultuous,
> somewhat random, interactive learning process linking a world-wide network of
> knowledge sources to the subtle unpredictability of customer's end uses.
>
> James Brian Quinn, 1985

18.1 DEFINING TECHNOLOGY MANAGEMENT

18.1.1 What is technology and innovation management?

Broadly defined, technology management covers all aspects of planning, organising, leading and controlling in technology-intensive business environments. Many other technical management disciplines, such as production and operations management, already describe the management of activities concerned with the *use* of technology (the blocks in the second column of table 18.1). Consequently, a narrower definition of technology management, focused mainly on the management of innovation and how innovation interacts with other disciplines, has emerged (the blocks in the third column of table 18.1).

Innovation management complements and expands on the other technical management disciplines by focusing on that body of knowledge concerned with the creation and commercialisation of new technologies.

Table 18.1 Management of technology typology

	Use of technology	Creation of technology
Rapidly changing technologies	Managing organisational response to the introduction of rapidly changing technologies	Managing highly skilled technologists to create and commercialise new technologies, products or processes
Slowly changing technologies	Developing organisations that make the most efficient and cost-effective use of established technologies	Managing the most efficient and cost-effective creation and commercialisation of new products and processes based on stable technologies

Innovation has two parts: (1) the generation of an idea or invention, and (2) the conversion of that idea into a business or other useful application. Thus:

Innovation = Invention + Commercialisation

This chapter deals with the management of these two aspects of innovation. The study of idea generation and invention is mainly concerned with what is often called research and development (R&D). Commercialisation is mainly concerned with issues such as the marketing and production of new products and processes.

18.1.2 The innovation process

The innovation process can be broken down into a number of steps that are common to most innovation activities. These are as follows.

▪ **Basic research:** This is usually based on one or more of the natural sciences and involves studies that involve the understanding of how the laws of nature regulate the world around us. An example of this is the human genome project, which identifies the genes in human cells.

▪ **Applied research:** This takes the research one step closer to a commercial product by looking at a potential application for the basic research. An example of this is perhaps the identification of the specific gene (or gene combination) that influences baldness. The output of applied research is often called a technology.

▪ **Idea generation:** At some point in the process, an idea for a potentially marketable product or process results from the research. An example of this is perhaps the idea of a product that treats baldness. In some cases, innovations require no formal research or use technologies that were developed many years previously and are simply new ideas or insights that can then be developed further.

- **Product and/or process development:** This is the development of a product up to the stage of being marketable and manufacturable. In our example, this would mean that a product that treats baldness has been formulated and tested. Specific drug delivery systems might have been developed, as well as special machines and processes to manufacture the product.
- **Market entry:** During this phase, the production and marketing of the product are commenced.

The innovation process is rarely simple and straightforward and often requires one to backtrack after following paths that turn out to be dead-ends. In particular, the research steps sometimes yield technologies that are never commercialised, or are used in unexpected ways many years into the future. Who would have guessed that research into quantum mechanics and the laser would play such an important role today in products like fibre optic cables and DVDs?

It is rare for one company to be involved in all stages of the innovation process. Research, for example, is often the domain of the universities, which will disseminate their knowledge through publications, lectures and the like. More commercially minded companies will take this research further to develop useful products, services or manufacturing processes. It is also unlikely that any one company will be involved in marketing the innovation and it is quite usual that a network of distributors or resellers will be used. It is clear that successful innovation is often more dependant on designing the right organisational model, selecting the right partners and managing the interactions than on the technology itself.

18.1.3 Classification of innovation

Many different classifications of technological innovation have been developed. Each depends upon the perspective of the user, and thus each is useful in describing innovation in different circumstances. A particular innovation may fall into a number of different innovation categories, depending on the perspective.

The first classification of innovation is that of being incremental innovation, radical innovation or transformational innovation.

- **Incremental innovation** occurs when small improvements are made to a product or the processes used in manufacturing a product. These changes generally extend the competencies of the innovator. An example of this might be temperature compensators on the balance wheels of wristwatches in order to enhance their accuracy.
- **Radical innovation** occurs when major improvements are made to a product. These changes often make the competencies involved in the old technologies obsolete and sometimes require new marketing channels to be developed. An example of this would be the quartz watch, which uses new electronic technologies and does not require a network of skilled watchmakers for after-sales service.
- **Transformational innovation** occurs when the innovation is of such a fundamental nature that it enables the development of many other innovations. Examples of these are recent innovations in computing and communications, biotechnology and polymeric materials.

Example 18.1 Classification dilemmas

Classifying technology as being incremental, radical or transformational is not always easy. What is a radical innovation for car brake manufacturers (when disk brakes replaced drum brakes) would be an incremental innovation for the car industry. One litmus test that is useful in this regard is based on the following distinction.

- **Incremental innovation** strengthens the *firm's* competitive position and entrenches the nature of the *industry* by making it more competitive.

- **Radical innovation** puts *firms* out of business and changes the nature of the *industry* in which the innovation occurs.

- **Transformational innovation** destroys whole *industries* and changes the nature of *society*.

Another useful classification of innovation is into the categories of product innovation, process innovation or service innovation.

- **Product innovation** results in new or improved products. An example of this might be a new type of razor blade that is sharper and lasts longer than previous blades.
- **Process innovation** occurs when the manufacturing processes are improved to make the production of existing products cheaper or when new processes are developed specifically for making a new or improved product. An example of this would be the development of a manufacturing technology to spray-coat razor blades with graphite for extra smoothness.
- **Service innovation** occurs when new ways of delivering services are developed. An example of this form of innovation would be the use of automatic teller machines (ATMs) in banks to replace human tellers.

Example 18.2 Innovation types

The distinction of these different types of innovation is not always clear and often depends on one's perspective. To a client of a bank, ATMs might represent service innovation; to the staff of a bank, it might represent process innovation; and to the supplier of the ATM, it might represent product innovation.

18.2 HOW TECHNOLOGIES AND INDUSTRIES EVOLVE

18.2.1 Why study technology evolution?

One of the keys to successful strategic planning is the ability to understand the environment in which an organisation has to function. Technology is an important component of that environment. Historically, technological change has been the graveyard for many an old company and an opportunity for growth for many a small start-up company. These so-called technological discontinuities often have a bigger impact on the future profitability of companies than other competitive forces.

Example 18.3 Technological discontinuities

Before 1948, most of the electronics industry was based on the use of thermionic valves for use in amplification and as logic gates. Very few of the makers of thermionic valves managed to make the switch to the semiconductor technology after its introduction. Most of the companies currently making integrated circuits were founded after the era of thermionic valves. All that remains of the thermionic valve industry are a few suppliers for niche markets such as up-market stereo systems.

Understanding the evolution of a technology represents one of the most important inputs into any strategic planning process. In order to understand the evolution process, one needs to look at it in two dimensions: that of the technology itself and then that of the industry in which the technology is used. Each gives different insights into the evolutionary processes.

18.2.2 Evolution of technologies

One of the most popular models describing the evolution of a technology is the 'S-Curve' proposed by R.N. Foster (Foster, 1986:31). Foster proposes that the performance of a particular technology increases at a very slow rate in its initial stages, much faster in later stages, and then slows down again as that particular technology reaches its technical limits. This is shown in figure 18.1. The process whereby performance increases along an S-Curve is one of incremental innovation, as performance bottlenecks are systematically removed.

The reader should be wary of confusing technology S-Curves (technological performance on the vertical axis) with product sales (diffusion) life cycle curves (cumulative sales in the vertical axis), for they describe two different, but interrelated, phenomena.

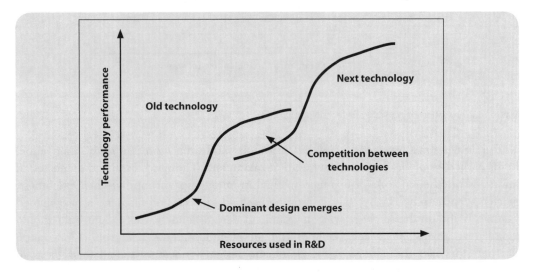

Figure 18.1 Foster S-Curves

The point at which the technology performance suddenly starts to increase rapidly generally coincides with the emergence of the dominant design. The dominant design is defined as that product or combination of components making up a product that is accepted into the marketplace as meeting most of the users' needs. Before the emergence of the dominant design, a lot of effort is spent in developing a variety of new product designs that may not gain wide acceptance in the marketplace. It is only when a dominant design emerges that efforts are focused on improving the performance and price of the technology.

After a dominant design has been established, the technology performance increases rapidly and the costs of the products based on the technology decrease. Thereafter a mature stage is reached where the fundamental technical limits to the technology are approached. It is at this stage that companies selling products based on the old technology are most vulnerable to radical innovations in that technology.

The new technology created by the radical innovation initially has inferior performance characteristics relative to the old technology. However, the same pattern emerges as the new technology establishes a dominant design and eventually overtakes and replaces the old technology. The Foster model proposes that these cycles continue indefinitely: as one technology matures, another replaces it.

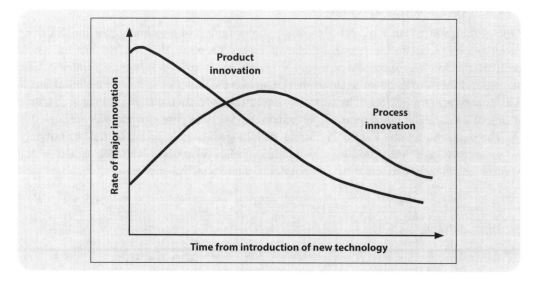

Figure 18.2 Innovation change over the technology life cycle

As technologies mature, the focus of the innovators shifts from *product* to *process* innovation. In the early days of incandescent light bulb development,[1] Thomas Edison and others tested hundreds of different sorts of filament material in an attempt to improve their brightness, efficiency and useful life. By 1884, Edison had settled on a carbon filament, a pear-shaped vacuum bulb and the screw base. This dominant design defined what a light bulb should look like and is instantly recognisable more than a century later. While product innovation did not stop with the advent of the dominant design, innovators shifted their focus to manufacturing and marketing the new product. Process innovations such as the Sprengel mercury pump and the paste-mould glass-blowing machine reduced the cost of manufacture ten-fold over the subsequent decade.

18.2.3 Evolution of industries

At the same time that a technology is moving up its S-Curve, the industry built around that technology also undergoes change (Utterback, 1994:90). This process of industry change is often more visible than the evolutionary changes in the technology itself and can serve as a reliable indicator of the maturity of a particular technology.

The most important changes are in the following areas.

- In the early stages, the *product line* tends to be varied, with products being custom built in small batches. As the industry matures, the product line becomes narrower, with only a few standard products. The first mass-produced car, the Model-T Ford, was reputed to have been available in 'any colour as long as it's black'.
- *Manufacturing processes* move from being based on generic machine tools (e.g. lathes, milling machines) to specialised equipment designed specifically for the manufacture of a specific product. Manufacturing processes move from being flexible 'job shops' to inflexible, but highly efficient, production lines. These days, motor manufacturers design whole plants around the production of one specific model of car.
- The *components* used in manufacture initially are generic 'off-the-shelf' items. As the industry matures, the components used in manufacture are specific to that product and are quite often sub-systems. Complex and product-specific components such as engines, headlights and windscreens are common in the motor industry today.
- *Organisational structures* start off being loose, flexible and supportive of freethinking and risk taking; and later become rigid, regulated and controlled. While this is a normal part of an industry's evolution, it often causes problems with the nurturing of new ideas and can potentially stifle innovation. For this reason, many companies in mature industries house their R&D labs in a separate organisational structure away from the mainstream activities.
- The *nature of competition* changes over the evolutionary cycle. In the early stages, products compete largely on features and functionality, and later this shifts largely to price competition.
- The *industry structure* changes dramatically. In the early stages, many small companies compete. As the industry matures, many of these companies exit the industry or are taken over until there are just a few giants left. The colourful history of the motor industry is now reflected in extinct companies such as Benz, Daimler, and the Olds Motor Vehicle Company, which are now just parts of brand names for the few global players that have survived.

18.3 TECHNOLOGY STRATEGY

The technology strategy of a company cannot be seen in isolation and needs to be tightly integrated into the financial, human resource, manufacturing and marketing strategies of a company. In particular, technology strategy involves choices between alternative new technologies, the criteria by which they are embodied into new products and processes, and the use of resources to accomplish this.

A technology strategy (Maidique & Patch, 1978) involves choices in six areas: the timing of market entry, the technologies forming the portfolio, the level of expertise in each technology, the source of the technology, the level of investment in R&D, and the organisational policies and practices. The last of these is covered in section 18.4.

18.3.1 The timing of market entry

At least three strategies can be followed by innovating companies: a first-to-market strategy, a fast-follower strategy or a cost-minimisation strategy. This strategic choice often sets the pattern for the rest of the strategy decisions that follow.

The *first-to-market strategy* aims to get new products to the market before the competition. It provides the advantage of a temporary monopoly while competitors are still developing their products. A first-to-market strategy requires a strong commitment to R&D. It is a high-risk approach, as it requires investments in products that may not gain acceptance in the market.

The *fast-follower strategy* aims to achieve early entry in the growth phase of the market by imitating the innovations of others. This approach also requires a strong commitment to R&D, but with an emphasis on building capacity that will enable the company to be fast at product development rather than being good at producing breakthroughs in the area of basic research. This is a lower-risk strategy, as only successful innovations are imitated and the failures are avoided.

The *cost-minimisation strategy* is essentially one of late entry into the market, with an emphasis on mass, low-cost production. This strategy is low risk in terms of product acceptance, as the dimensions of consumer choice are well known at this stage. The commitment to R&D is relatively low, but an emphasis is placed on process knowledge and on the implementation of new manufacturing technologies.

18.3.2 The technology portfolio

The selection of the technologies that are to form part of the portfolio is probably the most important question facing the strategic planner. The possibility of technological discontinuities makes this choice extremely difficult. The choice of technology is particularly important to those firms who have chosen a first-to-market strategy, as the choices have to be made long before there is clarity as to what technology will gain market acceptance.

Few organisations start with a 'clean sheet' of competencies, and the existing technology portfolio may very well be the starting point for strategic planning. It is obviously easier to build on existing competencies and use these in new products than to develop new competencies from scratch. This is the basis of what Prahalad calls 'core competencies' (Prahalad, 1993), suggesting that competencies can be leveraged by using them across several products and markets. It is important to be careful that the pursuit of the economies and advantages of core competencies do not blind the organisation to new, disruptive technologies.

18.3.3 The level of expertise

The level of competence an organisation chooses to adopt is tied to its market entry strategy. A first-to-market strategy probably implies that the company will aim to have state-of-the-art knowledge of the core technologies in its field. The organisation that chooses a fast-follower strategy will also probably need to have at least some state-of-the-art knowledge in core technologies. In both cases, non-core knowledge would probably be held at a lower level of competence.

Example 18.4 Technology strategy

The first commercial Internet browser to appear on the market was called Mosaic, and later became Netscape. In this case, Microsoft had a fast-follower strategy, successfully introducing its browser product after Netscape.

More recently, Microsoft has sourced technologies in unfamiliar markets (at least to Microsoft) through the acquisitions of Groove Networks (collaboration) and Giant (spyware). In familiar markets (operating systems and office productivity suites) Microsoft sources new technology mainly through internal development.

Organisations competing with a cost-minimisation strategy generally find it appropriate to keep only a superficial level of technology competence in-house and these organisations would mostly buy or license technology when appropriate. Their source of competitive advantage is in process management and in adapting technologies to make efficient use of them. Competencies in these areas would be maintained at a high level.

18.3.4 The sourcing of technology

A number of options may be available to companies in order to get access to a particular technology or competence. The first of these is to develop the technology internally. This is a slow approach, and is generally appropriate when the technology is very new or closely held by a limited number of competitors. Another way of acquiring technology is by licensing. This is simply the purchase of the right to use a particular technology. In complex technologies, the licence agreement would include access to technical backup and training to help with the transfer of the technology.

Firms with large cash resources may simply buy a company that has the desired technology and competencies. These are often small companies with good technologies or products, but that lack the resources to set up marketing channels. These purchase agreements normally bind the essential personnel to the company for a period in order that their competencies can be transferred.

Finally, a company may buy out the essential personnel from a rival firm. This works better when acquiring competencies rather than technologies, as technologies that are protected by a strong patent regime will still have to be licensed.

18.3.5 The R&D investment level

The aggregate level of R&D investment in a firm is highly dependent on the maturity of the firm, the nature of the industry and the market entry strategy. Small new technology-based firms may have limited sales and the spending on R&D may be very high as a percentage of sales.

Industries that are driven by advances in science, such as biotechnology and electronics, tend to spend a greater percentage of their budget (usually measured as percentage of sales) on R&D. In particular, where the innovation is difficult to copy (e.g. due to patent protection on new drug formulations) and the process very risky, spending on R&D can be as high as 30 to 40 per cent of sales. Firms in mature industries, where competition is based on cost, may spend less than 1 per cent of sales on R&D.

A first-to-market strategy requires a heavy investment in (broad) R&D – which may often not be as focused as that of fast followers. A fast-follower strategy will generally require a lesser, more-focused investment. A cost-minimisation strategy will require a very small R&D investment, as most technology is licensed in.

18.4 THE INNOVATIVE ORGANISATION

Innovative activity in an organisation is difficult to enhance and easy to stifle. A structure that is too tight and focused on the short term will prevent risk taking and discourage staff from putting forward new ideas. A structure that is too loose or has a 'blue sky' time frame will result in unfocused R&D that has no commercial application, or takes so long to develop that market opportunities are missed. This balance between tight and loose, rigid and free is one of the main themes of organisational theory in innovative environments. Organisations in different industries or in different phases of the innovation or technology cycle will find different balance points between these extremes. This section explains some of the areas that need to be considered in managing this balance.

18.4.1 Vision, leadership and the will to innovate

As we have seen, the technology environment is inherently one in which risks are high. For this reason, the owners and top management of a company must understand the business they are in and act accordingly. They must understand that any money invested will not yield immediate returns. Top management also need to keep their eye on the technological environment and be prepared to react swiftly to signs of change. It is common for top management of these technologically innovative companies to have an understanding of the underlying technologies and have the ability to make technology strategy decisions personally. Management that are remote from the technology, or that manage by financial statements alone, cannot be successful in the innovation environment.

18.4.2 Appropriate structure

No single organisational structure can meet all the needs of innovative companies. Generally, tasks that are capable of being formalised or programmed are best performed in rigid organisational structures. Tasks that are unstructured, complex and dynamic are generally best performed in loose organisational structures. It follows therefore that organisations managing technology where product development and radical innovation are predominant should generally have a looser organisational structure. Organisations managing technology in the later stages, where cost-reduction pressures are predominant, should generally have more rigid organisational structures. Modern organisations need to be both rigid and flexible at the same time as they manage existing businesses and future opportunities. This can be achieved by means of the so-called 'ambidextrous organisation' (Nel, 2005:297).

18.4.3 Key individuals

Over and above the engineering and scientific roles one finds in a technologically innovative company, five other critical roles have been identified (Roberts & Fusfeld, 1982). While multiple roles may be performed by a single person, each of these roles is critical to the successful execution of R&D projects. Management should deliberately constitute teams with people who can fill these roles. The first critical role is that of the *idea generator*. The idea generator is generally an expert in his/her field and enjoys solving problems and being creative. The second critical role is that of the *champion*. The champion is characterised by his/her energy, enthusiasm and tenacity in promoting an idea. The champion is often not an expert in any field and may be a 'jack-of-all-trades'. The third critical role is that of the *project leader*. The project leader is a pragmatist who focuses on order, discipline and clear planning. He/She is good at making compromise decisions and balancing the requirements of the various disciplines, and is adept at using organisational structures to get the job done. The fourth critical role is that of the *gatekeeper*. The gatekeeper provides the communication channel between the members of the team and the rest of the organisation, and into other organisations. Gatekeepers are technically up to date, and maintain their knowledge base by reading journals, attending conferences and talking to others. The last critical role is that of the *sponsor*. The sponsor is normally a senior person in the organisation who can guide a project through organisational politics, can authorise additional resources if necessary and can give sage advice.

18.4.4 Effective teamwork

Most development work is multi-disciplinary in nature and requires the co-operation of various individuals in order to be a success. Thus, successful team functioning is a vital part of innovative organisations. Successful teams do not happen by accident. They result from careful selection, spending time on team building and role clarity, and balancing the attention paid to team outputs with that paid to team processes. More detail on team dynamics is given in chapter 5 ('Managing People and Teams').

18.4.5 Innovative climate

Organisational climate refers to the patterns of shared values, beliefs and norms that shape organisational behaviour. Behaviour, or 'how we do things around here', and the supporting procedures and processes are the outward symptoms of the deeper values, beliefs and norms. It follows that environmental factors can destroy an innovative climate. Too much top-down authority has a negative effect on innovation. Freedom and an emphasis on horizontal relationships do not mean that top management can neglect to give leadership and direction, and apply controls when innovation becomes unfocused. Lacks of tools and resources, or rigid budget constraints, also have a negative effect on innovation.

18.4.6 The learning organisation

Because innovation is in effect the application of knowledge, the effective acquisition, storage, retrieval and application of knowledge is vital to the innovative organisation. The ability to build on internal knowledge is an important characteristic of an innovative organisation. Evidence suggests that many organisations repeat mistakes and fail to learn from earlier successes and failures. Innovative organisations should have the ability to 'unlearn' behaviours and routines that do not work or have become inappropriate.

18.4.7 Customer focus

It is essential that management focus on understanding the *real* needs of existing *and* future customers and use this information to shape the strategic direction of the company. Where existing technologies and markets are used, traditional market research methodologies can be used in the development of new products and services. In cases where new technologies are used in new markets, traditional market research can give erroneous results, as it is very hard to get people to enunciate latent needs. You will find more on the marketing of new products in chapter 11.

18.5 DEVELOPING NEW PRODUCTS

Developing new products is seen by many, for good reason, to be an inherently inefficient process. In many of the other areas of management science (e.g. manufacturing) the accumulated research and experience over many decades has made these processes extremely efficient, with little wastage. In most companies, the product development process is almost at the opposite end of the scale, with huge inefficiencies and wasted resources. While the process of developing new products does contain inherent risks and will probably never be as efficient as some other processes, the introduction of new-product development best practices can easily double the process efficiency.

18.5.1 Key factors for product development

Successful product development is far more than just the technical development of the product that meets the specified functional criteria. Success in this context cuts across a number of areas and disciplines and is concerned with aspects such as acceptance in the market and effective use of development resources, in addition to the technical functionality of the product. The key factors (Cooper *et al.*, 1998) that contribute to this success are as follows.

- **Differentiated, superior products:** The main success factor is delivering a product that has unique customer benefits and value.
- **Thorough up-front homework:** New-product projects that move from the idea stage right into development with little or no up-front homework are doomed to failure. Enough time and resources must be devoted to the activities that precede the design and development of the product.
- **Emphasis on the voice of the customer:** Successful new-product projects are a result of a slave-like dedication to the voice of the customer. This covers all phases of the product development cycle, from idea generation and product design right through to launch.
- **Sharp, stable and early product definition:** Defining the product in detail before development begins is a major contributor to new-product success. In addition, this contributes enormously to an on-time and on-budget development process. This definition goes beyond features and functionality and includes a marketing concept and target market definitions.
- **Planning and resourcing the market launch early:** A strong market launch (see the section on new product marketing in chapter 11) underlies the success of any product. Make the launch planning part of the new-product process and develop a launch plan even before development proceeds.

- **Building tough Go/Kill decision points into the process:** Successful processes start off with many ideas that are evaluated and the poor and marginal ideas are then killed off before they incur substantial costs. Most product development processes are log-jammed with too many marginal projects and too few resources. It is therefore important to have tough Go/Kill decision points and the will to enforce them.

There are a number of best-practice methodologies such as Stage-Gate™, Systems Engineering, Agile Development and many others. Each has found favour in certain industries (e.g. consumer products, the military and software) and most try to address the above success criteria in ways that makes sense in that particular industry. These methodologies contain common themes such as idea management, project management and portfolio management, each of which are described in more detail below.

18.5.2 Sourcing and generating ideas

One of the key criteria for successful product development is to source/generate as many initial ideas as possible. These ideas are then screened to weed out ideas that clearly have no commercial potential or are not aligned with the company's strategy. The ideas that remain (and it might take a second screening round to come to a manageable number) are developed further by doing a concept and/or feasibility study before a small number are selected for product development.

There are a number of techniques that can be used to source/generate ideas. For practical reasons, not all of them can be used simultaneously, but it is important that sources of new ideas are reasonably diverse. The main categories of idea generation are as follows.

- **Suggestion schemes:** While suggestion schemes can play a role in generating new ideas, the success thus far has been mixed. In order to work, they need to be designed properly with a transparent adjudication process. Aspects such as rewards for successful suggestions and feedback on rejected ideas also need to be thoroughly thought through before implementation, otherwise the scheme will lose credibility and will just become a way of venting frustrations within the company.
- **Brainstorming:** This remains a popular and useful process. It is important to involve a diverse range of participants such as marketing, technical and representatives, as well as clients, industry experts and even people from other industries to enrich the process. A number of techniques are available to enhance the brainstorming process, including the Theory of Inventive Problem Solving or TRIZ (for those interested in further information, a search for TRIZ on the Internet will yield many informative sites).
- **The voice of the customer:** This involves asking existing and potential customers for feedback and ideas that could lead to product improvements or new products. Techniques such as interviews, focus groups and questionnaires can be used. Unfortunately, customers will generally use their existing frame of reference. This method may therefore be a useful source of information for incremental innovations, but will not elicit ideas for more radical innovations. By carefully choosing the questions to get to the real needs, one can overcome this limitation somewhat. A number of formal techniques, such as Quality Function Deployment, can be used to elicit and give some definition to ideas.
- **Customer observation:** Also called ethnography, this is an incredibly simple way of gathering data about how users interact with products. It involves the observation of

products and how they are used in everyday life. Users accept many product shortcomings almost unconsciously and will often work around them without even thinking that the product could be improved. A keen observer will be able to identify these shortcomings quickly and easily as part of a customer observation process.

■ **Lead users:** Lead users are a special class of the innovator segment in the product life cycle adoption curve. They have a good sense of the current 'state of the art', have user needs that lead other customers, and often have technical expertise that enables them to develop or modify existing products to address these needs. This combination makes them an extremely valuable source of ideas.

18.5.3 Portfolio management

The second area of management attention in a product development process is that of managing the product portfolio. A product portfolio is the collection of current products, product enhancements and future products that a company is currently marketing or intends to market. Managing a portfolio means taking deliberate steps to enhance its long-term value. Decision making regarding the selection of products that are to form part of the portfolio are the subject of continual refinement. For this, it is important that there is a clear set of values and criteria that can guide the decision-making process. The generic criteria for deciding on what constitutes a good portfolio are: the financial value of the portfolio, the alignment of the portfolio with the company strategy, the balance between the various projects and the resource loading. Company-specific criteria need to be developed for each of these for practical decision-making purposes.

The *financial value* criterion requires that the financial return from the project portfolio should be maximised. Many companies use familiar discounted cash-flow methodologies such as net present value (NPV). Using NPV involves making a decision on the discount rate to be used, which, given the risks involved in product development, is at best a guess. See chapter 16 for more information on this topic.

The *strategic alignment* criterion measures each project in the portfolio against the strategic direction of the company. This requires that the company has enunciated a clear strategic direction.

Formulating criteria to ensure portfolio *balance* is more complex, as the concept of balance is different for each company. Indeed, it is likely that any one company would have a number of balance criteria that it would need to juggle. Typical balance criteria are:

■ balance between high-risk (radical innovation) and lower-risk (incremental innovation) types of projects;
■ balance across the product range to ensure customers can buy a product that matches their functional needs; and
■ balance across geographic, demographic and other market segments, etc.

Lastly, one needs to look at the *resource requirements* to support the planned development programme – i.e. across all projects. One of the truisms of product development is that almost all product development departments are overcommitted and underresourced. There is a significant body of reliable research that says that by applying the right time management process and deliberately *over*resourcing the product development team, overall efficiencies can be improved significantly. While most project scheduling software packages also include some form of resource planning tool, a simple spreadsheet-based calculation is often more than adequate.

18.5.4 Project management

The generic project management principles are equally applicable to product development projects and are not repeated in this chapter (see chapter 13 for more information on project management). However, it is important to note some specific project management best practices that are used extensively in new-product development. Almost every single product development methodology uses a phased development and decision-making process. The most common of these is the Stage-Gate™ process, where the stages refer to activities such as *feasibility, development, testing, launch,* etc. Companies should develop standard lists of the minimum activities for each stage that project managers are obliged to complete. The concept of 'gates' refer to hold points in the process that stop the movement from one stage to another unless there is a formal reapproval of the project. Typically, gate criteria are as follows.

- Is the quality of the work in the preceding stage complete and correct? The time pressures of product development often lead to short cuts being taken. Are these acceptable and if not, what should be done about them?
- Does the project still fit the original merit criteria? The viability of a project changes with time and events in the environment in which the project is located. If it is no longer viable, can it be put back on track by a simple change, or should it be stopped completely?
- How does this project rank against other projects in the portfolio? Is there still balance? Is the value being optimised? Is it still aligned with the company strategy?
- Are there adequate resources for the project? Is this project a marginal project that is diverting resources away from more promising projects?

Based on the answers to these questions, an individual project can be stopped completely, delayed or speeded up. It is a natural tendency not to want to stop or delay a project and one always needs to guard against the default decision of 'letting the project through' when it should really be stopped.

18.6 MANAGING KNOWLEDGE AND INTELLECTUAL PROPERTY

One of the greatest challenges to innovators is to be able to protect and enjoy the benefits of the revenue stream that results from innovation. This is the so-called appropriability problem. Over and above patent protection, various other means can be used to appropriate these benefits. In fact, patents are, in certain circumstances, a poor way of protecting innovation, as they require the innovator to disclose publicly the nature of the innovation. This allows shrewd competitors to invent around the patent, or at least to gain market intelligence about forthcoming new products.

18.6.1 Non-statutory mechanisms

There are many highly effective, non-statutory mechanisms available to the innovator to appropriate the benefits of innovation. The most common is simply to use *secrecy* as a way of keeping the technology from others. This normally involves the use of non-disclosure agreements, restraints of trade and other contractual agreements, as well as measures such as disclosing technology to staff on a need-to-know basis. In the absence of other measures, secrecy is not a very effective strategy.

More commonly in the technology field, *tacit knowledge* and *product complexity* are used to appropriate the benefits of technology. Tacit knowledge is the knowledge embedded in the people and processes in the company that is difficult to codify and transfer to others. Anyone who has tried arc welding, or riding a horse, will know that these are difficult competencies to learn from written instructions! Protecting the benefits from innovation by means of tacit knowledge is most useful when complex products or processes are involved.

Another strategy is to use first-mover advantage to establish *industry standards* based on your particular architecture. The innovator's familiarity with the standards so established gives him/her a window of opportunity while others struggle to understand and adapt to the new standards. This is most useful if the innovator's technology offering is so compelling, or if the innovator has sufficient marketing power to influence the creation of standards in this manner.

Another way of appropriating technological advantage is to use the *learning curve*[2] to keep ahead of the competition. By having first-mover advantage, the learning curve should ensure that the innovator always retains cost advantage through his/her cumulative experience.

Lastly, the innovator can use the control of complementary assets to ensure that the returns can be appropriated. Complementary assets are those things that are vital to the manufacture or marketing of a product and could include things like raw materials, marketing and distribution networks. Control of complementary asset that are both *scarce* and *important* often determines who will reap the benefits from innovation.

Example 18.5 Patent protection vs secrecy

Pharmaceutical products such as new drugs tend to be protected almost exclusively by patents. The exact chemical composition of the active ingredient in the drug is described in the patent. It is then very difficult for a competitor to copy the drug without infringing the patent holder's rights. On the other hand, the formulation for a new brand of cola is very difficult to patent. An imitator just has to change the ratio of the constituents slightly and he/she will not infringe the rights of the patent holder. For this reason, the innovator might opt for using secrecy to protect the cola formulation.

18.6.2 Statutory protection mechanisms

In addition to patent protection, there are a number of other legal mechanisms to protect your intellectual property. These include copyright, registered designs and trademarks. When considering any form of legal protection mechanism, it is advisable to obtain the advice of a specialist in the field, such as a patent attorney. More information can be obtained from the Companies and Intellectual Property Registration Office (CIPRO) (see www.cipro.cp.za) or from the Design Institute of the SABS (www.designinstitute.org.za). In summary, the following protections are available.

■ **Patents** protect any non-obvious invention that shows clear novelty. One needs to be very careful here, for even making one's own idea public before it is patented, e.g. at a trade show or in a newspaper article, puts the idea in the public domain and prevents anybody (including the inventor) from patenting that idea. Patents protect the owner by preventing any other person (without permission, which is normally called a 'licence') from making, selling or even using the product. Certain classes of products such as software, games, business methods and the like cannot be patented – however, the rules in this regard are different in each country. The patenting process normally starts with provisional patent, which is valid for one year, after which the holder must apply for patent protection in each country in which he/she wants protection. South Africa is a signatory to the Patent Co-operation Treaty (PCT), which can extend the validity period of the provisional patent and result in some cost savings when patenting in a number of countries. Patents generally provide protection for a period of 20 years.

- **Copyright** is the simplest and easiest form of protection. Originally intended to cover literary and musical works, it is now used to protect software and engineering drawings. Copyright (other than films in South Africa) does not require formal registration and subsists automatically by law. It is usual to identify the copyright holder by placing the author's name, the copyright symbol © and the date of publication on the document. Copyright is valid for 50 years from publication or the death of the author (this varies by product type and legal jurisdiction). It is important that when a copyrightable work is commissioned from any supplier or even an employee the assignment of the copyright is clearly addressed, as the author may be the holder by default.

- A **registered design** is similar to a patent in than it prevents others from making, selling or using the article described by the design without permission. Two types of registered design are recognised: an aesthetic design with features judged solely by the eye, and a functional design, which has to do with the functions an article has to perform. The protection afforded by a registered design is 15 years and 10 years respectively, both of which are subject to annual renewal. One has to apply for design registration in each country in which one requires protection.

- **Trademarks** are another form of protection that applies to product names and other identifying symbols such as logos. This prevents others from using your registered trademarks or even confusingly similar marks. The application is specific to each country and one needs to make application in each country in which one requires protection. Trademarks can be renewed every 10 years indefinitely. It is important to note that the registration of a company name or Internet domain name does not give one any preference or rights regarding the registration of trademarks.

Self-assessment

18.1 Describe the innovation cycle.

18.2 Describe the different types of innovation. Why would one want to classify innovation?

18.3 Explain why it is important to study technological evolution.

18.4 Define 'technological discontinuity' and 'dominant design'. Explain the relevance of these two concepts to a firm's competitive strategy.

18.5 Compare the characteristics of firms and industries where technologies are young with those where technologies are mature.

18.6 List the organisational factors that support the creation and maintenance of an innovative organisational environment. Give a brief description of each.

18.7 Describe the elements of a technology strategy. Explain how a first-to-market or fast-follower strategy would influence each of these technology strategy components.

18.8 Briefly explain the link between business strategy and technology strategy.

18.9 Describe the ways in which the benefits of innovation can be appropriated. When would one use patent protection in preference to other means?

18.10 Briefly explain how a Foster S-Curve is constructed. Explain why the evolution of technology may follow such a pattern.

Mini-projects

A. The watchmaking industry was turned on its head when electronic movements replaced the old mechanical movements. Describe how this technological change changed the distribution channels for watches.

B. It is often assumed that small companies are more innovative than large companies, but is this in fact true? What advantages and disadvantages does each have when innovating? What can each do to try to overcome these disadvantages?

C. Innovation is dependent on an enthusiastic and creative workforce. What human resource practices and techniques can be used in an R&D environment to enhance motivation and creativity?

References

Cooper, R.G. Edgett, S.H. & Kleinschmidt, E.J. 1998. *Portfolio Management for New Products*. New York: Perseus Books.

Foster, R.N. 1986. *Innovation: The Attacker's Advantage*. New York: Summit Books.

Maidique, M.A. & Patch, P. 1978. *Corporate Strategy and Technology Policy*. Harvard Business School Case 9-679-033. Boston: Harvard Business School Press.

Moore. G. 1999. *Crossing the Chasm*. New York: Harpur Collins.

Nel, W.P. 2005. 'Technology and OD.' In Meyer, M. & Botha, E. (eds). *Organisation Development and Transformation in South Africa*. 2nd edition, pp. 283–306. Durban: LexisNexis Butterworths.

Prahalad, C.K. 1993. 'The role of core competencies in the corporation.' *Research Technology Management*, November-December.

Roberts, E. & Fusfeld, A. 1982. 'Critical functions: Needed roles in the innovation process.' In Katz, R. (ed.). *Career Issues in Human Resource Management*. Englewood Cliffs: Prentice-Hall.

Rogers, E.M. 1995. *Diffusion of Innovations*. New York: Free Press.

Tidd, J., Bessant, J. & Pavitt, K. 2001. *Managing Innovation: Integrating Technological, Market and Organisational Change*. Chichester: John Wiley.

Utterback, J.M. 1994. *Mastering the Dynamics of Innovation*. Boston: Harvard Business School Press.

(Footnotes)

1 See Utterback (1994) for a summary of how light bulbs were developed.

2 The learning curve predicts that the higher the *cumulative* production of a given product, the lower the unit production costs will be.

19 An Overview of Environmental Management and Sustainable Development Concepts for Management Practices

Alan Brent

Study objectives

After studying this chapter, you should be able to:

- understand the concept of sustainable development, and specifically the underlying South African legislation that enforces it, and its influence on industry and business practices and technology strategies
- recognise the changes that are, accordingly, required in business practices, especially the necessity of a holistic life cycle management (LCM) approach, which integrates the three life cycles that are important to managers in industry:
  - □ project LCM;
  - □ asset (or process) LCM; and
  - □ product LCM
- describe the following organisation- and product/process-level sustainability tools that are available to managers, engineers, technologists and scientists:
  - □ environmental management systems (EMSs);
  - □ environmental auditing;
  - □ environmental costing and accounting;
  - □ environmental risk assessment (ERA);
  - □ life cycle assessment (LCA);
  - □ life cycle engineering (LCE); and
  - □ environmental impact assessment (EIA), together with social impact assessment (SIA)
- describe various strategies that organisations can use to reduce or eliminate unwanted sustainability impacts
- explain the role of engineers, technologists and scientists in improving sustainability performances through technological innovations
- recognise opportunities for industries to fund sustainability-oriented technologies. The example of the Clean Development Mechanism is used

The aim of this chapter is to introduce readers to the subject of environmental management and sustainable development, and its integration with normal management practices.

19.1 INTRODUCTION

In 1966 the prevailing open system economic model that viewed increasing production as good for the economy and whose success is measured by the amount of inputs (raw materials and energy) that can be converted to outputs (reservoirs of pollution) was questioned. The problem with this model is that the impact on the environment is not considered. Kenneth Boulding, a professor of economics, therefore offered an alternative model that is known as the 'spaceship economy', i.e. an economy that operates within a closed system (Reid, 1995:25). According to this theory, the Earth is seen as virtually a closed system, with mainly only sunlight and gravitational forces acting on it from the outside, and the overall influences of economic activities are assessed in terms of this closed system in which these activities take place.

Today it is generally recognised that humankind is using resources at a faster rate than the growth in its numbers in the closed system, and the scale of use results not only in local or regional impacts, but also in global impacts. More and more studies are showing that humankind has reached the point where the feasibility and desirability of continued industrial growth should be questioned.[1] Various scientific studies highlight that the long-term viability of traditional forms of 'progress' may not be possible in the future. During the last decades scientists have discovered new ways of how humankind may impact on nature, e.g. ozone depletion in the earth's upper atmosphere and global warming. However, much has still to be learnt about how big a threat such impacts may have on the survival and future prosperity of the people of this planet.

All of this has left humankind with the challenge of beneficiating minerals and generating energy but using and disposing of associated processed products in such a way that the natural environment is not significantly degraded. Similarly, the direct social impacts of economic activities on society at large are increasingly being considered.

19.2 SUSTAINABILITY CONCERNS AND THE MOVE TOWARDS SUSTAINABLE DEVELOPMENT

Environmental concerns can be traced as far back as the 15th century. For example, in the 1500s Georgius Agicola (1950:8) wrote about the effect of mining and metallurgical operations on the environment. However, the first major scientific concerns about the sustainability of development were published in the early 1960s (Carson, 1963; Dubos & Ward, 1972). These suggested a link between development and human activities and damage to biological species and human health. In the same decade, concerns about a global population explosion and its impacts on the environment and social structures were raised (Ehrlich & Ehrlich, 1990). By the early 1970s, these concerns were translated into a call for the integration of environmental and development strategies. This approach was emphasised at the United Nations Conference on the Human Environment (1972), which stated that:

> ... although states have a right to exploit their own resources pursuant to their own environmental policies, they nevertheless have a responsibility to ensure that activities within their borders do not cause damage to the environment of other states or areas beyond their limits of national jurisdiction.

---

[1] These questions are most often posed by non-governmental organisation (NGOs), e.g. Ecosystem Valuation (www.ecosystemvaluation.org), Earthday Network (www.earthday.net) and Rocky Mountain Institute (www.rmi.org).

The end of the 1970s saw the move to link environmental and economic aspects, with the International Conference on Environment and Economics, held in 1984, concluding that the environment and economics should be mutually reinforcing (SEDAC, 2003). Debates and work in this field continued throughout the 1980s, when the United Nations World Commission on Environment and Development (WCED) finally coined the concept of 'sustainable development' in the now-famous *Brundtland Report* (later published as a book, *Our Common Future*) (WCED, 1987): 'Sustainable Development is development that meets the needs of the present without compromising the ability of future generations to meet their own needs.'

19.2.1 The introduction of sustainable development in South African legislation

Chapter 2 of the Constitution of the Republic of South Africa Act 108 of 1996 subordinates all other legislation pertaining to the environment (see figure 19.1) (Sampson, 2001). Section 24 (a) of the Act states that: 'everyone has the right to an environment that is not harmful to their health or well being ....' Furthermore, section 24 (b) states that: 'everyone has the right to have the environment protected for the benefit of present and future generations ....' The Act empowers communities to defend their own or the public interest. Any person has the right to take a company to court for any infringement of the law.

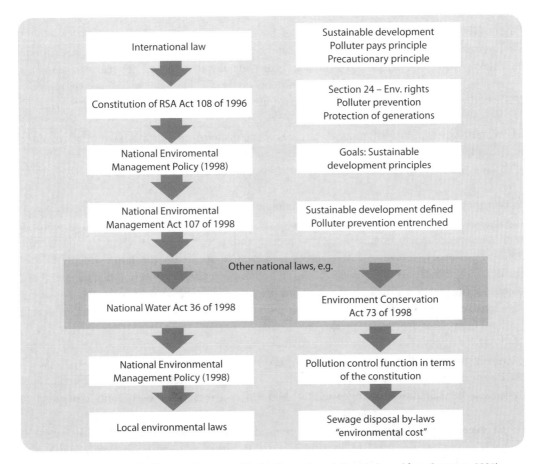

Figure 19.1 Growth of South African environmental legislation and regulations (adapted from Sampson, 2001)

The National Environmental Management Act (NEMA) 107 of 1998 dictates the right of individuals and communities with respect to the environment. Section 28 of NEMA states that companies have a positive duty to take reasonable measures to prevent pollution or degradation occurring, continuing or recurring, or where it cannot be avoided, to minimise and rectify pollution and its effects. This applies to every person, but particularly to landowners, controllers of land and those with the right to use land or premises. It also applies to historic pollution or degradation, or where there is a threat that pollution is likely to occur. However, a court of law must be able to deem it significant pollution or degradation, which can be open to different interpretations.

Section 24 (7) of NEMA holds past and present directors, who held positions at the time the company committed an offence, liable under schedule 3 laws, i.e. similar to fraud. Thereby, the individual persons are responsible for proving that 'all reasonable steps' were taken to prevent pollution and degradation of the environment. Reasonable measures, as defined by Section 28 of the act, include:

- ceasing, modifying or controlling any activity or process that may impact negatively on the environment;
- containing pollution;
- eliminating the source of pollution; and
- remedying the negative effects that an organisation may have on the environment.

Under NEMA, certain administrative duties have to be undertaken by companies.

- Environmental impact assessment (EIA) authorisations, including social impact assessments (SIAs), have to be obtained from local, provincial and national government agencies, and in some cases international governments, as applicable for a specific development.
- Scheduled process certificates in terms of air pollution have to be obtained from local, provincial and national government agencies.
- Water licences, trade effluent permits and waste disposal site permits have to be obtained from local, provincial and national government agencies.
- Other industry-specific (e.g. mining and forestry) requirements have to be met.

It is important to note that the environmental regulations and legislation applicable to management practices are quite vast and are consequently not dealt with in a comprehensive manner here. Rather, engineers, technologists and scientists are advised to consult the literature in order to ensure that all aspects of the law are dealt with appropriately from a sustainability perspective, e.g. the publication of Sampson (2001). Most notably, all engineers, technologists and scientists must understand the following NEMA principles.

- **Cradle-to-grave:** Organisations should consider, investigate and analyse all aspects in the life cycle of a product, process or service.
- **Polluter pays:** Anyone whose activities cause or are likely to cause damage to the environment shall bear the cost of full preventive or restorative measures.
- **Precautionary principle:** Where there are threats of serious or irreversible damage, lack of full scientific certainty shall not be used as a reason for postponing cost-effective measures to prevent degradation.
- **Environmental justice:** Adverse environmental impacts shall not be distributed in such a manner as to unfairly discriminate against any person, particularly vulnerable and disadvantaged persons.

- **Waste prevention and minimisation:** Resource usage and production wastages within operations shall be minimised to reduce cleanup requirements, i.e. the concept of cleaner production.
- **BPEO:** The principle of best practicable environmental option (BPEO) for a given set of objectives is the option that results in the least damage to the environment as a whole, at acceptable cost, in the long term as well as the short term.

19.3 SUSTAINABLE DEVELOPMENT AND ITS INFLUENCE ON BUSINESS AND INDUSTRY

The United Nations World Commission on Environment and Development (WCED) *Brundtland Report* is viewed as a major political turning point for the concept of sustainable development (Mebratu, 1998). Since then the influence of the concept has increased extensively and it features more and more as a core element in policy documents of governments and international agencies (Mebratu, 1998). The World Summit on Sustainable Development in 2002 highlighted this growing recognition of the concept by both governments and businesses at a global level (WSSD, 2002).

The concept of sustainable development is nevertheless inherently vague (Daly, 1990). Its exact definition is therefore still disputed, as are its underlying aspects. There are currently over 100 definitions of sustainability and sustainable development, but most agree that the concept aims to satisfy social, environmental and economic goals. These goals are also referred to as the three pillars or objectives of sustainable development (Azapagic & Perdan, 2000; Sillanpää, 1999; IISD, 2005). The interaction of the three pillars, as shown in figure 19.2 on the next page, contributes to sustainable development (Labuschagne *et al.*, 2005). Thus, meeting the needs of the future depends on how well these interconnected economic, social and environmental objectives or needs are balanced during current decision-making and management practices (World Bank Group, 2005).

Although the concept of sustainable development is understood, intuitively it remains difficult to express it in concrete, operational terms (Briassoulis, 2001). Business and industry, together forming one of the three pillars of society (the other two being government and civil society) (Wartick & Wood, 1998), have a responsibility towards the whole of society to actively engage in the sustainability arena (Holliday *et al.*, 2002). The pressure is therefore mounting for businesses and industry to align operational processes with the three objectives of sustainable development (Keeble *et al.*, 2003) and accordingly measure the sustainability performances of operational activities (Labuschagne *et al.*, 2005). Four different types of drivers for the incorporation of sustainability in business practices have been identified (Goede, 2003). An adaptation of the identified drivers is illustrated in figure 19.3 on the next page.

Merkel (1998:336) explains sustainable development as follows: 'Sustainable Development means nothing more than using resources no faster than they can regenerate themselves, and releasing pollutants to no greater extent than natural resources can assimilate them.' Brune (1997:687) states that:

> ... implicit in the sustainability criterion is the fact that renewable resources must be used at their rates of regeneration and this is also true for the capacity of the environment to assimilate waste. Waste cannot be accumulated in the environment without changing, usually adversely, its quality. The rapid use of non-renewable resources is, by definition, not sustainable. Sustainability requires that these resources are used at a rate decided by technical progress and their substitution for renewable alternatives.

405

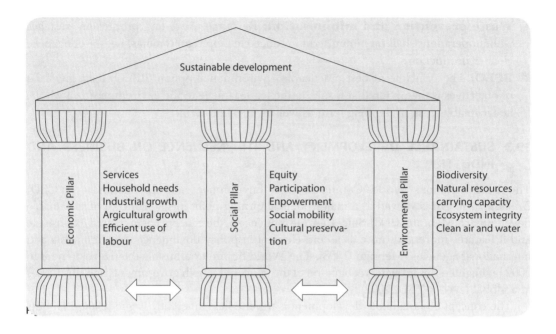

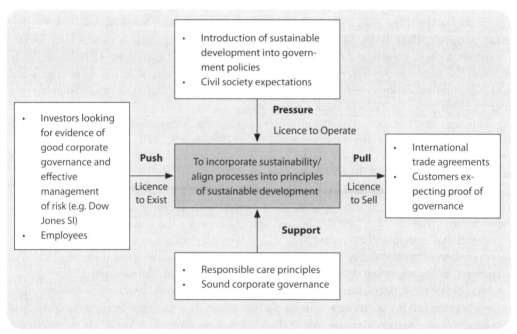

Figure 19.3 Drivers for the incorporation of sustainable development into business practices (adapted from Goede, 2003)

In order to assist business and industry, the International Institute for Sustainable Development (IISD) has suggested the following definition of sustainable development for the business community: 'For the business enterprise, sustainable development means adopting business strategies and activities that meet the needs of the enterprise and its stakeholders today, while protecting, sustaining and enhancing the human and natural resources that will be needed in the future' (IISD, 2005).

These definitions of sustainable development pose the complex challenge of sustaining economic growth without harming the environment. The challenge of sustainable development can be addressed to various degrees of success through:

- lifestyle changes: reducing waste, less materialistic attitudes, smaller families, re-use of objects and scarce resources where possible, i.e. the concept of sustainable consumption;
- more efficient environmental legislation and its implementation, e.g. legislation that enforces rehabilitation and recycling;
- the development of more-advanced and cost-effective end-of-pipe environmental technologies, e.g. catalytic converters and others;
- the development of technologies and green products that will result in no pollution or much less pollution than current products and technologies, e.g. cleaner production options, nanotechnologies that will use less material, fuel cell cars, biodegradable products and genetically engineered crops that will eliminate the use of pesticides;
- the development of technologies that will ensure greater efficiency in the use of scarce, non-renewable resources, e.g. internal combustion engines that are much more fuel-efficient and that will therefore ensure that oil reserves are depleted at a lower rate;
- the development of technologies that use abundant resources rather than scarce resources, fewer resources or alternative resources to ensure that adequate volumes of all resources will be available for future generations, e.g. replacement of steam-driven electricity generation with biomass- and wind-driven electricity generation;
- the development of renewable resources, e.g. biomass and fish farming; and
- environmental education and awareness.

The above illustrates that the engineer, technologist and scientist have a big role to play in the development of sustainable technologies and products (see section 19.12). Furthermore, as previously discussed, the management practices of the organisations in which these engineers, technologists and scientists operate must be aligned with the objectives of sustainable development.

19.4 CHANGES IN BUSINESS PRACTICES TOWARDS MEETING THE SUSTAINABILITY CHALLENGE

Three levels within an organisation have been identified that can be subjected to change namely: the strategic level, the process or methodological level, and the operational level (Du Toit, 2003). In order for sustainability to manifest itself within a company, change needs to take place on all three levels. This is, however, not currently the case. On the strategic and operational levels there is evidence of the integration of sustainability into the business environment. Some companies have started to define sustainable development for their businesses, while others endorse international agreements or include the principles of sustainable development in their companies' vision and mission statements. The majority of emphasis has fallen on the operational level, where companies implement environmental management systems (EMSs) and report on the sustainability of their operations in annual sustainable development reports. Companies also tend to place an increasing emphasis on corporate social responsibility and corporate philanthropic projects.

However, the *Sustainability Survey Report* of PricewaterhouseCoopers (2002) revealed that of the 101 Fortune 1 000 companies that were interviewed, 72 per cent of the respondents did not include the risk and/or opportunities of sustainability in their project, investment and transaction evaluation processes. Research by the Institute for Economy and the Environment at the University of St. Gallen (IWOe-HSG) has further revealed that traditional business management systems are solely geared towards financial performance and therefore exclude environmental and social sustainability aspects (Bieker *et al.*, 2001). Figure 19.4 illustrates the statistics from the 2002 PricewaterhouseCoopers survey. The statistics show that further change is required at the business processes and methodologies level to ensure an overall sustainability focus.

The traditional top-down and bottom-up approaches to incorporating sustainability within organisations have not seemed to be effective to a large extent. Practical tools that systematically include sustainability within the evaluation processes are described below to align business methodologies with the principles of sustainable development (Gladwin *et al.*, 1995) and to incorporate economic, environmental and social performances in the policies, culture and decision-making processes of business and industry (Brent *et al.*, 2002). These performance objectives manifest in three operational focal points that are fundamental to industry (Brent & Visser, 2005).

- **Projects:** The concept of sustainable development must be integrated in the planning and management over the life cycle of projects.
- **Assets:** The life cycle of assets must be optimised and managed in terms of sustainable development performance objectives of the production or manufacturing facility.
- **Products:** The influence of products on economies, environments and society as a whole must be considered, i.e. the concept of product stewardship (US EPA, 2003).

A comprehensive life cycle management (LCM) approach is subsequently required, which assures that the operational processes are consistent and that there is effective sharing and co-ordination of resources, information and technologies (ISO 2002; Brent & Visser, 2005). Such a holistic LCM approach would require an effective integration of the three life cycles (see figure 19.5 on the next page) (Labuschagne & Brent, 2005). Management practices must adhere to sustainability performances in all three of these life cycles.

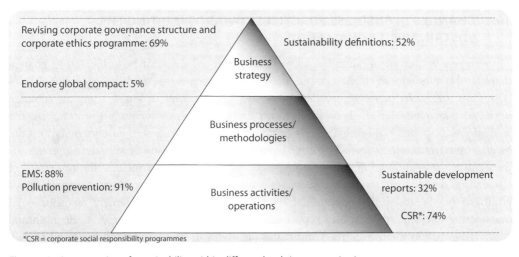

Revising corporate governance structure and corporate ethics programme: 69%

Sustainability definitions: 52%

Business strategy

Endorse global compact: 5%

Business processes/ methodologies

EMS: 88%
Pollution prevention: 91%

Business activities/ operations

Sustainable development reports: 32%

CSR*: 74%

*CSR = corporate social responsibility programmes

Figure 19.4 Incorporation of sustainability within different levels in an organisation

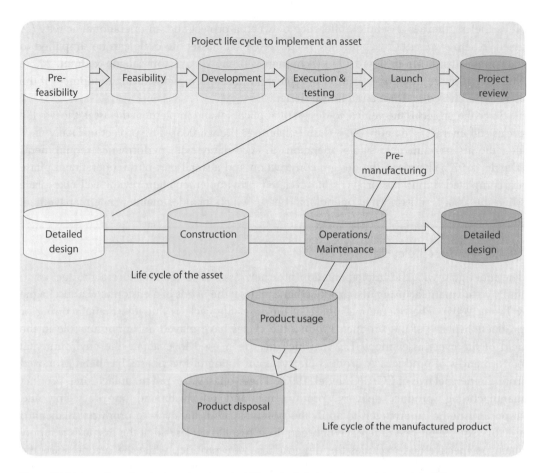

Figure 19.5 Integration of project, asset and product life cycles in industry to meet the sustainability challenge (Labuschagne & Brent, 2005)

19.4.1 Project life cycle

The project life cycle is discussed in both chapters 1 and 13. Since a project is a temporary endeavour undertaken to create a unique product or service, the project itself is likely to have minimal economic, environmental and/or social consequences. It is more likely that the 'product' or deliverables of the project will have these consequences and impacts.

Various project life cycle approaches have been defined, e.g. control-oriented models, quality-oriented models, risk-oriented models, and a fractal approach to the project life cycle, as well as some company-specific project life cycles (Morris, 1994:245–8; Cooper, 2001; Bonnal *et al.*, 2002). The simplified project life cycle shown in figure 19.5 is a practical model that has been introduced (Buttrick, 2000), which is based on primary research and theory development in the field of project management, and is currently used in the South African process industry.

19.4.2 Asset life cycle

The project life cycle and asset life cycle are often viewed as one life cycle, due to the fact that the two life cycles contribute to the same value chain (Labuschagne & Brent, 2005).

Nevertheless, there is a definite difference between a project and an operational activity (or asset) (Kliem *et al.*, 1997; Schuman & Brent, 2005). The asset life cycle can be simplified to four phases (as shown in figure 19.5) (Schuman & Brent, 2005; Labuschagne & Brent, 2005). The design phase of an asset can be the selection phase of manufacturing equipment if the asset is purchased and not an in-house design.

Since the project is the vehicle to design (if applicable) and implement the asset, the two life cycles still interact (see figure 19.5) (Labuschagne & Brent, 2005). The project normally ends after the asset commences stable operations in accordance with performance requirements (Dingle, 1997). Therefore, the design, construction and a small part of the operational phase are completed during the project's life cycle. A post-implementation review will take place when the asset is in its normal operational phase, which includes maintenance management practices.

19.4.3 Product life cycle

The main goal with the implementation of a new asset is to manufacture a product or to improve the manufacturing of a product that can meet the needs of a customer (Labuschagne & Brent, 2005). The operational phase of the asset life cycle is thus the manufacturing or production phase of the product. Product life cycles have played an important role in the field of life cycle assessment (LCA), which has been used to evaluate the environmental performances of products. A product life cycle consisting of five phases has been proposed from the perspective of LCA (Graedel, 1998). These phases are: pre-manufacturing, product manufacturing, product delivery, product use and refurbishment, and recycling and disposal. Another approach is to apply the generic systems life cycle to products (Blanchard & Fabrycky, 1998). The difference between these two life cycles is that the first uses a supply chain perspective and excludes the design phase of a product, while the second starts the life cycle of a product with the need identification, and considers supply chain activities as part of the production phase. A simplified supply-chain-focused product life cycle is used to describe the interaction between the product and asset life cycles in figure 19.5 (Labuschagne & Brent, 2005).

19.4.4 Sustainability tools for management support

In summary, companies are increasingly accountable for the impacts of an implemented project on society, the environment and the economy long after the project has been completed, i.e. beyond the normally considered project life cycle. Furthermore, in industry, the asset life cycle resulting from the project and the subsequent product life cycle resulting from the asset, or, in some cases, the product life cycle resulting directly from a project, have economic, social and environmental consequences, which are in turn associated with an implemented project (see example 19.1). Aligning project management methodologies, which are a core business methodology of most companies, with sustainable development principles therefore requires that the sustainability consequences of these asset and product life cycles must be considered during the project life cycle. A number of tools have been developed and introduced to assist managers to integrate these life cycles through a holistic management approach and to optimise business practices in terms of sustainability performances, some of which are highlighted in figure 19.6 on page 412. These tools are described in greater detail in table 19.1 on page 412. Although figure 19.6 uses the project/asset life cycle management perspective to illustrate the use of these tools, the tools can also be specifically applied to other life cycle management focus areas, e.g. technology management, and production, operation and maintenance management.

Example 19.1 Project/Asset/Product life cycle interaction example: Mittal Steel South Africa – Saldanha Steel operations

The Saldanha Steel Project (SSP) was a joint venture between the then Iron and Steel Industrial Corporation (ISCOR) and the Industrial Development Corporation (IDC), now Mittal Steel South Africa (http://www.iscor.co.za). SSP announced its plans to build a steel plant to the public in December 1994. The R6.8 billion Saldanha Steel development, situated on the Cape west coast roughly 10 km from Langebaan lagoon's ecologically sensitive wetlands, has been designed to produce 1.25 million tons of hot-rolled carbon steel coil per year for the international market. The mill was finally commissioned, after much controversy, in 1998. Many different environmental (and social) impact assessments were undertaken (http://www.lib.uct.ac.za/govpubs/EIAs.htm), the validity of which are still debated, and many associated project-specific compliance criteria had to be addressed and/or demonstrated. These included, but are not limited to, the following factors.

■ *The life cycle of the physical plant asset:* Plans had to be in place to manage and minimise the effects of construction; the normal operations must have no detrimental effects on the surrounding communities and the natural environment, e.g. the operating plant has to be aesthetically pleasing and management plans must be in place for large maintenance cycles; and current operations must allocate resources to develop local communities and provide for the adequate management of the plant's end of life through rehabilitation and socio-economic activities.

■ *The life cycle of the plant product:* Infrastructure had to be developed, with minimal detrimental effects on the natural environment and society in general, to supply the plant with the necessary resources, e.g. the railway line from Sishen for the ore was upgraded and Eskom power lines from the Highveld were constructed; and after-gate management of the coil product must be sustainable, e.g. the Saldanha Bay port's cargo quay was upgraded and good transport management practices of the shipping routes have to be in place.

Compliances with such project-specific criteria are ongoing more than half a decade after the project was successfully implemented and normal operations were reached. This example illustrates that the environmental and social implications of the life cycles of physical assets and the related products often extend beyond the life cycles of the responsible projects, and must be considered as part of the earlier phases of project management models and practices where project changes with minimum cost and risk implications are possible.

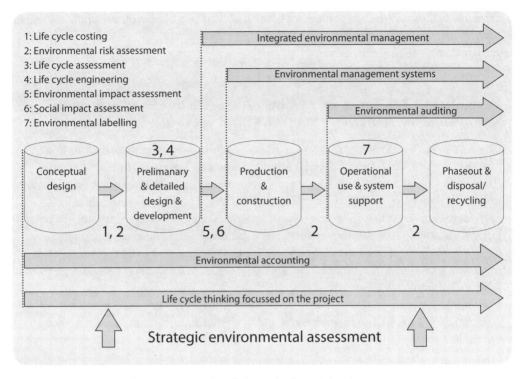

Figure 19.6 Incorporation of sustainability tools in the life cycle of a typical project

Note:
▧ Arrows indicate tools that are organisationally focused.
▧ Numbers indicate tools that are focused on products and processes.

Table 19.1 Definitions of the different sustainability tools

Sustainability tools		Definitions and descriptions
Organisationally focused	Integrated environmental management (IEM)	A philosophy that describes a code of practice for ensuring that environmental considerations are fully integrated into all stages of the development process in order to achieve a desirable balance between conservation and development (DEAT, 1992)
	Environmental management system (EMS)[a]	A system that provides a structured process for continual improvement and enables an organisation to achieve and systematically control the level of environmental performance that it sets itself. In general, this is based on a dynamic cyclical process of 'plan, implement, check and review' (DEAT, 2000).
	Environmental auditing[a]	A management tool comprising a systematic, documented periodic and objective evaluation of how well environmental organisation, management and equipment are performing, with the aim of helping safeguard the environment by facilitating management control of environmental practices and assessing compliance with company policies, which would include meeting regulatory requirements (ICC, 1989)

Sustainability tools		Definitions and descriptions
Organisationally focused	Environmental accounting[a]	Concerns the definition, assessment and allocation of environmental costs and expenditures for the purposes of cost and resource management, compliance reporting and capital budgeting, planning, and operational decision making. Environmental accounting can further be divided into three main areas: financial environmental accounting, managerial environmental accounting and strategic management environmental accounting (Gale & Stokoe, 2001).
	Life cycle thinking	A holistic consideration of product, process or activity systems. Life cycle thinking addresses the associated life-cycle-generated impacts through the use of different tools aiming at minimising them such as: life cycle assessment (LCA), life cycle costing (LCC) and life cycle engineering (LCE); collectively referred to as life cycle management (LCM). The concept of life cycle thinking integrates existing consumption and production strategies, preventing a piecemeal approach. Life cycle approaches avoid problem-shifting from one life cycle stage to another, from one geographic area to another and from one environmental medium to another (UNEP, 2003).
	Strategic environmental assessment (SEA)	A process of integrating the concept of sustainability into strategic decision making OR a process to assess the environmental implications of a proposed strategic decision, policy, plan, programme, piece of legislation or major (regional) plan (DEAT, 2000)3
	Life cycle costing (LCC)	An assessment of all costs associated with the life cycle of a product, process or activity that are directly covered by any one or more of the actors in the life cycle (supplier, producer, user/consumer, end-of-life actor), with complementary inclusion of externalities that are anticipated to be internalised in the decision-relevant future (Rebitzer & Hunkeler, 2003). Externalities refer to costs as per the environmental accounting tools and environmental economics and resource economics methods (Leiman & van Zyl, 2004; Blignaut et al., 2004; De Wit et al., 2004).
	Environmental risk assessment (ERA)[a]	Involves the examination of risks resulting from natural events (flooding, extreme weather events, etc.), technology, practices, processes, products, agents (chemical, biological, radiological, etc.) and industrial activities that may pose threats to ecosystems, animals and people. Environmental health risk assessment addresses human health concerns and ecological risk assessment addresses environmental media and organisms. ERA is predominantly a scientific activity and involves a critical review of available data for the purpose of identifying and possibly quantifying the risks associated with a potential threat (Five Winds International, 2004).
	Life cycle assessment (LCA)[a]	A decision-making tool to identify environmental burdens and evaluate the environmental consequences of a product, process or service over its life cycle from cradle to grave. LCA has been standardised by ISO and forms the conceptual basis for a number of management approaches that consider a product across its life cycle, covering resource acquisition, product manufacturing, use and end of life (Five Winds International, 2004).

Sustainability tools		Definitions and descriptions
Product/process focused	Life cycle engineering (LCE)[a]	A tool for system analysis and evaluation, which combines the tools of LCA and LCC with technical data; therefore, with the LCE methodology, decision support is given from the point of view of environmental, economical and technical aspects (IKP, 2004); also known as design for environment (DfE) (Minnesota Office of Environmental Assistance, 2001).
	Environmental impact assessment (EIA)[a]	A detailed study of the environmental consequences of a proposed course of action. An environmental assessment or evaluation is a study of the environmental effects of a decision, project, undertaking or activity. It is most often used within an integrated environmental management (IEM) planning process, as a decision-support tool to compare different options (DEAT, 1998).
	Social impact assessment (SIA)	Includes the processes of analysing, monitoring and managing the intended and unintended social consequences, both positive and negative, of planned interventions (policies, plans, programmes, projects) and any social change processes invoked by those interventions. Its primary purpose is to bring about a more sustainable and equitable biophysical and human environment (IAIA, 2003).
	Environmental labelling	Includes a number of activities, ranging from business-to-business transfers of product-specific environmental information to environmental labelling in retail marketing. The overall goal of environmental labelling (or eco-labelling) is to encourage the demand for, and supply of, those products and services that are environmentally preferable through the provision of verifiable, accurate and non-deceptive information on environmental features of products and services (Five Winds International, 2004).

[a] These tools, which engineer, technologist and scientist managers typically apply directly from an environmental perspective, are discussed in greater detail in the sections below.

19.5 THE IMPORTANCE OF ISO 14000 AND EMS STANDARDS FOR NORMAL MANAGEMENT PRACTICES

Distinguishing the sustainability tools in two main focus areas (see figure 19.6 on page 411 and table 19.1 on the same page) is primarily an outcome of the work carried out by Technical Committee (TC) 207 of the International Organisation for Standardisation (ISO), which, in turn, is responsible for developing and continually improving the ISO 14000 family of standards (ISO, 2005) (see figure 19.7). ISO 14000 is a response to the identified need of industry to incorporate EMSs into existing business practices due to the trend in globalisation and increased competition (refer to the abovementioned drivers in figure 19.3). The increased pressure experienced by companies to demonstrate improved environmental stewardship and the associated burden of the related accountability resulted in the need for an international EMS standard. The consequence was the development and publication of the ISO 14000 family of standards within a period of two years by the ISO (Tibor & Feldman, 1996), and this is today the most commonly used EMS standard in industry. Table 19.2 summarises the most important standards in the series.

ISO 14000 aims to achieve standardisation in the field of environmental management and thereby guides and assists organisations to implement and maintain an EMS. In the same way, ISO 14000 also contributes to sustainable development. However, the often complex EMSs and practices that are incorporated in an organisation's existing structure through the ISO 14000 standards deal with environmental management solely. Human, or social, and socio-economic aspects do not receive much attention in the current standards' documentation. Organisations need to incorporate these aspects into formal management systems through the available tools described in table 19.1 to ensure that all objectives of sustainable development are met. Also, it must be noted that the ISO has recognised this deficiency and is continually improving the standards accordingly. Furthermore, efforts are being made to officially integrate the different standards. For example, the ISO 14000 environmental standards are being merged with the ISO 9000 quality standards to form the ISO 19000 group of standards.

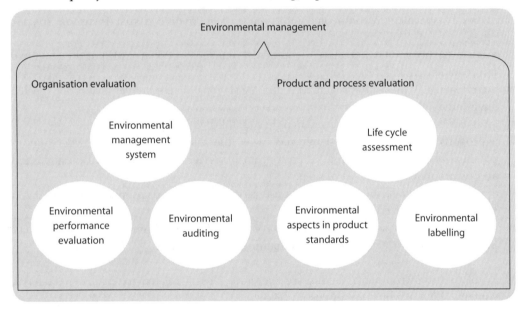

Figure 19.7 Focus areas of TC207 environmental tools and standards

Table 19.2 Main ISO standards pertaining to environmental management systems (Mohr-Swart, 2003)

Standard*	Description
ISO 14001	Environmental management systems – specification with guidance for use
ISO 14004	Environmental management systems – general guidelines on principles, systems and supporting techniques
ISO 14010	Guidelines for environmental auditing – general principles
ISO 14020	Environmental labels and declarations – general principles
ISO 14030	Environmental management – environmental performance evaluation – guidelines
ISO 14040	Environmental management – life cycle assessment – principles and framework

*Only the main standard groups are shown; refer to the ISO documentation for the sub-

sections. These can be obtained from Standards South Africa (http://www.stansa.gov.za). All of the other tools that are discussed form part of environmental management systems.

19.6 ENVIRONMENTAL AUDITING

Environmental auditing marked the growing shift from companies merely complying with regulations to companies developing forward-looking sustainable environmental management strategies. This tool can also be economically justified, as practice has shown that it can lead to more favourable insurance rates and improvement in a company's public image, and can also assist companies in avoiding legal action for environmental damage. Environmental audits are divided into two broad categories:

- industrial or private sector corporate environmental audits; and
- local authority or higher-level government environmental audits.

Six types of specific environmental audits have been identified that focus on specific activities:

- a site or facility audit, where an environmental management plan (EMP) or the continual improvement of an EMS is revisited;
- a compliance audit, where regulations and legislations dictate the verification of specific environmental release loads;
- an issues audit, where a risk assessment has established a possible concern;
- a property transfer audit, also referred to as due diligence risk audits, where future potential risk associated with a property from historical operations must be established and incorporated in the cost of sale;
- a waste audit, where the life cycle of waste streams must be determined and associated risks established or confirmed; and
- a life cycle audit, where life cycle tools have established environmental performances of product, process or activity life cycles that must be confirmed.

Many of these environmental audits are therefore closely related to environmental risk assessments (ERAs), which are discussed below.

19.7 ENVIRONMENTAL ACCOUNTING

Table 19.1 on page 412 shows, from a corporation level, that environmental accounting can be divided into financial environmental accounting, managerial environmental accounting and strategic management environmental accounting (Mohr-Swart, 2003) (see table 19.3). In addition, at a government level, natural resource accounting is also often performed, whereby macro-economic measures are used and the focus is typically the nation. Indicators include gross domestic product, taking environmental costs into consideration. Very often, the different accounting procedures are based on a full-cost accounting (FCA) approach (Matthews & Lockhart, 2001), which incorporates cost-benefit analysis (CBA) methods (Leiman & Tuomi, 2004) with environmental economics and resource economics methods (Leiman & Van Zyl, 2004; Blignaut et al., 2004; De Wit et al., 2004).

Table 19.3 Differences among management accounting, financial accounting and strategic management accounting (adapted from Mohr-Swart, 2003)

Factor	Management accounting	Financial accounting	Strategic management accounting
Regulations	Prepared for internal use, no external regulations	Prepared according to GAAP	Prepared for internal use, no external regulations
Range and detail of information	May encompass very detailed and highly aggregated financial, non-financial and qualitative information	Broad based, lacks details, provides an overview of position and performance of an organisation over a period of time	Emphasises the cohesive and consistency of macro- and micro-level activities and focuses on position relative to competitors
Reporting intervals	Needs of users dictate intervals	Produced annually but may be semi-annually or quarterly	Needs of users dictate intervals
Time period	May include historical and current information	Provides information on the performance and position of the past period, i.e. backward looking	Future oriented
Nature of reports	Specific purpose of reports for specific managers making a particular decision	General purpose reports for shareholders and broad-based stakeholders	General purpose reports for strategic decision making

19.7.1 Financial environmental accounting

Financial environmental accounting refers to the estimation and public reporting of liabilities and costs related to environmental matters. The assessment and reporting of environmental risks and liabilities, capitalisation for environmentally related expenditures and the treatment of environmental debt all fall into the stream of environmental accounting. Generally Accepted Accounting Principles (GAAP) are the basis for this reporting.

19.7.2 Managerial environmental accounting

Managerial environmental accounting focuses on internal management and decision-making processes through various techniques of cost allocation, performance measurement and business analysis. Managerial environmental accounting can involve data on costs, production levels, inventory and other vital aspects of business.

The interdisciplinary nature of managerial environmental management accounting promotes contributions from scientists, economists, engineers and policy advisors in identifying internal and external environmental costs, while the management accounting profession utilises its expertise in allocating the costs within existing and emerging environmental and sustainability accounting frameworks.

19.7.3 Strategic management environmental accounting

Strategic management environmental accounting has an internal focus and is not controlled by any external regulations. It focuses on establishing the position of the enterprise relative to its competitors. The needs of the users dictate the frequency of production of strategic management environmental accounts for decision making.

19.8 ENVIRONMENTAL RISK ASSESSMENT

Environmental risk assessment (ERA) is the process of assigning magnitudes and probabilities to adverse effects of human activities (including technological innovation) or natural catastrophes. This process involves identifying hazards and using measurements, testing, and mathematical or statistical methods to quantify the relationship between the initiating event and the effects. ERAs are defined in two broad groups: human health risk assessment and ecological risk assessment. The detailed processes of these two types of ERAs are described elsewhere (Five Winds International, 2004). The output of the ERAs is typically used as input to define a risk management process, which, in turn, provides the input for further ERAs (see figure 19.8).

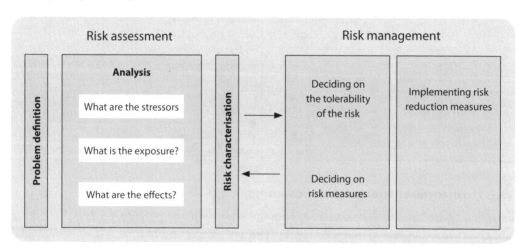

Figure 19.8 Relationship between risk assessment and risk management

19.9 LIFE CYCLE ASSESSMENT

The primary custodian of technical developments in the field of life cycle assessment (LCA) as a tool in the ISO 14000 series (see section 19.5 and Table 19.2) is the Society of Environmental Toxicology and Chemistry (SETAC) (2005). A number of workshops sponsored by SETAC documented LCA as a quantitative procedure to assess environmental burdens associated with the life cycle of an activity (product, process or service) by (Fava *et al.*, 1991):

- identifying and quantifying energy and materials used and wastes released to the environment;
- assessing the potential impact of those energy and material uses and releases to the environment; and
- identifying and evaluating opportunities to effect environmental improvements.

The LCA tool is, however, criticised in that subjective results may be obtained for the routine analyses of products due to (SETAC, 2005):
- the subjective basis and limitations in the collection and analyses of data; and
- variations in the temporal scale, spatial scale and locale, and assignment procedures of values to different environmental impacts.

These deficiencies must be considered when the LCA procedure is applied. A complete life cycle includes raw material extraction (including water), processing, transportation, manufacturing, distribution, use, re-use, maintenance, recycling and final waste disposal (Consoli *et al.*, 1993). The main objectives of executing an LCA study are to (Allen *et al.*, 1997):
- provide a profile (as complete as possible) of the interactions of an activity (product, process or service) with the environment;
- contribute to the understanding of the overall and independent nature of the environmental consequences of human activities; and
- provide decision makers with information that quantifies the potential environmental impacts of activities and identifies opportunities for environmental improvements.

In this respect, companies have used LCAs to fully comprehend the overall environmental consequences of products and changes in production processes (Baumann, 1996; Grotz & Scholl, 1996; Broberg & Christenen, 1999). This, in turn, has introduced the concept of product stewardship as a business decision mechanism (US EPA, 1997), whereby responsibility is accepted for the environmental practices upstream (suppliers) and downstream (customers or clients) of a company's activity, i.e. the 'cradle-to-grave' concept. Internally, industry has used LCA for cleaner production purposes (UNEPTIE, 2005).

LCA is also increasingly used as a tool for policy development by regulatory authorities that influence business decisions (Allen *et al.*, 1997; Troge, 2000). Options for possible waste management practices have been good examples of using LCA results for policy purposes (Tukker, 1999).

As LCA results are often used for company in-house and policy decisions, the formal ISO LCA procedure supports the main phases of theoretical decision-making and analytical processes (Seppälä, 1999), i.e.:
- structuring of the problem;
- construction of the decision/preference model; and
- sensitivity analysis.

Accordingly, the framework for executing an LCA study is well documented in the ISO publications and is illustrated in figure 19.9 (STANSA, 1998). In general, a complete LCA study must consist of four phases (Nietzel, 1996):
- **goal and scope definition,** which describes the application or specific interest and indicates the target group. A detailed description of the system to be studied is included, providing a clear delimitation of scope, periods and system boundaries;
- **life cycle inventory (LCI) analysis,** which quantifies the environmentally relevant inputs and outputs of the studied system, which is essentially a mass and energy balance of each unit, or smaller, process within the larger system. The ISO has provided a general framework for the inventory analysis (ISO 14041);

■ **life cycle impact assessment (LCIA),** which quantifies the environmental impact potential of the inventory data. An LCIA procedure has been developed for application in the South African context (Brent, 2003; Brent; 2004; Brent & Visser, 2005); and

■ **interpretation and improvement analysis,** whereby options are identified and evaluated to reduce the environmental impacts of the studied system.

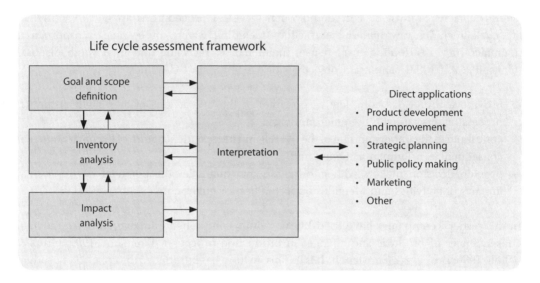

Figure 19.9 Standardised phases of the LCA tool (STANSA, 1998)

19.10   LIFE CYCLE ENGINEERING

The design for environment (DfE) concept has resulted in certain modifications that have been made to the environmental LCA procedure to adapt to DfE applications (Curran, 1999). The consequent life cycle engineering (LCE) approach (Cooper & Vigon, 1999) evaluates the environmental implications of a product, process or service in the design phase, i.e. it is proactive rather than reactive, as is the case with many assessment tools. LCE further incorporates data of economic and environmental aspects, together with an evaluation of the designed technology, as a combined decision-support system in the design phase (see figure 19.10) (IKP, 2004). Economic aspects are assessed through the total cost assessment (TCA) (AICHe, 2000) and life cycle costing (LCC) tools (Fava & Smith, 1998) and environmental aspects through the conventional LCA tool and related life cycle impact assessment (LCIA) procedure (ISO, 1998). The comprehensive integration of economic and environmental impact assessments for typical life cycle evaluations of products is illustrated in figure 19.11 on the next page (Sevitz *et al.*, 2003).

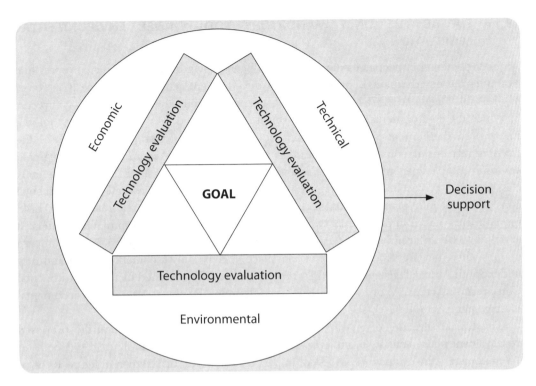

Figure 19.10 The decision support mechanism of life cycle engineering (IKP, 2004)

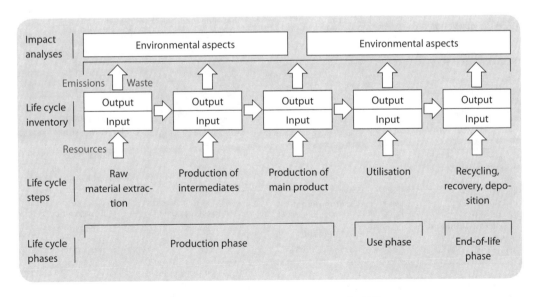

Figure 19.11 The life cycle approach for 'cradle-to-grave' analysis (Sevitz *et al.*, 2003)

Life cycle management (LCM) is an extension of the LCE concept, i.e. the life cycles of products, processes and services are managed beyond the design phase. Thereby, the total potential liabilities for a manufacturer of a product are evaluated (Rebitzer *et al.*, 2001).

19.11 EIA AS THE ONLY MANDATORY TOOL UNDER ENVIRONMENTAL LEGISLATION

An environmental impact assessment (EIA) has been shown to be a systematic process that examines the environmental consequences of development actions in advance (Glasson *et al.*, 1995:3). Fuggle (1992:764) defines an EIA as: 'the administrative or regulatory process by which the environmental impact of a project is determined.' As a management tool, the EIA process therefore assists in determining how a new proposed project will impact on the environment during construction, operations and final closure. According to the integrated environmental management (IEM) framework for South Africa, an EIA is consequently classified as a project-level assessment tool (DEAT, 1998) that incorporates environmental considerations into planning and decision-making processes and is a reaction to EMS requirements, i.e. to improve quality of decisions with respect to environmental matters. It attempts to be an accurate, critical and objective assessment tool, and is used to compare options where new developments are to take place. It is the only tool mandatory for industry under South African legislation. The EIA process is currently governed by the Environmental Conservation Act (ECA) of 1989, but will soon fall under the newer National Environmental Management Act (NEMA) (see section 19.11.2).

The typical structure of an EIA is shown in figure 19.12. In the figure, the proposed development phase consists of multiple attributes, which are shown in figure 19.13 both on the next page – the evaluation process of impact sources is, in turn, shown in figure 19.14

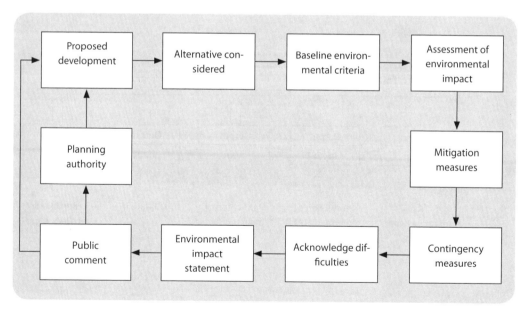

Figure 19.12 The typical structure of environmental impact assessments

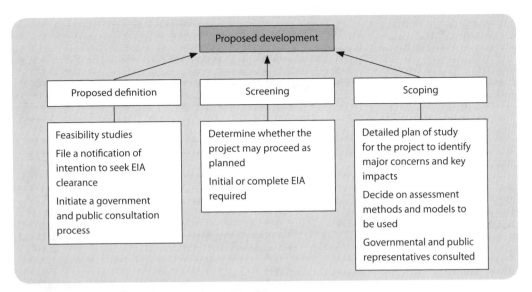

Figure 19.13 The proposed development phase of the EIA process

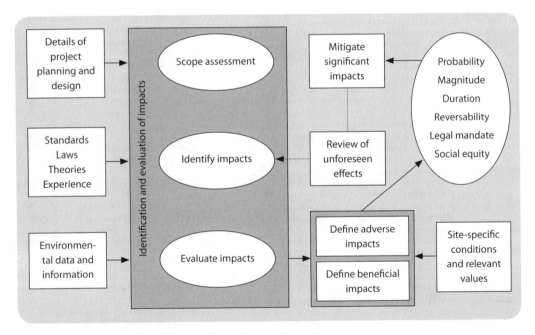

Figure 19.14 Inputs and outputs relating to the evaluation of impact sources

19.11.1 Existing EIA process followed in South Africa

The application procedure or EIA process that must be followed in order to obtain authorisation to commence with a listed activity can be divided into three phases that follow on each other (see figure 19.15) (DEAT, 1998):

■ Phase 1: Screening/Application for authorisation/plan of study for scoping
■ Phase 2: Scoping
■ Phase 3: Environmental impact assessment.

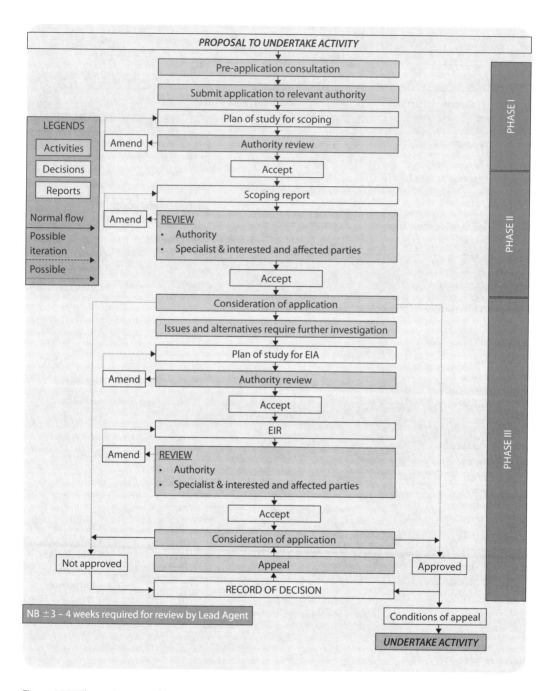

Figure 19.15 The environmental impact assessment process in South Africa

Review by various role players take place throughout all of the three phases and the EIA process is concluded by a final decision of the lead authority for the proposed project. According to Glasson *et al.* (1995:141), the purpose of the review of the EIA proposal is to evaluate its strengths and weaknesses and to provide the basis for deciding whether it should be approved or not. When an EIA is conducted for a proposed activity, various reports (e.g. a scoping report, an environmental impact report (EIR), etc.) need to be prepared and distributed to various parties for perusal. These parties include interested and affected parties (I&APs), specialists (when required), relevant authorities, etc. The cognisance of the various reports in the different phases of an EIA therefore provides a basis for decision making by the authorities as to whether a proposed activity is authorised.

The relevant authorities should provide a record of decision (ROD) to the applicant of a proposed project, whether or not the proposal is approved. An explanation of how environmental considerations were taken into account and weighed against other considerations must be documented in the ROD. The ROD must reflect any conditions of approval and should be made available on request to any I&APs (DEAT, 1992:8). According to the DEAT (1998:30), the relevant authority will decide to issue a ROD, which gives either an authorisation with or without conditions, or rejection of the application.

Phase 1: Screening/Application for authorisation/plan of study for scoping

According to Wood (2000:71), screening incorporates the decision on whether the nature of the action and its possible impacts are such that it should be submitted to environmental assessment. Glasson *et al.* (1995:3) state that screening narrows the application of EIA to those projects that may have significant environmental impacts. Screening may be partly determined by the EIA regulations operating in a country at the time of assessment. The Consultative Forum on Mining and the Environment (2002:14) explains: 'the level of effort in environmental assessment and public participation required for a project is a function of a combination of the following: the scale of anticipated impacts, and the scale of public sensitivity.'

In South Africa the activities that require an EIA in terms of the Environment Conservation Act (ECA) are listed in Schedule 1 of the regulations. These activities are referred to as the listed activities. The problem with the list of activities is that it does not capture all activities that can have a potential impact on the environment. Using the listed activities as the only basis for screening has, accordingly, led to concerns.

Nevertheless, as part of the screening phase, an application for authorisation and a document laying out a plan of study for scoping must be prepared and submitted to the relevant authority for review. The relevant authority must approve this plan.

Phase 2: Scoping

Most of the EIAs in South Africa end after this phase, meaning that all the identified environmental aspects or impacts are not only identified, but also addressed and resolved, i.e. mitigation measures are in place. DEAT (1998:7) defines scoping as: 'the process of identifying the significant issues, alternatives and decision points which should be addressed by a particular EIR, and may include a preliminary assessment of potential impacts' (see box 19.2).

Example 19.2 EIA scoping process example: Best practices of Sasol

The Sasol Quality Management System (Sasol, 2002a & 2002b) specifies the following best practices, which are not necessarily required by the ECA regulations.

1. Appoint an independent environmental consultant, if he/she was not appointed before or during the screening and application phase. It often happens that the applicant appoints an independent consultant to assist with the preparation and submission of the application for authorisation and the plan of study for scoping. The applicant may, however, submit these documents to the relevant authority.

2. Conduct an internal scoping exercise (not requested in terms of the regulations – good practice), which consists of:
    ■ identifying internally environmental aspects with regard to the proposed project; and
    ■ compiling an aspect register, which lists all the identified aspects applicable to the proposed project.

3. Prepare a background information document (not requested in terms of the regulations – good practice).

4. Conduct the public participation process.

5. Capture all identified environmental aspects and add to the aspect register.

6. Determine the significance of potential environmental impacts.

7. Prepare a draft scoping report and distribute to I&APs for comment.

8. Prepare a final scoping report and submit to the relevant authority and other consulting authorities for review. Also make a copy of the final scoping report available to I&APs for their cognisance.

After a scoping report has been submitted, the relevant authority may decide, through the ROD, that either:

■ the information contained in the scoping report is sufficient for the consideration of the application without further investigation; or

■ the information contained in the scoping report should be supplemented by an environmental impact assessment that focuses on the identified alternatives and environmental issues identified in the scoping report.

Phase 3: Environmental impact assessment

The authorities can require that a proposed project should go through the third phase, which is known as the environmental impact assessment phase, in other words that a full EIA be done. This will require the applicant, or the consultant appointed by the applicant, to (Sasol, 2002a:14):
■ submit a plan of study for EIA to the relevant authority;

- conduct specialist studies on specific identified environmental aspects, or on social aspects, i.e. a social impact assessment (SIA);
- conduct public participation;
- submit an environmental impact report (EIR) including the relevant specialist studies to the relevant authorities;
- obtain a ROD from the authorities and ensure that the ROD conditions are met; and
- advertise the ROD (I&APs can appeals against the said development).

19.11.2 Proposed new EIA process to be followed in South Africa (draft legislation)

The proposed new EIA process will still have the three phases described above. Similar review and decision making will also apply. However, according to the draft update EIA legislation, some specifics of the proposed EIA process will, amongst others, differ from the existing EIA process as follows.

- The proposed new EIA process will provide an updated list of activities and also a list of areas for which an EIA would be required.
- Projects will be screened under different categories and, depending on the category, different EIA processes will be followed.
- Small- or low-impact projects will only undergo a 'screening' process where an initial assessment report would have to be submitted for review. A decision on whether the proposed project can be implemented or not will be based on this report.
- Large- or high-impact projects need to go through the normal scoping process, as well as the assessment phase of the EIA. No project will be approved after the scoping phase of the EIA anymore.

19.12 CORPORATION AND SECTOR TECHNOLOGY STRATEGIES TO ADDRESS CURRENT AND FUTURE SUSTAINABILITY CHALLENGES

The Industrial Revolution brought about the belief that technology could solve most problems, and was characterised by the following changes in society's behaviours.

- Natural power, such as wind, water, people and animals, was replaced by steam, fuel and electricity, which increased the dependency on non-renewable resources such coal and crude oil.
- More materials such as iron, glass and chemicals were used.
- Machines were increasingly used to complement or replace the labour of humans.
- The focus was on the division of labour, increased productivity through mass production, and standardisation.
- The idea of taming or even conquering nature was established.

Some of these changes have resulted in the realisation that such practices are unsustainable. Apart from adapting business practices, various technology strategies were subsequently evolved to address the abovementioned sustainability concerns and challenges. For example, one such strategy is that of recycling. Recycling results in the usage of less virgin material, and less waste material is generated for treatment and disposal. The limitation of recycling is that commodities such as coal and oil cannot be recycled. However, the amount of such non-renewable resources that are used can be reduced by more efficient energy conversion technologies, e.g. hybrid automobiles (see box 19.3).

If consumption patterns do not adopt such new technologies, i.e. if they do not diffuse in big volumes, then society may never achieve the general objectives of sustainable development. Governments can, however, also play an important role in the early adoption of new technologies, e.g. by using fuel cell buses for public transport. These technologies can then be improved and costs reduced so that they can eventually compete with incumbent technologies that have gone through many years or decades of continuous improvement. Furthermore, environmental labelling schemes[2] (see table 19.1) can promote and provide consumers with access to environmentally friendly products. For examples, green products in the building and construction industry, especially paints, have received much attention.

Example 19.3 Technology strategy example: The challenge of personal sustainable transportation

Before 1969 it was believed that the only way to reduce the impact of internal combustion engines on the environment was by means of end-of-pipe technology. Car manufacturers widely believed that there was a trade-off among the various pollutants emitted from internal combustion engines (Heisler, 1999:625), which could only be solved by means of the add-on process of catalytic conversion. Honda, however, designed the CVCC (compound vortex controlled combustion) engine during the years 1969 to 1971. With this engine, Honda engineers tried not to produce pollutants in the first place, thereby reducing the need for later clean-up. In a similar way, they overcame the traditional trade-off between fuel economy and engine power by means of their VTEC (variable valve timing and lift electronic control) technology (De Witt, 1998:893–911).

The Honda Insight is the first gasoline-electric hybrid vehicle to be sold in the United States. The heart of the hybrid system is Honda's innovative Integrated Motor Assist (IMA™), which couples a one-litre, three-cylinder engine with an ultra-thin electric motor for outstanding performance and efficiency. The IMA system, combined with a rigid aluminium body structure and world-class aerodynamic design, gives the Insight the ability to achieve 68 miles per gallon of gasoline (US EPA highway estimate). It meets California's stringent ultra-low-emission vehicle (ULEV) standard, which makes it one of the world's cleanest, most fuel-efficient gasoline-powered cars. Because its primary motive power comes from its VTEC™-E gasoline engine, the Insight drives just like any other automobile. It has a 10.6-gallon petrol tank and no external power supply is needed for recharging. Batteries are recharged by means of regenerative braking. The Insight's lightweight aluminium body and reinforced frame are 47 per cent lighter than a comparable steel body and have superior bending and torsional rigidity. Its highly aerodynamic body has one of the lowest coefficients of drag (0.25) of any mass-produced automobile (http://www.honda.com).

---

[2] Many of these schemes may be an entrance barrier for products into the marketplace, e.g. products that are exported to the European Union must often comply with its eco-labelling scheme: http://europa.eu.int/comm/environment/ecolabel/index_en.htm.

19.12.1 Technological innovation as the solution to sustainable development challenges

Many economists, business leaders and strategists acknowledge the important role of innovation, and more specifically technological innovation, in the global economy. During the greater part of the industrial era many concerns about the ability of future generations to meet their own needs were rebuffed by technological optimism, i.e. the belief that future technological innovations would automatically make each generation better off than previous ones (Pirages, 1996:4). Publications such as the *Brundtland Report* (see section 19.2) and that released by the Club of Rome (*Limits to Growth*) have again emphasised the concept of sustainable development and have put a question mark behind the ability of current technological innovation practice to make a contribution. Technology is sometimes perceived as a controversial solution to the challenge of sustainable development, as it may be argued that technology itself is the cause of the problem. In other words, technology is responsible for bringing material wealth and lifestyles that are no longer sustainable. However, Van der Wal (1994) argues that a philosophical analysis and explanation of the phenomenon of technology is necessary if the presumption may be made that there is a special relationship between the present sustainability crisis and modern technology.

This trend towards an unsustainable society through technological innovation is shifting. Currently there is greater adoption of technologies that make use of renewable resources. Some of these technologies, such as photovoltaic cells, however, still have a long development path ahead before they can compete with conventional electricity generating technologies. Consequently there is much scope for and money to be made from greener technologies and therefore much research and development is done in this area (see box 19.4). In this regard, each market segment requires different business and technology strategies to achieve sustainable development.

- In the consumer economy, companies must decrease their environmental footprint by replacing mature technologies.
- In the emerging economy, companies have the challenge of meeting rapidly growing demands without repeating wasteful, outdated practices.

Example 19.4 Green technologies example: A competitive edge for the sugar industry as it becomes a producer of bio-plastics

Bio-plastics are just one of a number of diversification opportunities for the sugar industry, according to researchers from of the Cooperative Research Centre for Sugar Industry Innovation through Biotechnology (CRC SIIB) (http://www.crca.asn.au/activities/2005/ScienceAction04.htm#contents):

> ... every household will be using bio-degradable plastic bags, bottles and containers, every car will have bio-degradable plastic dashboards and fittings, fine clothing will be crafted from these biopolymers to replace petrochemical plastic and nylon with bio-nylons and bio-fabrics, all made from renewable resources.'

Research being carried out by the CRC is building on an already proven process for making plastics from sucrose that needs only a minor shift in economics for the process to become a market leader. The sugar industry produces just four products: sugar, molasses, alcohol and bagasse, the last of which is normally considered as waste matter. However, in the future, sugar will be seen as a carbon source for materials and energy production. Australia, and many other countries, can capitalise on this only if the sugar industry survives the period of transition.

This transition phase will take a decade or so to optimise the science and build the infrastructure to get the industry from a sucrose-based industry to a sugarcane industry producing lots of new products. The production of ethanol could create the breathing space to make this transition. According to the CRC:

> ... in the future we could be producing not just sucrose, ethanol, energy (burning ba-gasse), and molasses, but a whole range of bio-polymers with a vast array of proper-ties, not the least of which is biodegradability. This will greatly reduce the impact of world price fluctuations on our producers.

It could also change rural economies. As the sugar industry starts producing new biopolymers, sugar mills will become the hub for a range of manufacturing industries making the new products to feed into growing international markets. Industries will be attracted to this source of raw materials and will convert them into the range of plastics required by the automotive, electronics, household goods, fabrics and even the carpet industries. These will be clean green industries using renewable resources.

These clean green technologies will help reduce society's dependency on petrochemi-cals, replace non-biodegradable plastics with biodegradable ones, and reduce green-house emissions.

One Dutch study determined that: 'ten- to twenty-fold eco-efficiency improvements will be needed to achieve meaningful reductions in environmental stress' (Weaver, 2000:7). Duchin and Lange (1994) state that recycling, increased fuel efficiency and other technological adjustments will not go far enough to provide for truly sustainable development in the long term. Furthermore, Hart (1999:24) states that: 'environmental management serves only to improve incrementally the performance of existing products and processes.'

A new body of theory and practice is subsequently required that should create greater consistency between the process of technological change and the societal objectives of sustainability (Weaver, 2000:9). The normal incremental nature of most innovation processes is therefore not good enough. A strong need for radical or transformational technologies exists. One problem with radical technologies is that they often take a long time to diffuse through society because of the lock-in of old technologies. Technology development and adoption is often integrated with cultural and social values, and getting sustainable technologies adopted by markets is an immense challenge. For example, huge volumes of genetically engineered crops are produced and consumed in the United States. However, in Europe there is a lot of resistance to the use of genetically engineered food. Such debates are plentiful and ongoing.

In summary, a sustainable technology will only be successful if it can take market share away from eco-inefficient solutions (Weaver, 2000:54).

19.13   FUNDING OPPORTUNITIES FOR SUSTAINABILITY-ORIENTED TECHNOLOGIES

The increasing trends towards more sustainable technologies have introduced opportunities to fund these technologies. The Clean Development Mechanism (CDM), which forms part of the Kyoto Protocol of 1997, is such a case (Brent *et al.*, 2005).

The purpose of the CDM is to assist developing countries (not included in Annex 1 of Article 12 of the Kyoto Protocol) in achieving sustainable development and in contributing to the ultimate objective of the United National Framework Convention on Climate Change, and to assist developed countries (included in Annex 1) in achieving compliance with quantified emission limitation and reduction commitments under Article 3 of the protocol (UN Framework Convention on Climate Change, 2003). The CDM specifically aims to lower the overall cost of reducing greenhouse gas (GHG) emissions released to the atmosphere in developing countries by forming a means for international trading of such emissions. Annex 1 countries can thereby purchase reduced GHG emissions in non-Annex 1 countries and the funds are allocated to reduce the implementation cost of the CDM-eligible project (or technology) in the host country (see box 19.5). The host country has to give a final approval for each CDM project through its Designated National Authority (DNA). The definition of criteria that lead to an approval or a rejection of a project proposal is the full responsibility of the host country.

This example emphasises the importance of engineering, technology and scientist managers taking cognisance of new international opportunities that may arise to financially support new first-world technologies in developing countries.

Example 19.5 CDM project example: South African technology opportunities in the international ferrochrome industry

South Africa holds the largest share of the world's chrome resources (Brent *et al.*, 2002), which are mainly used in the steel manufacturing industry. Whilst the international chrome price has stagnated, it is still estimated that a new production facility is constructed in South Africa every second year with a minimum capacity of 100 000 tonnes per annum (tpa) of ferrochrome (Morrison & Singh, 2001). The conventional technology (used world-wide) utilises a pelletising-sintering treatment of the chrome ore before smelting. This conventional technology (for a 100 000 tpa plant) uses approximately 240 GWh of electricity, which in turn is primarily generated from coal in South Africa. A new technology has been developed by the CSIR in South Africa (Morrison & Singh, 2001), which saves in the order of 30 per cent of the process energy (coal). Consequently, a large reduction in greenhouse gas emissions (138 000 tpa CO_2) can be obtained through the introduction of the technology in the South African ferrochrome production industry, and the technology is therefore potentially eligible for CDM funding; according to the Executive Board (EB) established by the Kyoto Protocol, to be eligible, a technology must reduce greenhouse gas emissions by at least 100 000 tpa compared to the baseline technology, termed certified emission reductions (CERs).

Once the technology has been demonstrated and an approved designated operational entity (DOE) has verified the CERs, the technology can be made available on the open market. In the case of South Africa, PricewaterhouseCoopers is an approved DOE. The current (2005) market price for CERs is US$3–5/tpa (http://www.das-ib.de/). Developed (i.e. Annex 1) countries can then purchase these CERs at the open market price. The funding flows through the EB to the project implementer to financially support the green technology. However, before the EB gives final approval, the DNA of the developing country where the technology will be implemented must also provide written approval that it deems the technology to comply with the sustainable development criteria of that country (Brent *et al.*, 2005). In the case of South African, the DNA is the CDM secretariat, which is situated in the Department of Minerals and Energy (DME).

Self-assessment

19.1 Describe the following concepts, principles and methodologies in the context of the South African (or your applicable country's) legislation:
- cradle-to-grave
- polluter pays
- precautionary principle
- integrated environmental management
- environmental management system
- environmental auditing
- environmental accounting
- life cycle thinking
- life cycle costing
- environmental risk assessment
- life cycle assessment
- life cycle engineering
- environmental impact assessment
- life cycle management.

19.2 Describe in a comprehensive manner, using an example for clarification purposes, the different phases of the following methodologies:
- life cycle assessment, using the ISO 14040, SETAC and the US EPA documentation as reference material; and
- environmental impact assessment, using DEAT documentation and other international reference material, e.g. the international principles of EIA prescribed by the International Association for Impact Assessment (IAIA). Specifically reference the EIA process in at least one country other than South Africa.

19.3 Define the concept of sustainable development in the context of business/industry and explain how business (and operational) practices must be changed to address sustainable development through a holistic LCM approach.

19.4 Define the concept of sustainable development in the context of society in general and explain how technological strategies and innovations may align business/industry in terms of the increasing trend of sustainable production and consumption. Describe the role of engineers, technologists and scientists in the process.

19.5 Identify additional opportunities, other than the CDM process of the Kyoto Protocol, to obtain direct funding or financial support for green technologies.

Mini-projects

Mini-project 19.1 (sustainable development reporting)

By consulting the sustainable development reports of ten multinational companies, identify the aspects that are similar between the policies and visions of these companies, and define a generic framework by which a South African company can formulate a sustainable development policy and vision according to international best practices. Then, by following the principles of the Global Reporting Initiative (GRI), develop a standard template for a comprehensive sustainable development report for a specific industry sector of your choice.

Mini-project 19.2 (aligning project management practices with EIAs)

From literature, describe the project life cycle phases that have been developed for different industry sectors for which EIAs are compulsory as stipulated in the DEAT EIA guidelines, i.e. as listed activities. For these different project life cycles, explain how you would align the project phases with the phases of the EIA process to optimise the time schedule of the project and minimise potential project risks that you have identified.

Mini-project 19.3 (performing comprehensive life cycle assessments)

Passenger vehicles are often offered to the customer with leather or velour fabric as interior upholstery-trimming options. By following the ISO 14040 (and other) guidelines, conduct a comprehensive life cycle assessment of the material options for a specific vehicle of your choice. From your analysis, with supportive arguments, identify which option would be your preference from an environmental impact perspective. Explain what the two value chains can do to reduce their respective overall impacts.

Mini-project 19.4 (environmental management systems)

Select a typical company in an industry sector of your choice. Evaluate and describe the environmental management system of the company. Specifically identify the aspect of the ISO 14000 family of standards that are addressed in the EMS. Furthermore, describe how the EMS is integrated with other internal management practices, e.g. quality, operations and maintenance, and procurement management. Where possible, also identify possibilities for the EMS to contribute towards 'greening' the supply chain.

Mini-project 19.5 (sustainability costing)

From available literature, specify the financial information of a development project that has been undertaken. Considering cost-benefit analysis (CBA) and economic and resource economics approaches, incorporate the externalities associated with the project into the financials, and verify whether the project would still be economically viable in the long term.

Mini-project 19.6 (cleaner production)

For a specific sector, e.g. textiles, food and beverage, automotive manufacturing etc., compile, from available literature, the typical cleaner production opportunities that have been identified internationally. Perform SWOT analyses of these opportunities in the context of South Africa (or your applicable country) and describe how cleaner production can be promoted in the country through the specific sector.

Mini-project 19.7 (green products/technologies)

More companies are realising that there is a growing demand for green products, as consumers become aware of environmental issues such as global warming, upper-layer ozone-layer depletion and other forms of environmental degradation. Choose any product, activity, process or packaging and describe how it can be designed to become more environmentally friendly. For example:

- hybrid electric vehicles (HEV);
- ozone-friendly deodorants; and
- recycling technologies.

References

Allen, D.T., Consoli, F.J., Davis, G.A., Fava, J.A. & Warren, J.L. (eds). 1997. *Public Policy Applications of Life Cycle Assessment.* Society of Environmental Toxicology and Chemistry (SETAC), SETAC Press.

American Institute of Chemical Engineers (AIChe). 2000. *Total Cost Assessment Methodology.* Arthur D. Little report for the Center for Waste Reduction Technologies. Available at http://www.aiche.org/cwrt/projects/cost.htm, accessed 19 July 2005.

Azapagic, A. & Perdan, S. 2000. 'Indicators of sustainable development for industry.' *Trans Ichem,* 78[B], pp. 243–61.

Baumann. H. 1996. 'LCA use in Swedish industry.' *International Journal of Life Cycle Assessment,* 1 (3), pp. 122–6.

Bieker, T., Dyllick, T., Gminder, C.U. & Hockerts, K. 2001. 'Towards a sustainability Balanced Scorecard linking environmental and social sustainability to business strategy.' In *Proceedings of the Business Strategy and the Environment Conference.* Leeds: University of Leeds.

Blanchard, B.S. & Fabrycky, W.J. 1998. *Systems Engineering and Analysis.* 3rd edition. Upper Saddle River: Prentice-Hall.

Blignaut, J., Döckel, M., Mirrilees, R., Van Aarde, R. & Wilson, N. 2004. 'Capturing the value of environmental services.' In Blignaut, J. & De Wit, M. (eds). *Sustainable Options: Development Lessons from Applied Environmental Economics.* Cape Town: UCT Press.

Bonnal, P., Gourc, D. & Lacoste, G. 2002. 'The life cycle of technical projects.' *Project Management Journal* 33 (1), pp. 12–19.

Brent, A.C., Rohwer, M.B., Friedrich, E. & Von Blottnitz, H. 2002. 'Overview: LCA in South Africa: Status of life cycle assessment and engineering research in South Africa.' *International Journal of Life Cycle Assessment,* 7 (3), pp. 167–72.

Brent, A.C. 2003. 'A proposed lifecycle impact assessment framework for South Africa from available environmental data.' *South African Journal of Science,* 99 (3/4), pp. 115–22.

Brent, A.C. 2004. 'A life cycle impact assessment procedure with resource groups as areas of protection.' *International Journal of Life Cycle Assessment*, 9 (3), pp. 172–9.

Brent, A.C. & Visser, J.K. 2005. 'An environmental performance resource impact indicator for life cycle management in the manufacturing industry.' *Journal of Cleaner Production*, 13 (6), pp. 557–65.

Brent, A.C., Heuberger, R. & Manzini, D. 2005. 'Evaluating projects that are potentially eligible for Clean Development Mechanism (CDM) funding in the South African context: A case study to establish weighting values for sustainable development criteria.' *Environment and Development Economics*, 10 (5), 631–649.

Briassoulis, H. 2001. 'Sustainable development and its indicators: Through a (planner's) glass darkly.' *Journal of Environmental Planning and Management*, 44 (3), pp. 409–27.

Broberg, O. & Christensen, P. 1999. 'LCA experiences in Danish industry.' *International Journal of Life Cycle Assessment*, 4 (5), pp. 257–62.

Brune, D., Chapman, D.V., Gwynne, M.D. & Pacyna, J.M. (eds). 1997. *The Global Environment: Science, Technology and Management.* Vols 1 & 2. Weinheim: VCH.

Buttrick, R. 2000. *The Project Workout: A Toolkit for Reaping the Rewards from All Your Business Projects.* London: Prentice-Hall.

Carson, R. 1963. *Silent Spring.* London: Hamish Hamilton.

Consoli, F.J., Allen, D.T., Boustead, I., Fava, J.A., Franklin, W., Jensen, A.A., De Oude, N., Parrish, R., Perriman, R., Postlethwaite, D., Quay, B., Séguin, J. & Vigon, B. (eds). 1993. *Guidelines for Life-Cycle Assessment: A Code of Practice.* Society of Environmental Toxicology and Chemistry (SETAC), SETAC Press.

Consultative Forum on Mining and the Environment. 2002. *Public Participation for Stakeholders in the Mining Industry.* Johannesburg: Chamber of Mines of South Africa. Available at www.bullion.org.za, accessed 19 July 2005.

Cooper, R.G. 2001. *Winning at New Products.* 3rd edition. New York: Perseus.

Cooper, J.S. & Vigon, B. 1999. *Life Cycle Engineering Guidelines.* Prepared for US-EPA Office of Research and Development National Risk Management Research Laboratory by Battelle Memorial Institute, contract no. CR822956. Available at http://www.epa.gov, accessed 19 July 2005.

Curran, M.A. (ed.) 1999. 'The status of LCA in the USA.' *International Journal of Life Cycle Assessment*, 4 (3), pp. 123–24.

Daly, H.E. 1990. 'Toward some operational principles of sustainable development.' *Ecological Economics* 2, pp. 1–6.

Department of Environmental Affairs and Tourism (DEAT). 1992. *Integrated Environmental Management Procedure.* Pretoria: Directorate: Environmental Planning and Impact Management, DEAT.

Department of Environmental Affairs and Tourism (DEAT). 1998. *Regulations; Implementation of section 21, 22 and 26 of the Environmental Conservation Act, 1989.* Guideline document. Pretoria: Directorate: Environmental Planning and Impact Management, DEAT.

Department of Environmental Affairs and Tourism (DEAT). 2000. *Strategic Environmental Assessment in South Africa.* Guideline document. Pretoria: Directorate: Environmental Planning and Impact Management, DEAT.

Dingle, J. 1997. *Project Management: Orientation for Decision Makers.* London: Arnold.

Dubos, R. & Ward, B. 1972. *Only One Earth: The Care and Maintenance of a Small Planet.* Harmondsworth: Penguin.

Duchin, F. & Lange, G. 1994. *The Future of the Environment: Ecological Economics and Technological Change*. New York: Oxford University Press.

Du Toit, J. 2003. *Advanced Planning Implementation Methodology*. Unpublished master's thesis, University of Pretoria.

Ehrlich, P.R. & Ehrlich, A.H. 1990. *The Population Explosion*. London: Hutchinson.

De Wit, M., Harinath, V. & Letsoalo, A. 2004. *Environmental Economics*. Integrated Environmental Management Information Series 16. Pretoria: Department of Environmental Affairs and Tourism.

Fava, J.A., Densison, R., Jones, B., Curran, M.A., Vigon, B., Selke, S. & Barnum, J. (eds). 1991. *A Technical Framework for Life Cycle Assessment*. Society of Environmental Toxicology and Chemistry (SETAC), SETAC Press.

Fava, J.A. & Smith, J.K. 1998. 'Integrating financial and environmental information for better decision making.' *Journal of Industrial Ecology*, 2 (1), pp. 9–11.

Five Winds International. 2004. *Environmental Risk Assessment*. Available at http://www.fivewinds.com/publications/publications.cfm, accessed 1 July 2005.

Fuggle, R.F. 1992. 'Environmental evaluation.' In Fuggle, R.F. & Rabie, M.A. (eds). *Environmental Management in South Africa*. Cape Town: Juta.

Gale, R.J.P. & Stokoe, K.P. 2001. *Environmental Cost Accounting and Business Strategy: Handbook of Environmentally Conscious Manufacturing*. Boston: Kluwer.

Gladwin, T.N., Kennelly, J.J. & Krause, T-S. 1995. 'Shifting paradigms for sustainable development: Implications for management theory and research.' *Academy of Management Review*, 20 (4), pp. 874–907.

Goede, F. 2003. 'The future of SH&E in the process industry with the focus on products.' Presentation given at the Department of Engineering and Technology Management, University of Pretoria.

Graedel, T.E. 1998. *Streamlined Life-Cycle Assessment*. Upper Saddle River: Prentice-Hall.

Grotz, S. & Scholl, G. 1996. 'Application of LCA in German industry: Results of a survey.' *International Journal of Life Cycle Assessment*, 1 (4), pp. 226–30.

Heaton, G.R., Banks, R.D. & Ditz, D.W. 1994. *Missing Links: Technology and Environmental Improvement in the Industrializing World*. World Resources Institute. Available http://www.wri.org, accessed 19 July 2005.

Heisler, H. 1999. *Vehicle and Engine Technology*. 2nd edition. London: Arnold.

Holliday, C.O., Schmidheiny, S. & Watts, P. 2002. *Walking the Talk: The Business Case for Sustainable Development*. Sheffield: Greenleaf .

International Association for Impact Assessment (IAIA). 2003. *Social Impact Assessment: International Principles*. IAIA Special Publication Series, No. 2. Available at www.iaia.org/Non_Members/Pubs_Ref_Material/pubs_ref_material_index.htm, accessed 1 July 2005.

IKP. 2004. *Life Cycle Engineering: A Concise Guide at a Glance*. Available at http://129.69.122.198/english/loesungen_lcalce_e.html, accessed 1 July 2005.

International Chambers of Commerce (ICC). 1989. *Environmental Auditing*. Available at www.iccwbo.org/, accessed 1 July 2005.

International Institute for Sustainable Development (IISD), Deloitte & Touche, & World Business Council for Sustainable Development (WBCSD). 1992. *Business Strategies for Sustainable Development: Leadership and Accountability for the 90s*. Available at www.iisd.org/publications/publication.asp?pno=242, accessed 19 July 2005.

International Organisation for Standardisation (ISO). 1998. *Code of Practice: Environmental Management – Life Cycle Assessment – Life Cycle Impact Assessment*. International Standard ISO 14042. Geneva.

International Organisation for Standardisation (ISO). 2002. *Systems Engineering: System Life Cycle Processes*. ISO/IEC 15288:2002 standard. Available at www.iso.org, accessed 19 July 2005.

International Organisation for Standardisation (ISO). 2005. *ISO 14000*. Available at www.iso.org/iso/en/iso9000-14000/iso14000/iso14000index.html, accessed 1 July 2005.

Keeble, J.J., Topiol, S. & Berkeley, S. 2003. 'Using indicators to measure sustainability performance at a corporate and project level.' *Journal of Business Ethics*, 44, pp. 149–58.

Kerzner, H. 2001. *Project Management: A Systems Approach to Planning, Scheduling and Controlling*. New York: John Wiley.

Kesler, S.E. 1994. *Mineral Resources, Economics and the Environment*. New York: Macmillan.

Kliem, R.L., Ludin, I.S. & Robertson, K.L. 1997. *Project Management Methodology: A Practical Guide for the Next Millennium*. New York: Marcel Dekker.

Kumar, N. & Siddharthan, N. 1997. *Technology, Market Structure and Internationalization: Issues and Policies for Developing Countries*. London: Routledge.

Labuschagne, C. & Brent, A.C. 2005. 'Sustainable project life cycle management: The need to integrate life cycles in the manufacturing sector.' *International Journal of Project Management*, 23 (2), pp. 159–68.

Labuschagne, C., Brent, A.C. & Van Erck, R.P.G. 2005. 'Assessing the sustainability performances of industries.' *Journal of Cleaner Production*, 13 (4), pp. 373–85.

Leiman, A. & Tuomi, K. 2004. *Cost Benefit Analysis*. Integrated Environmental Management Information Series 8. Pretoria: Department of Environmental Affairs and Tourism.

Leiman, A. & Van Zyl, H, 2004. 'Economics in impact assessment: The role of environmental and resource economics.' In Blignaut, J. & De Wit, M. (eds). *Sustainable Options: Development Lessons from Applied Environmental Economics*. Cape Town: UCT Press.

Mebratu, D. 1998. 'Sustainability and sustainable development: Historical and conceptual review.' *Environmental Impact Assessment Review*, 18, pp. 493–520.

Mathews, M.R. & Lockhart, J.A. 2001. *The Use of an Environmental Equity Account to Internalise Externalities*. Birmingham: Aston Business School Research Institute, RP0104.

Minnesota Office of Environmental Assistance (MOEA). 2001. *Design for the Environment: A Competitive Edge for the Future*. DfE toolkit and guide book. Available at www.moea.state.mn.us/pubs.cfm, accessed 1 July 2005.

Mohr-Swart, M. 2003. 'Environmental management accounting in the South African mining industry: Setting the scene.' Proceedings of the Mining and Sustainable Development Conference, Sandton, South Africa.

Morris, P.W.G. 1994. *The Management of Projects*. London: Thomas Telford.

Morrison, A.M. & Singh, V. 2001. *An LCA Evaluation of Ferrochrome Production Processes*. CSIR Process Technology Centre Report 8600/86DD/HTE20. Pretoria: CSIR.

Neitzel, H. 1996. 'Principles of product-related life cycle assessment: Conceptual framework/memorandum of understanding.' *International Journal of Life Cycle Assessment*, 1 (1), pp. 49–54.

Pirages, D.C. (ed.). 1996. *Building Sustainable Societies*. New York: Sharpe.

PricewaterhouseCoopers. 2002. *Sustainability Survey Report*. Available at www.pwcglobal.com, accessed 19 July 2005.

Rebitzer, G. & Hunkeler, D. 2003. 'Life cycle costing in LCM: Ambitions, opportunities and limitations.' *International Journal of Life Cycle Assessment*, 8 (5), pp. 253–6.

Rebitzer, G., Fullana. P., Jolliet, O. & Klöpffer, W. 2001. 'Advances in LCA and LCM.' *International Journal of Life Cycle Assessment*, 6 (4), pp. 187–91.

Reid, D. 1995. *Sustainable Development: An Introductory Guide*. London: Earthscan.

Sampson, I. 2001. *Introduction to a Legal Framework to Pollution Management in South Africa*. Deloitte & Touche and South African Water Research Commission Report TT 149/01. Pretoria: Deloitte & Touche.

Sasol. 2002a. *Quality Management System 918: Progressing an Environmental Impact Assessment*. South Africa: Sasol.

Sasol. 2002b. *Quality Management System 919: Environmental Verification Inspections on Projects*. South Africa: Sasol.

Schuman, C. & Brent, A.C. 2005. 'Asset life cycle management: Towards improving asset performance.' *International Journal of Operations and Production Management*, 25 (6), pp. 566–79.

Seppälä, J. 1999. 'Decision analysis as a tool for life cycle impact assessment.' LCA Documents 4. Landsberg: Ecomed.

Sevitz, J., Brent, A.C. & Fourie, A.B. 2003. 'An environmental comparison of plastic and paper consumer carrier bags in South Africa: Implications for the local manufacturing industry.' *South African Journal of Industrial Engineering*, 14 (1), pp. 67–82.

Sillanpää, M. 1999. 'A new deal for sustainable development in business.' In Bennett, M. & James, P. (eds). *Sustainable Measures*. Sheffield: Greenleaf .

Society of Environmental Toxicology and Chemistry (SETAC). 2005. *Life Cycle Assessment*. Available at http://www.setac.org/lca.html, accessed 1 July 2005.

Socioeconomic Data and Application Center (SEDAC). 2003. *International Conference on Environment and Economics: Conclusions*. Available at http://sedac.ciesin.org/, accessed 1 July 2005.

Standards South Africa (STANSA), 1998. *Code of Practice: Environmental Management – Life Cycle Assessment – Principles and Framework*. SABS ISO 14040: 1997. Pretoria.

Tibor, T. & Feldman, I. 1996. *ISO 14000: A Guide to the New Environmental Management Standards*. Chicago: Irwin.

Troge, A. 2000. 'Life cycle assessment and product-related environmental policy.' *International Journal of Life Cycle Assessment*, 5 (4), pp. 195–200.

Tukker, A. 1999. 'Life cycle assessment for waste, part I.' *International Journal of Life Cycle Assessment*, 4 (5), pp. 275–81.

United Nations Conference on the Human Environment (UNCHE). 1972. *Environment and Development: The Founex Report on Development and Environment*. New York: Carnegie Endowment for International Peace.

United Nations Environmental Programme (UNEP). 2003. *Life Cycle Thinking as a Solution*. Production and Consumption Branch, UNEPTIE. Available at www.uneptie.org/pc/sustain/lcinitiative/background.htm, accessed 1 July 2005.

United Nations Framework Convention on Climate Change (UNFCCC). 2003. *Clean Development Mechanism (CDM)*. Available at http://cdm.unfccc.int/, accessed 19 July 2005.

United States Environmental Protection Agency (US EPA). 1997. *Pathway to Product Stewardship: Life-Cycle Design as Business Decision-Support Tool*. Pollution Prevention and Toxics, EPA742-R-97-008.

United States Environmental Protection Agency (US EPA), 2003. *Product Stewardship*. Available at www.epa.gov/epaoswer/non-hw/reduce/epr/, accessed 19 July 2005.

Wartick, S.L. & Wood, D.J. 1998. *International Business and Society*. Malden: Blackwell.

Weaver, P., Jansen, L., Van Grootveld, G., Van Spiegel, E. & Vergragt, P. 2000. *Sustainable Technology Development*. Sheffield: Greenfield.

Van der Wal, K. 1994. 'Technology and the ecological crises.' In Zweers, W. & Boersema, J.J. (eds). *Ecology, Technology and Culture: Essays in Environmental Philosophy*. Cambridge: White Horse Press.

WCED (World Commission on Environment and Development). 1987. *Our Common Future*. Oxford: Oxford University Press.

Wood, C. 2000. 'Screening and scoping.' In Lee, N. & George, C. (eds). *Environmental Assessment in Developing and Transitional Countries*. London: John Wiley.

World Bank Group. 2005. *Sustainable Development*. Available at www.worldbank.org/wbi/sustainabledevelopment/, accessed 19 July 2005.

Journals that may be referenced on the topic matter

Environmental Impact Assessment Review: www.elsevier.nl/inca/publications/store/5/0/5/7/1/8/.

Environmental Management and Health: fernando.emeraldinsight.com/vl=6579500/cl=58/nw=1/rpsv/emh.htm.

International Journal of Life Cycle Assessment: www.scientificjournals.com/sj/lca/welcome.htm.

Journal of Cleaner Production: www.elsevier.nl/locate/jclepro/.

Journal of Industrial Ecology: mitpress.mit.edu/catalog/item/default.asp?ttype=4&tid=32.

World Wide Web resources

American Institute of Chemical Engineers (AIChe). Total Cost Assessment: www.aiche.org/cwrt/projects/cost.htm.

Automobile ecology information: msl1.mit.edu/esd123_2001/.

Biodiversity Support Program (BSP): BSPonline.org/.

Centre for Science and Environment: www.oneworld.org/cse.

Danish Environmental Protection Agency. Environmental Impact of Packaging Materials: www.mst.dk/waste/Packagings.htm.

Earth Day Network: www.earthday.net/about/.

Economic Input-Output Life Cycle Assessment: www.eiolca.net/.

Ecosystem Valuation: www.ecosystemvaluation.org.

Electronic Green Journal: egj.lib.uidaho.edu/egj12/weintraub1.html.

Global Reporting Initiative: www.globalreporting.org/.

International Association for Impact Assessment: www.iaia.org.

International Organisation for Standardisation (ISO). The ISO 14000 Family of International Standards: www.iso.org/iso/en/prods-services/otherpubs/iso14000/index.html.

LCA links: www.life-cycle.org/.

LCA engineering links: www.bmacnulty.tripod.com/lcaengineeringlinks.html

LCA resources at S&P, CML, Leiden: www.leidenuniv.nl/interfac/cml/.ssp/projects/lca2/index.html.

Minnesota Office of Environmental Assistance. Design for the Environment Toolkit: www.
    moea.state.mn.us/p2/dfetoolkit.cfm.
SETAC LCA site: www.setac.org/lca.html.
SETAC. Streamlined Life Cycle Assessment: www.setac.org/htdocs/what_intgrp_lca.html.
UNEP Global Life Cycle Initiative: www.uneptie.org/pc/sustain/lca/lcini.htm.
US EPA LCAccess: www.epa.gov/ORD/NRMRL/lcaccess/.
World Business Council for Sustainable Development: www.wbcsd.org.

20 Entrepreneurship

Jopie Coetzee

Study objectives

After studying this chapter, you should be able to:
- explain the nature and characteristics of an entrepreneur and intrapreneur
- explain the importance of innovation and creativity in entrepreneurship
- identify and evaluate a new business venture
- identify various entry options into a market
- describe the key building blocks of a new venture, i.e.
  - various forms of a business enterprise
  - legal considerations
  - various finance options;
- describe the annual budgeting and business planning cycle
- design the layout of a business plan for a start-up business venture and for a company already in operation
- understand the key issues and steps to take when starting a new venture
- network within the business sector of South Africa

The aim of this chapter is to introduce the culture of entrepreneurship and the process of converting dreams into business ventures.

20.1 INTRODUCTION

Just imagine the world today if it were not for those unknown inventors and developers of the wheel and fire, which introduced the Iron Age!

What would you have been studying today if the printing press, the steam engine, the rolling steel mill or the humble spinning wheel had not been developed and commercialised during the Industrial Revolution of the 18th century? And what about the modern Information Age without the cordless telephone, computer technology and satellites circling the Earth?

Fortunately, our world today is the beneficiary of these and many other splendid examples of entrepreneurship. As in today's world, these early entrepreneurs were driven by innovation and creativity and pursued their dreams with boundless energy, all believing they could make a difference to their circumstances and, perhaps, even humankind. They believed in themselves and in the value of their work. They did not consider the wealth they accumulated of prime importance. Making a difference gave them a sense of achievement as their prime reward.

Today, the world is experiencing a new wave of emerging entrepreneurs, perhaps as a consequence of the inefficiencies of large corporations, the liberalisation of socialist countries and the availability of technical, financial and marketing information. Globalisation, the increasing sophistication of consumer requirements and the mobility of capital has opened new business opportunities that were not available to previous generations. The 'Entrepreneurship Revolution' manifests itself from the flower market in Cape Town to street vendors in Jakarta to California's high-tech Silicon Valley as an alternative development mechanism to socialism and central government planning. It is a bottom-up revolution that is embraced by all cultures around the world. Whether they are Chinese, Indian, Brazilian, German or South African entrepreneurs, the culture of entrepreneurship has a common objective, namely to achieve self-actualisation and financial independence despite bureaucracy and other socio-political or economic obstacles (Berger, 1991:3).

The USA is a good example of the benefits of this silent revolution, which manifests itself in its buoyant economy as evidenced by the following examples.

- By the year 2000 there were some 30 million business enterprises in the USA compared to 18 million during 1990 – that's 1,2 million new business enterprises per year! Over the past 20 years, 36 million new jobs have been created, whereas Europe and Africa have seen a net loss of jobs.

- Small businesses and expanding firms accounted for some 70 per cent of all the new jobs. Since the Second World War, 50 per cent of all innovations and 95 per cent of all radical innovations came from new and smaller companies, e.g. the microcomputer, the pacemaker, overnight express delivery services, fast food etc. (Timmons, 1994:5).

In South Africa, the consensus view among leading businesspeople and politicians is that the culture of entrepreneurship is the only sustainable long-term option that can bring prosperity, security and happiness to the country's citizens. The task is daunting considering that the unemployment rate is some 35 per cent (ideally, it must be less than 10 per cent), a GDP per capita of $3 100 (compared to Australia's $16 000 and Mozambique's $80), a literacy rate of some 50 per cent (compared to Cuba's 95 per cent), rampant crime and increasing levels of corruption.

However, the overall trend of a growing culture of entrepreneurship in South Africa is encouraging, namely:
- the number of micro-businesses: some three million and growing; and
- the number of small- and medium-sized businesses: some one million and growing.

There are many governmental and private sector initiatives to assist small businesses to start up and prosper.

20.2 WHAT IS AN ENTREPRENEUR?

Simply put, an entrepreneur is someone who is his/her own boss. However, for the purposes of this textbook the following definition applies: An entrepreneur is a person who spots a gap in the market, and conceptualises and evaluates a business idea to fill that gap. If proven viable and the timing is right, then he/she implements the business idea by bringing together capital, technology and know-how to convert input materials into goods or services in order to fill that gap in the market in a profitable and sustainable manner.

Entrepreneurs from all cultures and from all walks of life have the following characteristics in common (Schöllhammer, 1979:10).

- **The need for achievement,** which can be defined as a want or a passion in a person that motivates behaviour towards accomplishment. This is about fulfilling a goal that challenges the individual's competence;
- **The preference for calculated risks:** Entrepreneurs prefer to set goals that require high levels of performance that they believe can be achieved. However, entrepreneurs are not gamblers;
- **The willingness to take personal responsibility for accomplishment:** They prefer to use their own resources in their own way and to be held accountable for results;
- **An ability to deal with conceptual ideas,** which are at times in conflict with each other or not yet fully developed or quantified. They possess the ability to think laterally and see opportunity where others only see problems and confusion;
- **A high perception of the probability of success:** Entrepreneurs have confidence in their ability to achieve success. They study the available facts and form realistic judgements. They rely on their intuition and proceed confidently when the facts may seem insufficient;
- **High levels of physical and intellectual energy:** Entrepreneurs are active, mobile and constantly looking for creative ways of getting a task done or improving their situation. As they are acutely aware of the passage of time, there is a sense of urgency in their work;
- **Future orientated:** By planning and thinking ahead, entrepreneurs anticipate future possibilities and threats. Although attention to detail is important, entrepreneurs never forget the bigger picture of what they are busy with;
- **An unusual skill in organising** and mobilising scarce resources towards achieving objectives; and
- **Money is considered as a means to achieve objectives:** Although wealth is important, the key driving force is to make a difference and to achieve self-actualisation.

Apart from the above, the following doing skills are also considered as essential for the successful entrepreneur: negotiation, report writing, making presentations, networking, deal making, conflict resolution, time management, influencing people and diplomacy.

What propels a person towards the risks and uncertainty of entrepreneurship? The following factors provide some insight (Dollinger, 1995:50):

- **negative displacement,** which is the marginalisation of individuals or groups from the mainstream of society, e.g. minority groups in a society, people who have been retrenched, immigrants, etc. have a need to get control over their circumstances and destinies;
- **finding oneself 'between things',** i.e. people who find themselves at the end of a major phase of their lives, e.g. the completion of full-time studies or completion of a major project are more inclined to do something new than those who are 'involved with things';

- **positive pull influences** from people they may hold in high regard, e.g. a partner, a family member, a customer, a mentor, etc.; and
- **positive push influences** like a promising career, a good education, appropriate experience, a dominant father/mother, etc.

Together with the above stimuli, the new entrepreneur also needs to have a favourable perception of the ultimate viability of the venture. Examples of success of similar ventures are helpful and, having a mentor who has 'been through the same experience' is a bonus. Perceptions of cultural desirability also provide a mental comfort zone that the new entrepreneur 'is doing the right thing'. Within a South African context, it is considered the right thing to start a new business for an Indian or Jewish person. A very supportive family network, even across country borders, would be available to guide and assist such people towards achieving success. They would also enjoy a high status within their communities.

Entrepreneurial activity essentially falls into two categories, namely *initiating* and *imitating*. Initiating entrepreneurs introduce new services, products, technologies, etc. Imitating entrepreneurs disseminate and copy the innovations of innovating entrepreneurs, e.g. introducing the USA fast food concept to South Africa.

Does the above mean that we should all resign from our full-time employment and start working for ourselves? Not necessarily! Many of today's large corporations started off as small entrepreneurial ventures like the Anglo-American Company, which was founded towards the end of the 19th century. Over the years many small companies have grown and own considerable assets today. In order to remain globally competitive, these companies had to shrug-off the burden of bureaucracy developed over time and they had to re-embrace the entrepreneurial spirit of their founders. This introduced a new management philosophy, which encourages employees to take risks, to tolerate mistakes (not too many, of course!) and to develop incentive schemes to reward individuals who improve productivity due to their own initiatives.

This led to the birth of the corporate *intrapreneur*. The intrapreneur has the same characteristics as the entrepreneur, except that he/she is accountable to the shareholders and has to work within the mission, objectives, resources and value system of the large corporation. Many of today's successful corporations have a corporate culture that embraces and rewards its intrapreneurs at all levels of the organisation (Kuratho, 1992:112).

An intrapreneur's thinking revolves around challenging convention, and exploiting new technology and opportunities. This kind of thinking leads to behaviour on four levels (adapted from Rwigema and Venter, 2004:77).

1. **New venture creation:** This involves new venture formation by identifying new markets for the company, or improving productivity to supply existing markets more effectively by way of increased production volume, lower costs or better quality.
2. **Innovation:** This involves striving for breakthroughs from research and development, as well as from personal creativity and risk taking (see chapter 18 for more information on technological innovation).
3. **Internal rejuvenation:** This involves the continual search for improved processes, product improvement and productivity improvements. This is done on an ongoing incremental manner, or even in quantum leaps of growth.
4. **Pro-activeness:** This involves planning change – intrapreneurs do not react to change being forced upon them. They lead change.

Today's successful businesses are innovative, nimble, close to the customer and quick to react to the market. Generally these companies are relatively small, highly focused and lean. This is leading to structural changes in the labour market arising from downsizing, outsourcing and sub-contracting. These changes on the one hand may lead to anxiety, but to those employees with an entrepreneurial mindset it may also lead to new opportunities and towards self-actualisation.

Why should an engineering student familiarise him-/herself with the fundamentals of entrepreneurship and intrapreneurship?

With the additional knowledge of management and entrepreneurship, engineers should be seen as *engineers of economic growth*. As such, the engineer will not only experience better career prospects, but his/her skills will be envied and sought after by blue-chip companies from around the world. Also, such additional knowledge is a *platform* from where to reach for the dream of financial independence, self-actualisation and making a difference as the ultimate reward.

Some thoughts on ENTREPRENEURSHIP

The usual tendency is to refer to a business owner as an entrepreneur.

Strictly speaking, that business owner does not continually act entrepreneurially. In fact, usually he/she acts administratively or managerially. The point is that entrepreneurship is that BEHAVIOUR that causes business formation and expansion. Elsewhere in this book the behavioural characteristics are discussed, but for me the key ones are:

- willingness to take risks;
- passion for his/her enterprise;
- a burning desire to succeed;
- innovation; and
- determination.

There are very many more such qualities, and other commentators may well identify others they consider the key ones.

Naturally, entrepreneurship can and is manifested by employees of large companies – particularly the winning, world class ones. That's what creates growth in these companies. Some people call this INTRAPRENEURSHIP.

It is essential to remember that a successful business requires ENTREPRENEURSHIP, DOING SKILLS (knowledge and experiences) and BUSINESS SKILLS. In order to run one's own business one MUST have vocational (doing) skills. It is also important to remember that ANY trade, profession or task (sophisticated or simple) can be done as an employee or as a SELF-EMPLOYED PERSON. The decision to rather work for oneself than to take a job is the manifestation of ENTREPRENEURSHIP.

Jo' Schwenke: Managing Director, Business Partners Limited

20.3 NEW BUSINESS OPPORTUNITIES

Identifying the correct business opportunity that matches the skills and resources available to an entrepreneur is of critical importance. The following key issues are therefore discussed in this section:

- the innovation and creativity process;
- the relationship between ideas and a viable business opportunity; and
- entry options into the market and the associated strategic posturing (also see the chapter 11 for more information on marketing).

20.3.1 The innovation and creativity process

In the previous section *innovation* and *creativity* have been identified as key drivers for the entrepreneur. The dynamics of these two processes determine the manner in which the entrepreneur goes about his/her task of identifying and developing new business opportunities.

The *creativity process* is driven by the internal orientation of the entrepreneur. Every person has knowledge and experiences, which are unique to him/her. This uniqueness serves as an instant reference to benchmark and evaluate new information. This process is also known as 'gut-feel' and is important in identifying and in the initial evaluation of new business concepts. This intuitive process is followed up with a more rational evaluation process. The creativity process demands extreme intellectual flexibility to deal with ambiguous and half-developed facts. It also requires restraint in order not to jump to conclusions and a maturity of mind to be willing to admit that initial conclusions may have been wrong. The creative act may from time to time occur as an 'illuminating flash' through the entrepreneur's mind. All 'loose ends' just fall into place in a logical and clear concept, short-circuiting the rational innovative process. This may provide the entrepreneur with a competitive advantage in terms of spotting a gap in the market, either where competitors fail to identify the gap or where it may take them much longer to come to a similar conclusion.

Example 20.1 Finding new uses for established technologies

One of the tricks that innovators use is to find new uses for old ideas and, if necessary, to adapt these to suit new applications. Well-known examples are Robert Fulton, who decided to use steam engines to propel boats 75 years after they had first been used in mines. Consider also the problem of experimental light bulbs that kept falling out of their fixtures – this was solved by one of Thomas Edison's staff by 'borrowing' the idea of a threaded cap that could be screwed onto a kerosene bottle (Hargadon & Sutton, 2000:158).

A more recent example is that of Standard Bank, which has teamed up with cellular phone operator MTN to achieve ubiquitous banking via the cellular phone (Needham, 2005:13).

The *innovative entrepreneur* has a specific set of skills that are being used, probably subconsciously, but in addition to his/her character disposition as outlined in section 20.1. These skills are as follows.

■ They are opportunity orientated and are always attuned to spotting a gap in the market.

■ They are constantly questioning the status quo.

■ They are strategists par excellence with well-defined plans and objectives, but with flexibility to adapt to new circumstances that may improve their end objective.

■ They are trend spotters, idea orientated and tenacious. They rely on intuition and are very creative and resourceful in the process of coming up with what's required to make things happen.

■ They have a healthy blend of optimism and realism with a positive frame of mind.

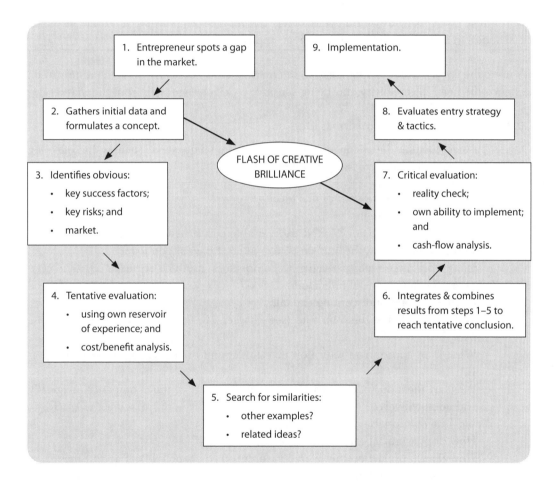

Figure 20.1 The entrepreneurial innovation and creativity process (Schöllhammer, 1979:31)

20.3.2 The origin of business ideas

Not every idea has the seeds of a profitable business opportunity. However, an abundance of varying ideas is the basis for lateral thinking, which may lead to a novel and highly profitable business venture. The golden rule is that a silly idea should not easily be discarded as folly; it just might trigger off a million dollar idea that nobody has thought of before!

Consider the following sources of business ideas:

- the printed media: newspapers, magazines, technical or special interest group publications, advertisements, etc.;
- the electronic media: the Internet, topical discussions on TV or radio, etc.;
- trade fairs and exhibitions not only for observing new products or services, but also for meeting new people and networking;
- government agencies for the promotion of small business and patent brokers;
- friends and family complaining about bad service, or 'wishing' that something was available to save them time or to take the hassle out of their everyday lives, saying 'would it not be nice if ...?' and
- within a going concern, employees being encouraged by way of financial incentive schemes to submit ideas that save costs or improve productivity.

It's up to the entrepreneur to spot a gap in the market and to conceive an idea on how to fill it, hence the need for creativity and being sensitive to 'signals' coming from the market.

20.3.3 How to evaluate a business idea

Once a business idea has been conceptualised, fundamental questions such as the following need to be answered with increasing confidence:

- Am I going to make money out of this idea?
    - What is the profit margin?
        - Cash cost of production? Fixed and variable cost structure (see chapter 14)?
        - What are the 'below-the-line' costs, e.g. interest payable on loans?
        - Is the sales price realistic relative to competitors' products/services?
    - What are the key financial indicators?
        - Capital: start-up and working capital?
        - What is a suitable discount rate (see chapter 16) to reflect the overall business risk?
        - What is an appropriate return on investment and break-even period (see chapter 16)?
        - What are the industry norms and how would the new venture compare strategically and competitively?
    - Would I be comfortable with the cash-flow profile?
        - How does it impact on my personal circumstances?
        - How does it impact on the overall cash flow (see chapter 14) of my business?
        - Is the cash flow optimal?

- Would I be able to sell the product or service?
    - What is the unique selling point? Why would a customer want to buy from me?
    - Customer profile and after-sales service?
    - Who are the competitors and would I be able to compete?
    - What are the quality (see chapter 8), environmental (see chapter 19) and hygiene requirements?
    - Is the locality of the business optimal?

- Advertising: where, how, when, cost, medium, profile of target group (see chapter 11)?
- Product life cycle: would I have to invest in ongoing research and development?

■ Would I be able to manage the key risks?
- Capital exposure and cash-flow variations (expected and unexpected)?
- Macro-economic influences, legal, technological and union matters (see chapter 4)?
- Availability of and retaining key personnel (see chapter 3)?
- Competitor counteractions: what are their options and would I be able to survive?
- How do I exit from this business and what would the cost be?

■ Am I going to enjoy this?
- How would this venture impact on my family and private life?
- What are the hours involved and will there be extensive travel involved?
- Do I have to quit my job and/or relocate to another town?
- Would I have to learn new skills and would I be able to utilise my strong points?
- Could I reconcile this venture and what comes with it with my culture and religion?
- Would I achieve a sense of achievement, pride and satisfaction?

■ What happens after this venture?
- Is this a one-off event or are there longer-term strategic benefits that may lead to further business opportunities?

The above questions are being addressed systematically by way of various techno-economic studies, which can be summarised as follows:

Table 20.1 Various levels of evaluation of a business idea (also see evaluation of projects – chapters 13 and 16)

Type of study	Level of accuracy	Status of assumptions and evaluation work
Scoping	~70%	■ Assumed, but based on experience and advice from experts ■ Bench-scale tests; line drawings
Pre-feasibility	~80%	■ Key issues researched; pilot scale tests ■ Received letters-of-intent to approve licences to operate, to supply input materials, to buy goods and/or services, to provide finance etc.
Feasibility	~90%	■ All issues fully researched; sensitivity analysis done ■ All permits and contract in order; engineering design completed

Although the success rate of ideas culminating into implementation is frustratingly low (in some cases as low as 100:1), it is important to have a full pipeline of new business ideas varying from the scoping study phase to the feasibility study phase. The entrepreneur should not be discouraged by this low success rate and the large volume of evaluation work, as it is at the heart of the competitiveness and sustainability of any company. Note that many new business ideas are evaluated and implemented by making use of project management methodology.

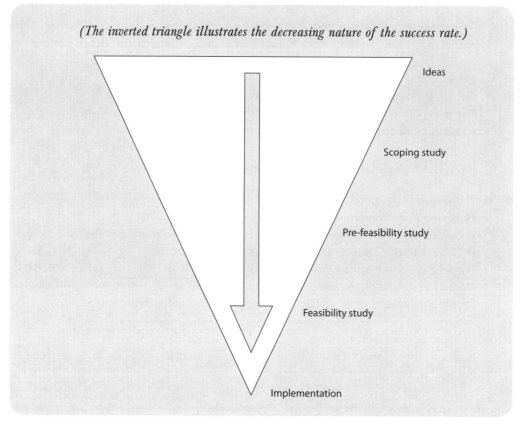

(The inverted triangle illustrates the decreasing nature of the success rate.)

Ideas

Scoping study

Pre-feasibility study

Feasibility study

Implementation

Figure 20.2 Success ratio and pipeline of new projects at various stages of development

20.3.4 Entry strategy options and strategic positioning of a new venture

How does an entrepreneur establish a business venture as a new entrant into an existing market? An unwelcoming and perhaps even a hostile reception from existing participants in the market may be awaiting! Hence, an entrepreneur should have a carefully designed entry strategy for the new venture, as competitors are not going to make it easy to move onto their turf.

As the new business venture will be at its most vulnerable during the entry phase (due to maximum financial risk with no means yet to generate profits), the following *entry strategy options* should be carefully considered.

1. Wedge into existing market sectors by way of exploiting market inefficiencies (see table 20.2 for sub-options).
2. Acquire new resources or increase the utilisation of existing resources (see table 20.3 for sub-options).
3. Identify and exploit unique opportunities relating to uncertainty and/or vulnerability during the life cycle of a specific product, a company or an industry (see table 20.4 for sub-options)

Table 20.2 Entry strategy option: Entry wedges

Sub-option	Brief description
Introduce new products & services	Relatively rare, but most potent if patented or difficult to imitate. A favourable customer perception of relative low product risk, an acceptable cost/benefit ratio and good after sales back-up add value to this option. This is usually supported by a technological leadership strategy – see chapter 18 for more information
Parallel competition	A 'me-too-but-better-than-you' strategy introducing the same product/service either at a lower cost, a better quality, slight innovative improvements or by targeting new customers. This is usually linked to a fast-follower technological strategy and capability (see chapter 18)
Franchising	Takes a proven formula for success and expands it into a new geographical area with support from the franchisor
Exploit slack capacity in the market	■ Existing production capacity being used to produce other products ■ Filling a product shortage by way of adding new production capacity ■ Geographic transfer of production capacity
Secure customer sponsorship	A letter-of-intent to purchase a product/services may be used as an impetus for a new venture and may also serve as collateral for finance
Parent company sponsorship to subsidiary, employees or strategic alliance partner	■ Providing licensed technology for restricted use ■ Joint venturing with a new entrant on the basis of a strategic trade-off ■ Relinquishing market share ■ Outsourcing of non-core business activities ■ Allowing management buy-outs ■ Sharing infrastructure to reduce the capital requirement

Table 20.3 Entry strategy option: Resource-based entries

Sub-option	Description
Secure economic rent	■ **Ownership rent:** Rents derived from acquiring, owning and controlling a valuable resource. See example 20.2 ■ **Monopoly rent:** Rents collected from government protection, collusive agreements or structural entry barriers to the industry ■ **Entrepreneurial rent:** Rents accrued from risk-taking behaviour due to special insights into complex or uncertain circumstances. This rent is of a diminishing nature when information enters the public domain ■ **Quasi rent:** Rents earned by employing firm-specific competencies to exploit a resource in a unique and cost-effective way
Utilise under-performing assets or provide assets during periods of peak output	A firm seldom fully utilises all its assets at full capacity all the time due to the ever-changing nature of the market: ■ physical assets; ■ technological and R&D assets; and ■ human, financial, intellectual and organisational assets

Table 20.4 Entry strategy option: Life-cycle opportunities (adapted from Dollinger, 1995:167)

Sub-option	Brief description
Emerging status	**Structural uncertainty:** No track record, standard operating procedures or customary practices in place yet **Strategic uncertainty:** How will the external environment react, e.g. competitors, government, communities, special interest groups, etc.? **Resource uncertainty:** Actual cost structure, quality variations, sustainability of output, ability of operators and management, etc.? **Market uncertainty:** Meeting product quality specifications, logistical supply chain working optimally, timely payment, etc.? **Investor uncertainty:** Will planned financial returns materialise?
Transitional status	Scarcity of resources to grow the business (e.g. physical or financial) Customer changes: more sophisticated, disillusioned, more demanding, sudden increase in demand, etc. De-bottlenecking of production resources and distribution channels Expanding market share, changing product mix or product characteristics
Mature status	Sustain production and maintain market share Optimisation of existing plant and resources Replacement of resources/equipment Creeping bureaucracy and systems getting outdated Industry responsibilities Complacency may lead to niche markets opening up
Declining status	Technological substitution Customers uncertain about long-term supply Changes to demographics in management and workforce Increased overheads Looking for new investments to sustain profitability

From a practical point of view, it is important that the chosen entry strategy should be fine-tuned and adapted to suit the specific practical circumstances before implementation. Usually it is very difficult and costly to change the strategic direction of a business venture once implemented.

20.4 FORMS OF BUSINESS OWNERSHIP AND MODALITIES OF START-UP

Selecting the optimal mode of entry and the form of business ownership are of critical importance to the success of a new venture. These decisions are influenced by the techno-economic and marketing realities of the new venture, the personality and financial capacity of the entrepreneur, and the external business environment. Forms of business ownership should also be understood within the context of different modalities of starting up different forms of businesses. See figure 20.3 on the next page.

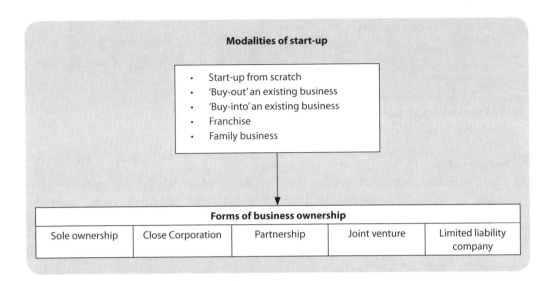

Figure 20.3 Forms of business ownership and modalities of start-up

20.4.1 Forms of business ownership

▓ **Sole ownership**

Because it is the simplest form of ownership, it is usually chosen where capital requirements are small.

❒ **Some advantages:**
  – simplicity: few legal requirements; and
  – tax benefits: business losses can be deducted from the owner's personal income.

❒ **Some disadvantages:**
  – unlimited liability, as the owner is responsible for all the debt and legal action; and
  – limited access to expertise, technology and finance and consequently, limited capacity to expand or to increase profitability.

❒ **Typical application:**
  Service-oriented industries such as any form of consultancy.

▓ **Close Corporation (CC)**

Since the Close Corporations Act was promulgated in 1984, the close corporation has become a popular form of business ownership in South Africa. The overriding intent was to introduce a simple, flexible and inexpensive legal form for up to ten persons engaged in business. The CC is a legal entity that can enter into contracts, own assets, operate a bank account and sue or be sued in court. The CC is owned and managed by its members, but exists independently of them.

❒ **Some advantages:**
  – The members are protected by having limited liability.
  – A CC is easy to establish due to simplified legal requirements and registration process.

☐ **Some disadvantages:**
- There can be a maximum 10 members, hence, growth may be limited. A CC can however be converted to a company. See disadvantages associated with a company.
- Only natural persons may be members.

☐ **Typical application:**
Service-oriented and small production-type businesses.

Partnership

The Companies Act limits a partnership to 20 members, except for partnerships of certain recognised professions such as accountants and attorneys. Apart from this restriction, a partnership is not subject to the requirements imposed on companies and close corporations by legislation.

No other statutory provisions govern partnerships. They are subject to the general principles of the law of contract and to various special principles of common law. They are not separate legal entities, and no registration formalities exist in relation to partnerships.

☐ **Some advantages:**
- easy to set up and manage; and
- limited up-front capital required.

☐ **Some disadvantages:**
- unlimited liability of the partners; and
- the actions of one partner may cause liability damage to the other partners.

☐ **Typical application:**
Ideally suited to knowledge-based, professional organisations such as legal and accounting firms.

Joint venture

A joint venture is in South African law nothing but a partnership. The term is normally used when an organisation is created by two or more independent organisations for a specific purpose over a specific period of time. Each partner in the joint venture has a pre-determined share of the ownership, capital contribution, management and the profit that the venture may generate.

☐ **Some advantages:**
- complementary skills, technologies, markets, etc., which make the venture viable; and
- reduced capital and business risk exposure to each partner.

☐ **Some disadvantages:**
- Corporate cultural differences may lead to unproductive management behaviour.
- The partners are dependent on each other for success, hence divergent views on major issues should be managed with skill and prudence to avoid financial loss.

☐ **Typical application:**
Joint ventures are often established when a project requires immense capital or each partner has something special to contribute to the overall success of the venture, e.g. technology, know-how, resource contributions or off-take agreements.

■ Limited liability company

The most common form of a business entity in South Africa is the limited liability company. Liability can be limited by guarantee or by shares.

Firstly, companies limited by guarantee are generally not-for-profit organisations that primarily promote religious, charitable, educational and certain other similar social interests. The names of companies limited by guarantee end with the words 'Limited (Limited by Guarantee)'.

Secondly, companies limited by shares may be public or private companies. The name of a public company ends with the word 'Limited', is usually listed on a bourse, and is therefore subject to the rules and regulations laid down by such a bourse (e.g. the Johannesburg Securities Exchange). The name of a private company ends with the words '(Proprietary) Limited' or Pty Ltd.

Typical applications:

Any organisation requiring significant amounts of capital or where it is deemed prudent to have the protection of limited liability.

The various forms of business ownership discussed in this section are summarised in table 20.5.

Table 20.5 A summary of the various business entity forms

Business entity	Key characteristics
Sole proprietorship	■ No statutory regulations
Partnerships (general)	■ Unlimited liability ■ Not more than 20 partners ■ No registration necessary ■ No detailed reporting, disclosure audit requirements ■ Every partner liable for all debts
Close corporation (CC)	■ Limited liability ■ Not more than 10 shareholders ■ Limited reporting and disclosure ■ No mandatory audit
Private companies (Pty) Ltd)	■ Limited liability ■ Not more than 50 shareholders ■ Limited reporting ■ Mandatory full-scope auditing
Public companies (Limited)	■ Limited liability ■ No restriction on the number of shareholders ■ Comprehensive reporting and disclosure ■ Mandatory full-scope auditing

20.4.2 Modalities of starting up a new venture

▥ **Family business (see example 20.2)**

The family business is one of the most popular forms of business in the world.

❑ **Some advantages:**
- – pooled expertise, capital, human resources and network; and
- – shared commitment, a support network and overhead costs.

❑ **Some disadvantages:**
- – There is an absence of professional management in a small business, which is often paternalistic or maternalistic.
- – Family politics may have a negative influence on efficiency, e.g. succession hierarchy, entry of new family members, compensation, allocation of responsibilities, etc.

❑ **Typical application:**

Entries into niche markets by small 'production' concerns that require limited capital initially. By 'production' one would include ventures as diverse as bakeries, farms, mines or the manufacturing of items.

Example 20.2 Family businesses in South Africa

In South Africa there are some very prominent examples of family businesses, e.g. the Oppenheimer family (De Beers and Anglo American companies), the Rupert family (Rembrandt company) and the Kunene brothers, who built a retail empire from scratch. These family businesses developed into world-class, multi-national companies and have managed to overcome the abovementioned potential disadvantages.

▥ **Franchise**

A franchise is a special type of business agreement, whereby a franchisor allows a franchisee to use the company's name, trademarks, technical data, recipes, etc. to establish a business in return for a royalty payment. The royalty payment is usually linked to a percentage of sales in the franchised operation. The Franchise Agreement typically provides for the following rights and obligations.

❑ **Franchisor:**
- – Provides market research, training, certain stock items and economies of scale relating to purchasing, distribution, advertising, etc.
- – Grants an exclusive right to trade within a certain geographical area.
- – Sets the standards of the service or product.

❑ **Franchisee:**
- – Manages the business on a day-to-day basis and optimises profitability within the exclusive geographical area allocated to him/her.

❑ **Some advantages:**
- – The franchisee benefits from the experience and know-how of the franchisor.
- – The franchisee does not have to go through the process of product design, development and launch.

☐ **Some disadvantages:**
  – A franchise limits creativity and the scope of expansion, unless the franchisor agrees.
  – Usually a large up-front payment is required for the 'name' of the franchisor.
☐ **Typical application:**
  Service industries, but particularly in the food and leisure industries.

■ **Other modalities of starting up a new venture**
  ☐ **Start-up from scratch:** High risk, but with tremendous potential for adding value to the up-front risk capital.
  ☐ **Buy-out of an existing business:** A detailed due diligence is required to ensure that there are no hidden fatal flaws. A premium is usually payable, as the business is a going concern and generating cash. Unless the incoming entrepreneur has a clear perception/action plan to increase profitability, return on investment is not high. The following is a list of various types of buying-out a business, either in a friendly or hostile manner:
    – an outright buy-out of 100 per cent of the business;
    – a management buy-out (MBO);
    – buy-out of the major shareholder and assume an active and controlling role; and
    – buy into the business as a minority shareholder and assume the role of 'trainee manager'. Once the new owner has learned the nature of the business, then the decision is made to buy out the major shareholder.

20.5 LEGAL CONSIDERATIONS FOR A NEW VENTURE

New business ventures require entry into one or more contracts, which may be interrelated.

A contract may be defined as any agreement between two or more parties and, if in compliance with the law it is legally binding (see chapter 6 – 'Engineering Contracts and Law'). It is good business practice that each party should appoint its own legal representative to protect its rights and to formalise the terms of an agreement. Each agreement should contain at least the following key clauses:
■ names and addresses of each party;
■ a description of the transaction and the associated rights and obligations of each party; and
■ arbitration procedures, force majeure and termination.

When starting a new venture the entrepreneur is obliged to register the business with the relevant government authorities and comply with all the applicable legislation, as illustrated in figure 20.4 on the next page.

20.6 SOURCES OF FINANCE

To select the most suitable source of finance is a 'best fit' among a multitude of parameters, such as:
■ the ownership structure, the projected cash flow and robustness of the new venture;
■ the technical, marketing and financial risk factors;
■ the track record of the entrepreneur and the management team; and
■ the collateral offered.

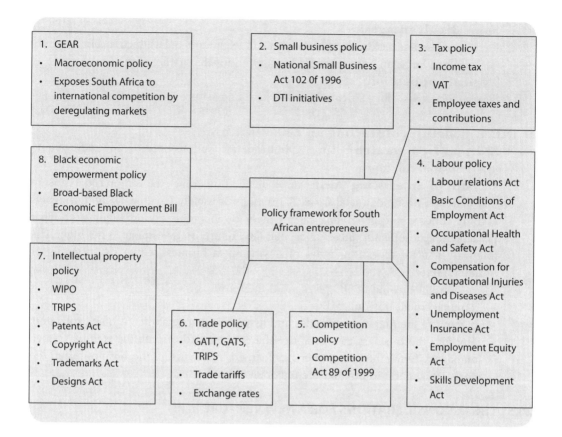

Figure 20.4 Policy framework for South African entrepreneurs (Rwemiga & Venter, 2004:313) (Note: see glossary at back of book for meanings of abbreviations.)

20.6.1 Financial institutions

At the beginning of the 21ˢᵗ century the financial services industry is going through a period of fundamental change with new product and technological innovations. Also, the boundaries between financial institutions have blurred due to competition. Hence, a prospective entrepreneur can go to any of the following institutions and can expect to be advised and assisted with a wide range of financial options, subject to their specific investment criteria. these financial institutions are commercial banks, merchant banks, private investment banks, private equity and venture capital funds, and development corporations.

For information on the above financial institutions, the following websites are worth accessing to begin your search for finance:
www.absa.co.za; www.standardbank.co.za; www,businesspartners.co.za; www.idc.co.za and www.allangray.co.za.

20.6.2 Johannesburg Securities Exchange

■ **Historical background:**

Benjamin Woolan founded the Johannesburg Stock Exchange (JSE) in 1887, a year after President Paul Kruger proclaimed as public diggings the nine farms that made up the

Witwatersrand goldfields, the richest of its kind in the world. A marketplace was required by the many mining and financial companies to buy and sell shares. Today the JSE is governed by the Stock Exchange Act and by its own rules and directives. After 108 years of open outcry trading, the JET (Johannesburg Equities Trading) System introduced the modern era of electronic trading in 1996.

- **Purpose and benefits of a stock exchange:**
  The JSE is structured without any bias towards any particular business or social community. A listing on the JSE enables large sums of capital to be raised for expansion of existing businesses, or to start up business ventures. The financial returns for the investor are in the form of capital appreciation of the shares and dividend payout.
- **Main listing categories:**
  The JSE operates two markets, namely the Main Board and AltX. The Main Board is for companies with a share capital of +R25 million and an audited profit level of +R8 million before tax at the time of listing. The AltX is for smaller, or start-up companies. The JSE's website at www.jse.co.za provides all information required for the prospective entrepreneur to consider this means for raising capital.

20.7 THE BUSINESS PLAN

20.7.1 What is a business plan?

A business plan is a written document that summarises a business opportunity, defines how management intends to capitalise on this opportunity and forecasts returns on shareholders' capital invested. The main uses of a business plan are to raise finance and to guide management towards achieving the company's strategic and operational business objectives.

During the preparatory stages of a business plan it is wise to spend more time on it than one would think necessary as it:

- forces you to arrange your thoughts in a logical order; and
- forces you to simulate reality and to anticipate pitfalls before they occur, thus avoiding costly mistakes.

A business plan is truly management's game plan to achieve the company's objectives. It enables management to assist in the planning, organising, leading and control of all strategic and tactical objectives throughout the company (see also chapter 2). It enables management, firstly to ensure the survival of and secondly to grow the company.

A business plan is the consolidation of various departmental budgets (see chapter 15 for more information on budgets). A budget is the detailed stipulation of the following on a monthly and annual basis to ensure the delivery of departmental objectives:

- output targets, such as production (see chapter 7 – 'Operations Management') and sales;
- quality assurance, such as operating costs, product quality, productivity ratios, SHEQ (safety, health, environment and quality), etc.;
- scheduling of input requirements, such as staff, start-up and working capital, electricity, etc.; and
- implementation planning, such as the allocation of who, what, when, where, how and MMMMM (i.e. men, money, machines, methods and material) to each project/task.

This can best be illustrated by the diagram below (figure 20.5).

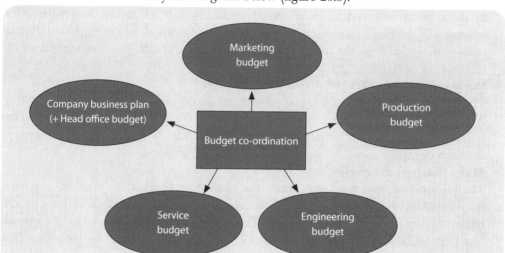

Figure 20.5 The integration of various departmental budgets into a business plan (see also chapter 15)

20.7.3 The annual business planning cycle of an operating company

The diagrams that follow describe the annual business cycle of an operating company.

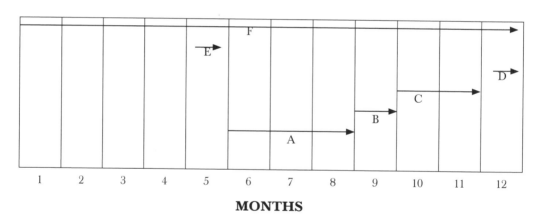

MONTHS

A: Collect and synthesise strategic information
B: Strategic planning
C: Business planning
D: Business plan approval
E: Business plan review
F: Monthly monitoring of progress relative to budget forecasts and corrective measures where appropriate.

A: Collect and synthesise strategic information

B: Strategic planning

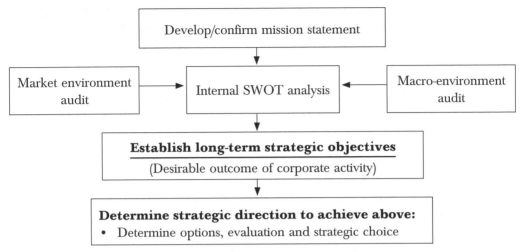

```
┌─────────────────────────────────────────┐
│      Develop/confirm mission statement    │
└─────────────────────────────────────────┘
                      │
                      ▼
┌──────────────────┐   ┌──────────────────────┐   ┌──────────────────┐
│ Market environment│ →│ Internal SWOT analysis │← │ Macro-environment │
│      audit        │   └──────────────────────┘   │      audit        │
└──────────────────┘             │                 └──────────────────┘
                                 ▼
         ┌───────────────────────────────────────────┐
         │  Establish long-term strategic objectives   │
         │    (Desirable outcome of corporate activity) │
         └───────────────────────────────────────────┘
                                 │
                                 ▼
         ┌───────────────────────────────────────────┐
         │  Determine strategic direction to achieve above: │
         │  • Determine options, evaluation and strategic choice │
         └───────────────────────────────────────────┘
```

C: Business planning

> **Agreement on:**
> - Tactical objectives
> - Key assumptions
> - Action plans
> - Responsibilities

> **Iterative departmental budgeting process:**
> - Synchronising inter-departmental action plans
> - Scheduling of activities & expenditure
> - Allocation of MMMMM + responsibilities
> - Financial analysis and sensitivity analysis

> **Compilation of overall business plan:**
> - Integrate financial results and report writing
> - Acceptance by management team

D: Business plan approval

> **Management presentations:**
> - Board and key shareholders to approve

E: Business plan review

> Formal strategic and operational reviews of progress;
> consider new strategic trends and how to capitalise

F : Monthly monitoring of progress relative to budget forecasts and corrective measures where appropriate.

20.7.2 Key success factors of a business plan

The business plan must be a credible anchor document that binds together the various divisions of the business into a coherent team effort, focused on achieving the company's strategic and operational objectives. Therefore, particular attention should be given to structure the business planning process for success by incorporating the following principles:

- an in-depth assessment of the macro- and market business environments, and outlining trends that may impact on the company – negatively and positively (see also chapter 17);
- an in-depth assessment of the company's relative competitive position, outlining the company's strengths, weaknesses, opportunities and threats (see also chapter 17); and
- an 'inclusive planning process' where management develops its own objectives and actions plans within the context of the strategic requirements by the board. As management is responsible for achieving the objectives of the business plan, it is imperative that they not only believe in it, but also that they align their personal and career objectives with the company's vision, values and objectives. Management also need to secure buy-in of all employees into the company's game plan to survive and grow.

The planning manager acts as the facilitator of the planning process. However, management remains involved at all times, as it is 'their' game plan for which they will be held responsible for delivery. The planning manager merely compiles the business plan document as a service to management.

20.7.4 Layout of the business plan of an operating company

Although the layout of business plans varies from industry to industry, and from company to company, the following layout will illustrate the key features common to most companies.

Paragraph	Main heading	Key features (see note 1, below)
	Executive summary	See note 2, below.
1	Review of past year	- Strategic and operational objectives achieved? - Reasons for performance or failure - Key achievements and challenges during the year
2	Analysis of the business environment	- Re-confirm vision, mission and values statement - Analysis of micro-, macro- and market environments - Strategic objectives for company over the next five years in all key performance areas
3	Setting tactical objectives for the planning period	- Statement of the company's operational objectives - Planning assumptions for the next year, e.g. sales targets, interest rate, inflation, productivity ratios, etc. - Departmental objectives for next year
4	Consolidated budget for the entire company	- Financial statements; capital budget - Project and finance plan for key initiatives

	Departmental budgets	
5	Marketing	▨ Scheduling of key outputs such as sales, production
6	Production	▨ Capital expenditure schedule
7	Engineering	▨ Operating costs
8	Services	▨ Key tactical action plans
9	SHEQ	▨ Allocation of resources to implement tactical plans
Appendices		

Note 1: For report writing, consider the information needs of the various stakeholders as illustrated below.

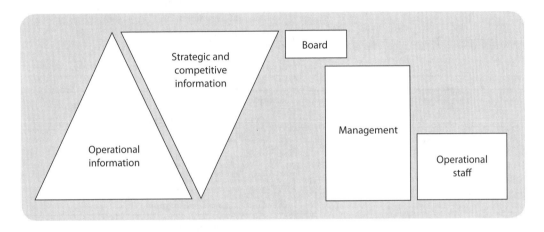

Figure 20.6 Typical information requirements from within the company

Note 2: The executive summary (1–2 pages):

The executive summary is compiled at the end of the planning process and contains all the essential features of the business plan. This section is most important as it is aimed at securing approval from the key decision makers. A decision maker does not read beyond this section if his/her interest has not been secured. Consequently, it should contain all the relevant supporting information required for executive decision making. The business case should be presented convincingly and enthusiastically. In particular, hard questions such as the following need to be answered.

▨ What is the validity and robustness of technical, marketing and financial assumptions?
▨ Are the objectives challenging, yet realistic?
▨ What is the break-even point and return on investment?
▨ What is the quality of management and the workforce?
▨ What is unique about the product and the company?
▨ How competitive is the company in the local and global market?
▨ What are the key risks and opportunities?
▨ What is the position regarding SHEQ?

- Are corporate governance systems in place?
- How sustainable is the company, given competition and unstable market conditions?

20.7.5 Layout of a business plan for a start-up new business

While using the same planning principles in the above, this type of business plan is focused on:

- *justifying* the business opportunity to new investors; and
- *explaining* how the business will be structured, implemented and managed.

Paragraph number	Main heading	Key features
	Executive summary	
1	Proposed venture	- The business concept - Viability statement and projected returns on investment - Implementation stages; location - Key risks and upside potential - What is unique about this venture?
2	Market analysis	- Target market; marketing gap; prices; supply/demand; competitor; client analysis
3	Key cornerstones of the business	- Form of business - Licences and permits obtained - Key contracts secured - Key service providers - Governance arrangements
4	Organisational design	- Ownership structure - Management team; organogram
5	Financial planning	- Capital schedule; finance planning - Financial reporting; cash flow
	Implementation planning	
6	Schedule of key events	- Key milestones; key dates
7	Start-up plan	- Site preparation; logistics supply; construction plan; commissioning
8	Operations plan	- Marketing plan; production plan; engineering plan; services plan; governance plan
	SHEQ plan	- Action plan to comply with all statutory and legal requirements
Appendices		

The following guidelines should be incorporated in this type of business plan.

- **Make a good first impression** by ensuring that the front page and contents pages are professional looking, business-like and eye-catching. It should contain the following information:
  - the name, and registration number of the business;
  - the names of the owners, management, solicitors and auditors;
  - the contact details of the business; and
  - the mission statement and logo of the business.
- **Start off with an interesting introduction** by being concise and stimulating interest in your proposition by giving:
  - a clear description of the product or service you intend to provide;
  - the reasons why you believe the venture would be viable;
  - the start-up capital required and the expected return on investment; and
  - the short-, medium- and long-term objectives of the new venture.
- **Provide a comprehensive description of the venture** by including:
  - the ownership and capital structure;
  - the form of the business;
  - compliance statement with legal, environmental and community requirements;
  - a technical description of the product or service, e.g. manufacturing process, equipment, technology to be used, floor plan layout, etc.;
  - a description of the location of the venture relative to key sources of input, competitors and markets;
  - the competence of the management and workforce, including the names, qualifications and experience of key individuals; and
  - a detailed SWOT analysis.
- **Comprehensively outline the proposed marketing strategy.** Keep in mind the golden rule of **No market – no business.** The new entrepreneur will have to convince potential investors that the product or service will be able to penetrate the market. The following are the key issues arising from detailed market analysis:
  - a description of the target market (who do you intend to sell to?);
  - the projected turnover and percentage market share to be captured;
  - sales price, cost structure and break-even analysis;
  - advertising and sales promotion action plan; and
  - a description of customer support and after sales service.
- **End off with a positive and motivating conclusion** (Fisher, 1997:4–11).

20.8 STARTING UP A NEW BUSINESS VENTURE

The golden rule is to obtain the services of independent specialist advisors, such as a commercial lawyer, a chartered accountant or a business advisor. Costly mistakes could be made by trying to do this yourself, especially if you are a new entrant. Should these services prove to be too expensive, then contact any financial institution or the Department of Trade and Industry (www.dti.gov.za) to assist you with mapping out the appropriate course of action to be taken.

However, in general terms the following action steps would apply upon completion of the business plan.

Phase 1: Ensure legal compliance:

- Secure all state, provincial and local permits, such EIA approvals and registration of the new business venture.
- Secure all commercial agreements, such as for marketing and the supply of all input goods/services.

Phase 2: Secure finance for the new ventures

Phase 3: Prepare to launch the new business:

- Make key appointments and muster management team.
- Get access to site and clear the site for construction.
- Get logistics and initial input goods/services in place, and on site.
- Commence construction and/or prepare the site for ongoing business.
- Promote and launch the business.
- Deliver first goods/services onto the market.

Phase 4: Ongoing business:

- The first focus is on the survival of the business – to reach financial break-even.
- The second focus is on sustainable and profitable growth.

20.9   THE BUSINESS NETWORK IN SOUTH AFRICA

There are essentially five categories of business networks, which interface informally and mostly on an 'as required' basis. Within each category there are various organisations that address the specific needs of its members. A few examples of each category is given below.

- Government/Industry interface, such as:
  - ❏ South African Chamber of Business (SACOB);
  - ❏ Afrikaanse Handelsinstituut (AHI);
  - ❏ National African Federated Chamber of Commerce (NAFCOC); and
  - ❏ Industrial Development Corporation (IDC).
- Specific interest groups, such as:
  - ❏ South African Institute of Mining and Metallurgy;
  - ❏ Institute of Mechanical Engineers; and
  - ❏ South African Society for Professional Engineers.
- South African/Foreign country interface, such as:
  - ❏ Southern African Development Community (SADC);
  - ❏ RSA/Australian Business Forum; and
  - ❏ Commonwealth Development Corporation (CDC).
- Personal business network:

  Each entrepreneur has his/her own business network consisting of a complex web of relationships of family, friends, business associates and colleagues. To build and maintain such a personal network over time is hard work. However, such a network may assist the entrepreneur through difficult times as well as open doors to new markets and technology, and perhaps even new ventures.

Self-assessment

20.1 List three examples of each form of business ownership.

20.2 List three examples of each type of financial institution.

20.3 Discuss the four possibilities or modalities of starting a new business, as well as the advantages and disadvantages of each.

20.4 Discuss the possible sources of entrepreneurial ideas.

20.5 Discuss why it is essential for engineers to be familiar with the fundamentals of entrepreneurship.

20.6 Define a business plan, also referring to the main uses of this document.

20.7 Discuss the annual planning cycle of an operating company.

20.8 Briefly describe the advantages and disadvantages of a company and a close corporation as forms of enterprises. Explain why most organisations involved in capital intensive businesses usually choose a company as a form of enterprise.

Mini-Projects

A. Set up an abbreviated business plan for a new small business, such as for a person selling fruit on the sidewalk (maximum of three pages).

B. Enquire from an entrepreneur in your neighbourhood about the following:
  - What made him/her decide to start his/her new venture?
  - How did he/she go about setting up this new venture?
  - What would he/she do differently when given the chance to start the venture again?
  - What are the joys and frustrations that he/she experienced from this new venture?

C. Compile an action plan to build an appropriate business network to add value to a business venture of your choice.

D. Make an appointment with any financial institution and enquire about the type of services they provide to entrepreneurs.

E. Make a list of all the entrepreneurship initiatives and support services in your town/region. What are the contact details and who is in charge of each organisation?

F. Visit two or more bookstores and assess how many books are available on entrepreneurship and starting up your own business.

G. Access your favourite Internet search engine and type in 'entrepreneur' or 'business plan' and see what happens!

H. Select a business venture of your choice and compile a business plan for presentation to prospective investors. Ask a merchant banker/commercial lawyer to give you a critique of your business plan.

References

Anderson, R.L. & Dunkelberg, J.S. 1990. *Entrepreneurship: Starting a New Business*. New York: Harper & Row.

Berger, B. 1991. *The Culture of Entrepreneurship*. San Francisco: Press Publishers.

Clarke, A.G. 1995. *Create Your Own Business: From Ideas to Action*. Cape Town: Human & Rousseau.

Clegg, G. & Barrow, C. 1984. *How to Start and Run Your Own Business*. London & Johannesburg: Macmillan.

David, F.R. 1995. *Concepts of Strategic Management*. 5th edition. Upper Saddle River: Prentice-Hall.

Dollinger, M.C.J. 1995. *Entrepreneurship: Strategies and Resources*. Homewood: Austin Press.

Gifford, P. 1985. *Intrapreneuring: Why You Don't Have to Leave the Corporation to Become an Entrepreneur*. New York: Harper & Row.

Guy, M. 1995. *Starting Your Own Business in South Africa*. 8th edition. Cape Town: Oxford University Press.

Hargadon, A. & Sutton, R.I. 2000. 'Building an innovation factory.' *Harvard Business Review*, May/June, pp. 157–66.

Kuratho, D.F. & Hodgets, R.M. 1992. *Entrepreneurship: A Contemporary Approach*. Fort Worth: Dryden Press.

Needham, C. 2005. 'Ringing the changes.' *Business Times Money*, 14 August 2005, p. 13.

Porter, M.E. 1980. *Competitive Strategy*. New York: Free Press.

Porter, M.E. 1985. *Creating and Sustaining Superior Performance*. New York: Free Press.

Rwemiga, H. & Venter, R. 2004. *Advanced Entrepreneurship*. Cape Town: Oxford University Press.

Schöllhammer, H. & Kuriloff, A.H. 1979. *Entrepreneurship and Small Business Management*. New York: John Wiley.

Timmons, J.A. 1994. *New Venture Creation: Entrepreneurship for the 21st century*. 4th edition. Boston: Irwin/McGraw-Hill.

Annexure A
Guidelines for Answering Mini-projects

Some guidelines for completing projects and the evaluation criteria that may be used by evaluators are as follows.

- Your completed mini-project should be 5–10 pages in length.
- The marker will focus on the content of the assignment. Make sure that your project is well structured and follows the accepted format for a report. Be sure to include an executive summary, introduction and other headings, and list your references at the end of the report.
- The following marking criteria may be used (these were adapted from criteria used by the Potchefstroom Business School)

Evaluation criteria that may be used by the evaluator:

		Max. mark (%)	Mark allocated (%)
1.	**Meeting the objectives of the assignment**		
1.1	The assignment has been understood and answered comprehensively.	20	
1.2	Independent work/thought is reflected.	10	
1.3	Insight into the topic is reflected.	10	
1.4	Logical, systematic thought and reasoning is demonstrated.	10	
1.5	Quality research (literature study or empirical work) has been done. Did the learner consult a number of industry experts?	10	
1.6	Conclusions are logical, meaningful and substantiated.	10	
	Sub-total	**70**	
2.	**Presentation and technical aspects**		
2.1	The content shows a logical and integrated development and forms a balanced, holistic whole.	10	
2.2	The executive summary reflects the content comprehensively and meaningfully.	5	
2.3	Style, language, layout and neatness are of an acceptable standard.	5	
	Sub-total	**20**	
3.	**General quality rating**		
	Evaluator's general evaluation mark of the assignment's quality – taking into consideration the above and other factors.	10	
	Total	**100**	

B
Annexure B
Glossary and Acronyms

A

accrued interest: The interest accrued on a bond since the last interest payment was made.

agility: The ability of an enterprise to manage the changing, unpredictable world of commerce and industry and survive in markets that demand rapid response to unexpected changes in customer demands, competitive challenges and technological breakthroughs.

amortise: To pay off a debt or cost. Normally this involves a gradual retirement of a debt with a series of payments.

annual report: A financial statement issued yearly by a corporation to its shareholders. It shows assets, liabilities, earnings, profits or losses incurred, etc.

annuity: A series of equal payments occurring at equal time periods that last for a finite length of time.

approved supplier: A supplier given preferential treatment in purchasing decisions by earning high ratings on quality, delivery, price and service.

artificial intelligence (AI): Enables computers to process information in the same manner as a human being.

assets: Everything that a company owns and that is due to it. Cash, investments, money due, materials and inventories, etc. are called current assets; land, buildings, machinery, etc. are known as fixed assets.

assignable cause: A source of variation in a process that can be isolated, especially when its significantly larger magnitude differentiates it from common (random) causes.

assignable variation: Variation caused by assignable causes.

attributes chart: A process control chart for attributes showing the number of defective items in each sample.

attribute inspection: Inspection requiring only a yes/no, pass/fail or good/bad judgement.

average chart: A process control chart that shows the variation of the averages of a number of samples.

B

backward integration: The strategy of a firm to acquire others that supply it with inputs.

balance sheet: Condensed financial statement showing the nature and amount of a company's assets, liabilities and owner's equity on a given date.

bill of materials: A list of the type and number of parts needed to produce a product.

bond: Basically an 'I owe you' agreement. A bond is evidence of a debt on which the issuing company usually promises to pay the bond holder a specific amount of interest for a specified length of time and to repay the loan on the expiration date.

book value: The value of a capitalised investment that has not been written off for tax purposes by depreciation, depletion or amortisation.

brainstorming: A decision-making or problem-solving technique used to improve group creativity through the spontaneous proposal of alternatives without concern for reality or tradition.

business plan: A formal document containing a mission statement, description of the firm's goods or services, a market analysis, financial projections and a description of management strategies for attaining goals.

C

CAD: Computer-aided design.

CAM: Computer-aided manufacture.

CAGR: Compound annual growth rate.

capital expenditure budget: A budget that indicates future investments to be made in buildings, equipment and other physical assets of an organisation.

capital gain/loss: A profit/loss from the sale of a capital asset.

cash: Cash at bank and on hand, and cash equivalents such as money market instruments.

cash flow: The reported net income of a corporation plus amounts charged for depreciation, depletion, amortisation, etc. that are book-keeping deductions and not paid out in actual rands and cents.

cash generated by operations: The difference between cash received during the period from customers, together with cash income from investments, and cash paid during the period for goods and services used in the operation.

CBA: Cost-benefit analysis.

CDM: Clean development mechanism.

CER: Certified emission reduction.

certification: Formal approval of a supplier as a source for purchased goods and services, bestowed after a supplier exhibits quality and process control, design and delivery standards, and other desirable traits.

change in working capital: The amount of cash utilised to finance an increase in the working capital or the amount of cash derived from a reduction in working capital. Working capital consists of stock, debtors and creditors.

commodity: Mass-produced, unspecialised product.

common cause: Causes of variation that are inherent in the process.

common variation: Variation remaining in the process output after all special variation has been removed; may be thought of as the natural variation in a process that is under statistical control.

competitive analysis: The process of analysing the business environment to collect relevant competitive intelligence and then integrating this information into new product development and delivery strategies.

compound interest: The interest that accumulates on unpaid interest as well as the original principal.

computer-integrated manufacturing (CIM): An integrated approach that combines CAD/CAM with the use of robots and computerised inventory management techniques.

concept: An abstract entity that can be characterised as a socially realised equivalence class of physical representations and is not identical to any particular physical representation in this class (a sedan car can be represented in a photograph, in speech, in written text, in a computer digital image, in CAD files, in a brain state and so on; the concept is the same, but the representation varies). A concept is therefore not the same as any physical configuration or entity.

concurrent engineering: An approach to new product development where the product and all of its associated processes, such as manufacturing, distribution and service, are developed in parallel.

conflict: Disagreement about the allocation of scarce resources or clashes regarding goals, values, etc.

constraints: Limitations on resources.

control: The process of ensuring that actual activities match planned activities.

co-ordination: The integration of the activities of the separate parts of an organisation to accomplish organisational goals.

cost of sales: All the costs incurred in getting the stock item to its saleable condition.

C_{pk} *index:* A widely used process capability index.

critical path: The longest path through a PERT network that identifies the maximum amount of time required for essential tasks.

CSI: Corporate social investment.

CSR: Corporate social responsibility.

current assets: Items owned by the business that are cash or will be converted into cash in the near future. current assets include stock, debtors and cash.

current liabilities: Includes money owed and payable by a company, usually within one year, and trade creditors.

cycle time: See 'design for assembly'.

D

data: Facts, words, numbers, names and figures that indicate activities and transactions to which they are applicable.

database: Data organised in a manner that makes it possible to access specific information.

decision making: The process of thought and deliberation that results in a decision. Decisions, the output of the decision-making process, are the means by which a manager seeks to achieve some desired state.

DEAT: Department of Environmental Affairs and Tourism

delegation: The act of assigning formal authority and responsibility for the completion of specific activities to a subordinate.

depreciation: A decrease in value because of age, wear and tear, or market conditions.

design: The aim of design is to give some form, pattern, structure or arrangement to an intended technological product so that it is an integrated and balanced whole that will do what is intended. Designing often begins with an idea in someone's mind and the designer has to be able to envisage situations, transformations and outcomes and model these in the mind's eye. Designing involves non-verbal, visual thinking and creativity. It is also about putting ideas together in new ways (*Microsoft Encarta 98*).

design for assembly: Making the product easier to assemble, thereby reducing cycle time during production.

design for disassembly: Designing new products so that they can be taken apart for processes such as recycling or disposal.

DfE: Design for environment.

dividends: An amount paid to shareholders out of profits as a reward for investing in the business.

DME: Department of Minerals and Energy.

DNA: Designated national authority.

DOE: Designated operational entity.

DTI: Department of Trade and Industry.

DWAF: Department of Water Affairs and Forestry.

E

early adopters: People who adopt an innovation quickly after it is introduced.

EB: Executive Board of the Kyoto Protocol.

ECA: Environmental Conservation Act.

EIA: Environmental impact assessment.

EIR: Environmental impact report.

EMS: Environmental management system.

engineering economics: The discipline that is concerned with the cost of alternative engineering solutions to a problem. It is a special branch of microeconomics.

entrepreneur: An individual who establishes and manages a business.

ERA: Environmental risk assessment.

ethics: The study of rights and who should benefit or be harmed by an action.

expenses: Costs incurred in selling stock items and administering the business.

F

FCA: Full-cost accounting.

financial accounting: Concerned with providing information to shareholders, creditors and other outsiders.

financial institutions: Organisations that act as intermediaries between individuals and other organisations that have a surplus of funds available and those that are experiencing a shortage of funds.

financial market: Reflects the total demand for and supply of funds and is, in a capitalist economy, usually sub-divided into a money market and a capital market.

financing costs: Interest paid on amounts borrowed from financial institutions to finance business operations.

fishbone chart: A chart resembling the skeleton of a fish in which the spine bone represents the major cause of quality problems and connecting bones contributing causes; reveals cause-effect linkages (also known as *cause-effect diagram* and *Ishikawa diagram*).

fixed assets: Items owned by the business that have enduring benefit to the business and form part of the infrastructure.

flexitime: A system that permits employees to arrange their work hours to suit their personal needs.

forward integration: The strategy of a firm to acquire other firms that are customers for its outputs.

G

GATT: General Agreement on Tariffs and Trade.

GATS: General Agreement on Trade in Services

goals: The desired future conditions that individuals, groups or organisations strive to achieve. The concept therefore includes missions, purposes, objectives, targets, quotas and deadlines in the wider sense.

gross profit: The difference between turnover and cost of sales.

H

human capital: An organisation's investment in the training and development of its members.

heuristic principles: A method of decision making that proceeds along empirical lines, using rules of thumb to find solutions or answers.

I

I&AP: Interested and affected party.

IEM: Integrated environmental management.

IISD: International Institute for Sustainable Development.

income: Units sold multiplied by unit price.

incremental cost: The cost that will be incurred as the result of increasing output by one more unit.

inflation: An increase in the money supply relative to goods and services that results in a continuing rise of prices.

information: Usable, relevant data that, in a processed form or even unprocessed state, is usable for decision making. Information involves the communication and reception of intelligence or knowledge. It appraises and notifies, stimulates, reduces uncertainty, reveals additional alternatives or helps eliminate irrelevant or poor ones, and influences individuals and stimulates them to action.

inputs (production inputs): Resources from the environment, such as raw materials and labour, that may enter any organisational system.

installment loan: This comes into existence when a person borrows a lump sum of money and agrees to make equal periodic payments until the loan is paid off. Everyday examples of installment loans are mortgages and vehicle loans.

intelligence information: Data on elements of the organisation's operating environment such as clients, competitors, suppliers, creditors and the government for use in short-run planning, and data on developments in the economic environment such as consumer income trends and spending patterns, and in the social and cultural environment for use in long-run strategic planning.

interest: A rental amount charged by financial institutions for the use of money.

interest period: The length of time after which interest is due.

interest rate: The rate of gain received from an investment.

Internet: The hardware consisting of cables or other hardware that links millions of computers connected to thousands of networks around the world.

interpersonal contract: An informal contract whereby two or more people agree to interact about and respect each other's goals, needs and expectations in a mutually beneficial relationship.

intranet: An in-house website that uses Internet technology within a corporate firewall. It is not accessible by the general public. Intranets are used by employees for communication within an organisation and may point to external resources if some intranet pages are linked to the Internet.

intrapreneur: A career manager or employee who creates shareholder value by utilising and managing the assets of the corporation as if he/she is the owner.

investment proposal: A single undertaking or project being considered as an investment possibility.

ISO: International Organisation for Standardisation.

K

kaizen: The Japanese word for continuous improvement.

L

late adopters: The last group of people in a market to adopt a new product.

LCA: Life cycle assessment.

LCC: Life cycle costing.

LCE: Life cycle engineering.

LCI: Life cycle inventory.

LCIA: Life cycle impact assessment.

LCM: Life cycle management.

liabilities: All the claims against a business. Liabilities include accounts payable, wages, salaries, dividends declared, and fixed or long-term liabilities such as mortgage bonds, debentures and bank loans.

liquidation: The process of converting securities or other property into cash. Upon dissolution of a company, the cash remaining after sale of assets and payment of creditors is distributed among the shareholders.

long-term liabilities: Monies owing that will be paid 12 months after the balance sheet date.

M

managerial accounting: Concerned with providing information to managers.

market: All the potential customers sharing a particular need or want who might be willing and able to engage in exchange to satisfy that need or want.

market penetration price strategy: The product is sold at a relatively low price to achieve high sales volume and high market penetration quickly.

mass customisation: To provide individual product configuration at mass-production cost levels.

mechatronics: The precise actuation and control of mechanical devices through electronics.

minimum rate of return: The rate of return at which an investor feels he/she has opportunities in which to invest available capital with reasonable risk. An investor normally seeks alternative investments that yield a rate of return better than this minimum rate.

mission statement: A broad organisational goal that justifies the organisation's existence.

model: A simplified representation of the key properties of a real-world object, event or relationship.

modem: A device that converts digital signals to analog signals and vice versa. It is used because computer hardware produces and receives digital signals, whereas most communication lines handle only voice or analog signals.

N

NEMA: National Environmental Management Act.

O

outputs (production outputs): Transformed inputs that are returned to the external environment as products or services.

overhead: A cost or expense inherent in performing an engineering, construction or manufacturing operation that cannot be charged to or identified with a part of the work done or product manufactured. Such a cost or expense must therefore be allocated on some arbitrary basis or handled as a business expense independent of the volume of production.

owner's equity: The amount of money the shareholders have invested in the business. It is made up of share capital, the original money provided by the shareholders and retained income.

P

patent: A patent confers the sole ownership of an invention on its holder.

PERT: Programme evaluation and review technique.

policy: A standing plan that establishes general guidelines for decision making.

principal: The amount of money borrowed.

problem: A situation that occurs when an actual state of affairs differs from a desired state of affairs.

procedure: A standing plan that contains detailed guidelines for handling organisational actions that occur regularly.

process: A process is a repetitive set of interacting activities that uses resources to transform a defined set of inputs into outputs that are of value to a customer.

process capability: In general, a statement of the ability of process output to meet specifications; inherent capability is the width (approximately six standard deviations) of the distribution of process output.

process control: A condition signifying that all assignable variation has been removed from process output so that only common or chance variation remains.

process control chart: Statistical control chart on which to record samples of measured process outputs. The purpose is to note whether the process is statistically stable or changing, so that adjustments can be made as needed.

process improvement team: A small work group that meets periodically to discuss ways to improve quality, productivity or the work environment.

product testing: The undertaking of test procedures to satisfy product specifications and customer requirements.

(project) budgeting: The process of anticipating project cash flow, including how much money will be spent and when.

project risk analysis: The process of determining possible risks to a project schedule and goals and the likelihood of such occurrences.

Q

quality assurance: In general, the activities associated with making sure that the customer receives quality goods and services.

quality characteristic: A process performance (output) property of a product or service deemed important enough to control.

quality cost: Costs of preventing defects, checking process output and paying for the results of defective output; more broadly, the loss to society of any deviation from target.

R

random events: Patternless occurrences for which there is no apparent cause.

range chart: A process control chart that shows the variation of the ranges of a number of samples.

rapid prototyping systems: Using systems that can create a physical prototype directly from a CAD representation.

R&D: Research and development.

RDD: Research, development and demonstration.

recognition agreement: This usually includes disciplinary, grievance and retrenchment procedures agreed upon by both an employer and a union through a collective bargaining process.

research and development (R&D): The entrepreneurial function that devotes organisational assets to the design, testing and production of new products.

retained income: That portion of the profit remaining at the end of each financial year after dividends have been declared. It belongs to the owners, but is not paid out to them as it may be used to finance new products, for example.

run chart: A running plot of measurements of some process or quality characteristic, piece by piece as the process continues.

S

science: Knowledge and understanding of the physical world and the underlying laws that govern it.

SEA: Strategic environmental assessment.

SETAC: Society of Environmental Toxicology and Chemistry.

SIA: Social impact assessment.

simple interest: Interest payment based only on the original principal amount that was borrowed and not unpaid interest.

sinking fund: Money regularly set aside by a company to redeem its bonds, debentures or preferred stock.

Six Sigma: Six Sigma is a statistical concept that measures a process in terms of defects. At the Six Sigma level there are only 3,4 defects per million opportunities. Six Sigma is also a philosophy of managing that focuses on eliminating defects through practices that emphasise the understanding, measurement and improving of processes.

solid modeling: See 'rapid prototyping systems'.

special cause variation: A type of variation in process output that can be traced to a specific cause such as a fault or malfunction, removal of which removes the variation. Also called assignable variation.

specification: Process output description commonly in two parts: the target (nominal) and the tolerances.

statistical process control (SPC): Collection of process analysis techniques including process flowchart, Pareto analysis, fishbone chart, run chart and control chart. These are all statistical techniques used to monitor, control and improve process performance over time by studying variation and its source.

synergy: A situation in which the whole is greater than its parts.

T

taxation: Amount paid to revenue authorities (e.g. SA Revenue Services) based on profits.

technology: A tool or systematic selection of tools and methodologies, often based on custom and practice, together with the application knowledge, designed to manipulate the materials of the physical world and create potential products in a reproducible and transferable way, or to effect change.

total quality management (TQM): A systematic, organisation-wide approach to quality that stresses continually improving all processes that deliver products and services, with the major outcome of exceeding the customers' needs. TQM is a management process that ensures that products and services are designed, developed, produced, delivered and supported to fully meet customer expectations.

trips: Trade-related aspects of intellectual property rights.

turnover: The total amount of sales made during a period for cash and on credit (items sold multiplied by their selling price).

U

UNEP: United Nations Environmental Programme.

US EPA: United States Environmental Protection Agency.

V

value engineering: A systematic approach to evaluating design alternatives that seeks to eliminate unnecessary features and functions and to achieve required functions at the lowest possible cost while optimising manufacturability, quality and delivery.

variables chart: A process control chart for variables along a scale showing the number of defective items in each sample.

variables inspection: A test in which measurements of an output characteristic are taken.

VAT: Value-added tax.

W

WCED: World Commission on Environment and Development.

website: Information unit containing information that is stored on a server. Access to this information can be obtained by entering the URL or website address into an Internet browser.

WIPO: World Intellectual Property Organisation.

WORLD VIEW:

- 'a set of presuppositions (or assumptions) which we hold (consciously or unconsciously) about the basic makeup of our world' (James Sire).
- 'a guide to life, a basic set of values that we acquire primarily from our culture. At the foundation of a worldview is a commitment to beliefs about our existence, our basic problems, and how to solve them' (R.T. Wright).

World Wide Web (WWW): Refers to the abstract cyberspace of information that 'flows' through the Internet.

WSSD: World Summit on Sustainable Development.

Z

zero defects: Proposed as the proper goal of a quality programme; an alternative to the past practice of setting an acceptable quality (defect) level.

Annexure C

Answers to Selected Self-assessment Questions (marked by '*' in the text)

CHAPTER 11

Answer 11.10

Examples of market segments:

Segment 1: Small building contractors that build or repair houses and DIY (do-it-yourself) home owners.

Segment 2: Big construction companies that build blocks of flats, factories, roads, bridges, reservoirs, dams, etc.

Product

Segment 1 will probably be interested in crusher sand and smaller (6,7 mm and 9,5 mm) aggregate. DIY homebuilders may want to buy a product that is packaged in small quantities (e.g. 40 kg bags). Packaging will be done by the quarry or by wholesale building suppliers and hardware shops.

Segment 2 may want a variety of aggregate sizes, e.g. 6,7 mm; 9,5 mm; 13,2 mm; 19,0 mm; 22 mm and 37 mm. This group may also require a variety of ready-mixed material. They may require SABS-certified aggregate to meet design specifications. They may have specific needs such as larger aggregate that may be required as railway track ballast.

Before you start your quarry you therefore have to determine whether the different properties of the aggregate meet the requirements of your target market.

Price

The price of your quarry's aggregate must be competitive. It is difficult to differentiate one quarry's product from another. By offering an SABS-certified product you can probably, however, ask higher prices than those quarries that don't. The offering of additional services such as delivery will also impact on the pricing strategy. Pricing may further depend on variables such as quantity ordered and product type.

Distribution

Aggregate is a low-priced product and is therefore usually sold to a market segment within a relatively short distance from the quarry. Due to transportation cost, an end user will probably find that the aggregate producer closest to him/her will be able to provide the product at the best price (transportation cost included). In the case of market segment 1, a quarry will have to deliver small quantities at a specific address or distribute the product through a hardware shop. If your target market is segment 2, then your customers will probably load the aggregate in big quantities at your quarry or you will have to deliver it per truck. Transport contractors may be used.

Marketing communication

The appointment of a sales representative may be the most effective way to generate new business if you target segment 2. When you target segment 1, it may however be more effective to make use of advertising in appropriate local newspapers and magazines (e.g. those that target DIY builders).

CHAPTER 14

Answer 14.14

Road	Probability of death	No of 'encroachments' per year	Potential lives saved per year (over 20 years)	Cost per life saved
1	1	0,006	0,006 x 1 x 23,3 = 0,14 (2,8)	R7m/2,8 = R2,5m
2	0,1	1,9	1,9 x 0,1 x 23,3 = 4,42 (88,54)	R7m/88,54 = R79 060

The money will be better spent on erecting guardrails along road number 2 (the four-lane highway).

CHAPTER 15

Answer 15.7

Table 15.3 extended: Relation between volume and surface area for a spherical container

Radius, r (m)	Volume, V (m³)	Surface area, A (m²)	V/A	A/V (3/r)
2,879	V_1 = 100 (or 100 000 litres)	104,2	0,96	1,04
3,628	V_2 = 200 (or 200 000 litres)	165,4	1,21	0,83
4,153	V_3 = 300 (or 300 000 litres)	216,7	1.38	0,72
4,571	V_4 = 400 (or 400 000 litres)	262,6	1,52	0,66
Comparison:	V_2/V_1 = 2	A_2/A_1 = 1,59		
	V_3/V_1 = 3	A_3/A_1 = 2,08		
	V_4/V_1 = 4	A_4/A_1 = 2,52		

The 400 000 litre container will probably cost more than 2,52 times the cost of the 100 000 litre container. If the same conditions as in example 15.3 apply then it will cost 2.64 [$4^{0,7}$] times as much.

Answer 15.13

Fixed cost of ABC plant = 0,4 x 180m = 72m.

Variable cost of ABC plant = 0,6 x 180m = 108m.

Cost estimate of variable cost component of new plant (before price escalation): 160 000 tons/140 000 tons x 108m = R123,4 million.

Cost estimate of total cost for new plant (before price escalation, therefore in 2005 terms) = 72 + 123,4 = R195,4 million (assume that the fixed cost components for both plants will be similar in 2005 terms).

Estimated total cost of new plant by 2008 = 195,4 x $(1 + 0,07)^3$ = R239,4m. (See chapter 16 for information on the future value of a single sum.)

CHAPTER 16

Answer 16.3

$FV = PV (1 + i)^n$; therefore, $PV = FV/(1 + i)^n = 20\ 000/(1 + 0,14)^5 = R10\ 387,37$.

Annexure D
How to Use the CD-ROM at the Back of the Book

Use Windows Explorer or 'My Computer' to have a look at the content of the CD (usually drive D) that you will find included with this book. Under the directory, 'CD_METS-2_ Tutor' you will for example find the 'Engineering Management tutor' which is an executable file called 'ENG_MGT'. Double click on the icon to open it.

Index

485

D

E

K

knowledge 248
   engineering 4-5
   leaders 26
   management 397-399
resource 297
tacit 398
Kyoto Protocol 431

L

LAN (local area network) 251
LCA (life cycle assessment) 410, 411, 413, 418-420
LCC (life cycle costing) 413
LCE (life cycle engineering) 411, 414, 420-421
LCIA (life cycle impact assessment) 420
LCI (life cycle inventory) analysis 419
LCL (lower process control limits) 174
LCM (life cycle management) 408, 421
labour
   cost estimating 328-329
legislation 60-86
   productivity 54-56
Labour
   Appeal Court 79
   Court 79
   Relations Act 65-57, 83-84, 91, 136
laggard 235
land 297
late majority 234-235
law see also legal, legislation
   and safety 183-185
   of comparative advantage 299
   of diminishing returns 297
layout, business plan 462-465
lead users 396
leader, manager as 14
leadership 22-27
   achievement-oriented 24
   benevolent autocratic 25
   characteristics of 22
   democratic 25
   directive 24
   exploitative autocratic 25
   innovative organisation 392
Likert's four systems 24-25
participative 24, 25
   supportive 24
   theories 23-27
   traits 22-23
leading, maintenance 204
learnership 85
learning
   curve effects 373
   curves 328-329, 398
   evaluation 50
   organisation 393
legal consideration, new venture 457-458

legislation
   draft, EIA process 427
   environmental 403, 422-427
   labour 60-86
   training 45-46
leniency 55
level
   of expertise 290
   of R and D 391
liabilities 302
liability
   administrative 184
   contracts 122
   criminal 184
   vicarious 183
liaison, manager as 14
life cycle
   assessment (LCA) 410, 411, 413, 418-420
   asset 409-411
   costing (LCC) 413
   engineering (LCE) 411, 414, 420-421
   impact assessment (LCIA) 420
   inventory (LCI) analysis 419
   iron law of 268
   management (LCM) 31-32, 408, 421
   of facility 7
   operational stage 32-33
   opportunities, entry strategy 452
   organisational 6-8, 31-32, 68
product 231-233, 409-411
project see project life cycle
   thinking 413
life plan, maintenance 214-215
Likert's four systems of leadership 24-25
limited
contractual capacity 116
liability company 455
line functions 19
list of deliverables and receivables 280-281
local area network (LAN) 251
lockouts 81-83
logic, errors of 55
logical relationships 285
long -term capacity planning 143
lose-lose strategy 94
lower process control limits (LCL) 174

M

MERSETA (Manufacturing, Engineering and Related Services SETA) 46
MIS (management information systems) 257
MRP (material requirements planning) 256
MRPII (manufacturing resource planning) 256
machine tools inventory 152
machines, process 166
macroeconomics 295
maintainability 208-209
maintenance